Dimension-reduction Representation of Stochastic Ground Motion and Engineering Applications

Zhangjun Liu • Zixin Liu • Xinxin Ruan

Dimension-reduction Representation of Stochastic Ground Motion and Engineering Applications

Springer

Zhangjun Liu
Wuhan Institute of Technology
Wuhan, China

Zixin Liu
University of Emergency Management
Sanhe, China

Xinxin Ruan
Xinyang Normal University
Xinyang, China

This work was supported by University of Emergency Management and Wuhan Institute of Technology

ISBN 978-981-92-0167-9 ISBN 978-981-92-0168-6 (eBook)
https://doi.org/10.1007/978-981-92-0168-6

This Springer imprint is published by the registered company Springer Nature Singapore Pte Ltd.
The registered company address is: 152 Beach Road, #21-01/04 Gateway East, Singapore 189721, Singapore

Preface

To guarantee the disaster-resilient safety and reliability of engineering structures, structural stochastic dynamic analysis serves as a crucial step. Efficient simulation of stochastic dynamic effects acting on engineering structures serves as both the foundation and prerequisite, and it also represents one of the primary bottlenecks in the comprehensive reliability assessment of structural disaster resistance. Earthquake ground motion is a typical stochastic dynamic action that displays pronounced non-stationary characteristics in both intensity and frequency, along with complex features, such as large-scale spatial distribution and multi-component nature. Accordingly, earthquake ground motion is generally characterized as a non-stationary stochastic process or a spatio-temporal stochastic field.

Although the number of strong motion records has increased significantly in recent years, the available recorded data remain insufficient to fully meet the practical demands of seismic analysis for engineering structures, due to the complexity of source mechanisms, variability in wave propagation paths, and differences in soil site conditions. With the continued advancement of structural dynamic time-history analysis methods and the increasing maturity of probabilistic design approaches grounded in random vibration theory, a systematic investigation of structural dynamic responses throughout the entire seismic process, along with their statistical significance, has become essential for seismic performance analysis and optimal structural design. Consequently, the thorough characterization of ground motion engineering properties and the development of rational stochastic seismic modeling have emerged as major research focuses over the past few decades.

Typically, stochastic earthquake ground motion simulations can be classified into two main categories: frequency-domain and time-domain analysis methods. Fundamentally, these two methods provide equivalent representations for both stationary and non-stationary stochastic processes, demonstrating their equal importance in engineering practice. At present, both methods are well-developed and extensively applied in simulating stochastic dynamic effects, such as earthquakes, strong winds, and waves. However, both approaches essentially belong to the category of conventional Monte Carlo methods and still suffer from two major limitations. On the one hand, they typically involve a large number of random variables and require an extremely large sample size, which poses severe challenges to the computational efficiency of nonlinear stochastic dynamic analysis for complex engineering structures. On the other hand, as random sampling techniques, conventional Monte Carlo methods struggle to acquire the complete probability information of the generated sample set, thereby restricting the refined dynamic analysis and reliability assessment of complex engineering structures.

To overcome these limitations, through more than a decade of sustained research efforts, the authors of this book have revealed the physical essence of dimension-reduction for high-dimensional stochastic dynamic systems and have proposed a novel dimension-reduction method for engineering stochastic processes and fields. By defining the standard orthogonal random variable set as a cluster of orthogonal functions with merely several elementary random variables, a unified dimension-reduction representation theory and an efficient simulation methodology for stochastic processes and spatio-temporal stochastic fields have been established, which are applicable to both frequency-domain and time-domain analysis methods. In essence, the dimension-reduction representation constitutes a full probability modeling approach and exhibits intrinsic consistency with "The Third Generation Structural Design Theory," such as the probability density evolution theory. Consequently, it can be further integrated with this theory to conduct refined nonlinear dynamic response analysis and reliability assessment for randomly excited complex engineering structural systems during their whole service lifecycle.

This book systematically introduces the theory and application of the proposed dimension-reduction simulation method for stochastic ground motion. The main contents are organized as follows: Chap. 1 introduces the fundamentals of probability theory and stochastic process; Chap. 2 addresses the fundamentals of earthquake ground motion; Chaps. 3 and 4 present the dimension-reduction representations of univariate stochastic ground motions using frequency-domain and time-domain analysis methods, respectively; Chap. 5 explains the dimension-reduction representation of multivariate stochastic ground motions; Chap. 6 expounds the dimension-reduction representation of stochastic ground motion fields; Chap. 7 describes the dimension-reduction modeling of stochastic ground motion via strong motion records; and Chap. 8 shows some representative engineering application of the dimension-reduction simulation method.

We would like to express our profound gratitude to Academician Jie Li of the Chinese Academy of Sciences and Professor Jianbing Chen from Tongji University for their long-term attention, encouragement, guidance, and warm help. Throughout our research, they have provided numerous insightful comments and valuable suggestions, which have played an indispensable role in deepening and refining the theoretical framework of this book.

Since 2008, postgraduates Yaolong Lei, Min Xiong, Xing Fang, Yong Wan, Bo Zeng, Linqiang Wu, Wei Liu, Zenghui Liu, Chenggao He, Kangxuan Lv, Yun Liu, Yunmu Jiang, Long Yan, Chunyuan Tan, Jianglin Zhou, Bohang Xu, Miao Liu, and others who were guided by Zhangjun Liu have made substantial contributions to the exploration, development, and refinement of this research, for which we are deeply grateful. Besides, Dr. Jing Hao, Dr. Jin Fan, and Dr. Xiaozhong Li from the Wuhan

Institute of Technology, Ph.D. candidate Xiaojiao Fu, and Master's students Shiwei Yin and Qifan Shu have provided invaluable support and assistance in completing this book, to whom we also express our sincere thanks.

The research presented in this book has been continuously supported by the National Natural Science Foundation of China (Grant Nos. 52478557, 52108444, 51978543, 51778343, 51278282, and 50808113), which is gratefully acknowledged.

We sincerely welcome readers to provide valuable comments and suggestions to facilitate the continued improvement of this book.

Wuhan, China
Sanhe, China
Xinyang, China
January, 2026

Zhangjun Liu
Zixin Liu
Xinxin Ruan

Contents

Chapter 1
Fundamentals of Probability Theory and Stochastic Process

Abstract This chapter primarily introduces fundamental concepts of probability theory and stochastic processes, providing the necessary mathematical foundation for the subsequent chapters. In presenting the basic notions of probability theory and stochastic processes, emphasis is placed on physical interpretation and intuitive understanding. The main topics covered include random events and probability; random variables together with their probability distributions and numerical characteristics; as well as the correlation, independence, and correlation-structure decomposition of random vectors. Particular emphasis is given to functions of random variables (vectors) and their associated probability distributions. In addition, the chapter discusses core concepts of stochastic processes, including time-domain and frequency-domain descriptions, and the spectral decomposition theory for both stationary and non-stationary processes.

Keywords Probability · Random variables · Random vectors · Function of random variables · Stochastic processes

1.1 Probability and Random Variables

1.1.1 Random Events and Probability

An activity carried out to achieve a specific objective can be termed an experiment. When a particular experiment is repeated numerous times under identical basic conditions, and the outcomes still cannot be precisely predicted but exhibit statistical regularity, such an experiment is called a random experiment. A collection of certain possible outcomes of a random experiment is referred to as a random event.

A random experiment has three fundamental features: (1) The basic conditions can be controlled and maintained consistently throughout the experiment; (2) The possible outcomes of the experiment, known as random events, can be identified, but the exact results cannot be precisely predicted; and (3) The probability associated with each possible outcome is known. The likelihood of a random event occurring is described by its probability. According to the classical frequency interpretation, when a random experiment is repeated many times under consistent conditions,

Z. Liu et al., *Dimension-reduction Representation of Stochastic Ground Motion and Engineering Applications*, https://doi.org/10.1007/978-981-92-0168-6_1

the outcomes exhibit statistical regularity, meaning the relative frequency of a specific outcome stabilizes and converges towards its theoretical probability. The randomness observed in such experiments arises from factors beyond our control; that is, despite maintaining consistent experimental conditions, uncontrollable influences still exist, which inherently cause the uncertainty in the experimental outcomes.

Consider the following random experiment involving the toss of a coin: repeatedly toss the coin in the same manner until it lands head-up on the table. Denote "the coin landing on the table with its head up" as T, and "the coin landing on the table with its tail up" as Q. Possible outcomes of this random experiment include:

$$\varpi_1 = T,\ \ \varpi_2 = QT,\ \ \varpi_3 = QQT,\ \ \cdots,\ \ \varpi_k = Q\cdots QT,\ \ \cdots$$

In ϖ_k, the first $k-1$ outcomes are all Q, with the final tossing (k-th) outcome is T. All possible outcomes constitute a set $\Omega = \{\varpi_1, \varpi_2, \cdots, \varpi_k, \cdots\}$, and each specific outcome ϖ of the experiment is an element within this set Ω. A particular outcome of this random experiment (for example, occurrence ϖ_k) is denoted as $\{\varpi = \varpi_k\}$, which is called a basic random event, and such an event ϖ_k is referred to as a sample point. The set Ω, comprising all sample points, is referred to as the sample space, where any subset thereof constitutes a random event. For instance, the random event defined as {"The number of coin tosses required upon termination of the experiment should be no less than 2 and no more than 5"} can be represented by the subset of $\{\varpi_2, \varpi_3, \varpi_4, \varpi_5\} = \{\varpi_k, 2 \leqslant k \leqslant 5\}$. Evidently, this subset is contained within the sample space $\Omega = \{\varpi_1, \varpi_2, \cdots, \varpi_k, \cdots\}$.

Suppose that the possibility of a coin landing on the table with its head up or tail up each time is the same, that is, the probability of each case is 1/2, and the outcome of each toss is entirely independent. Then, according to the multiplication rule, the probability of the random experiment outcome ϖ_k is:

$$p_k = \Pr\{\varpi = \varpi_k\} = \underbrace{\frac{1}{2} \times \frac{1}{2} \cdots \frac{1}{2}}_{\text{the first } k\text{-1 outcomes are all } Q} \times \frac{1}{2} = \frac{1}{2^k} \tag{1.1}$$

where $\Pr\{\cdot\}$ denotes the probability of the random experiment. It is straightforward to verify that:

$$\sum_{k=1}^{\infty} p_k = \frac{1}{2} + \frac{1}{2^2} + \cdots + \frac{1}{2^k} + \cdots = 1 \tag{1.2}$$

In other words, the sum of the probabilities of all basic random events equals 1, satisfying the compatibility condition of probability. Furthermore, the basic random events are mutually exclusive, meaning that any two distinct events cannot occur simultaneously.

In mathematics, $(\Omega, \mathcal{F}, P)$ is defined as a probability space, where Ω denotes the sample space, $\mathcal{F}$ represents the collection of events (including both basic random events and composite random events formed from basic events), and P is a probability measurement of a basic or a composite random event in $\mathcal{F}$. By defining the probability measure for basic random events, the probability of composite random events can be derived based on the rules of probability operations. Broadly speaking, the core objective of elementary probability theory, stochastic process, and stochastic dynamics is to derive the probability distribution of composite or equivalently transformed random events based on the probability distribution of basic random events.

1.1.2 Random Variables

Define a real value for each sample point ϖ (representing a basic random event) in the sample space, thereby constituting a mapping from the sample space Ω to the real number field X, say $\Omega \to R$, and the real number X is called a random variable:

$$X = X(\varpi) \tag{1.3}$$

Since $\{\varpi = \varpi_k\}$ is a random event in the sample space, the corresponding event $\{X = X(\varpi_k)\}$ is also a random one. Moreover, the probabilities are the same, namely:

$$\Pr\{X = X(\varpi_k)\} = \Pr\{\varpi = \varpi_k\} \tag{1.4}$$

For the aforementioned example of tossing a coin, define:

$$X = X(\varpi) = k \tag{1.5}$$

when $\varpi = \varpi_k$, according to Eq. (1.1), it can thus be obtained:

$$\Pr\{X = k\} = \Pr\{\varpi = \varpi_k\} = \frac{1}{2^k} = p_k \tag{1.6}$$

Equation (1.6) presents the probabilities for all possible values of the random variable $X(\varpi)$ (typically abbreviated as X), as defined in Eq. (1.5). This is referred to as the probability distribution of random variable X, with the concept of "mapping" playing a pivotal role herein. It is noteworthy that random variables serve as deterministic functions of sample points in the probability space, with their randomness entirely stemming from the randomness of the underlying basic random events.

For the coin-tossing example in Eq. (1.5), it can be observed that the basic random events can be enumerated (they can establish a one-to-one mapping relationship with the set of natural numbers, hence thus countably infinite). Consequently, the associated random variables can only take discrete values. Such random variables are termed discrete random variables. In contrast, many science and engineering problems involve random variables that can take continuous values, referred to as continuous random variables. Examples include the wave height of the sea surface, daily maximum temperature, and the peak acceleration of seismic ground motions.

Generally, the distribution function of a random variable X can be defined as:

$$F_X(x) = \Pr\{X \leq x\}, \quad -\infty < x < \infty \tag{1.7}$$

This function must satisfy the following conditions:

1. Regularity: $F_X(-\infty) = 0$;
2. Compatibility: $F_X(\infty) = 1$;
3. Monotonicity: $F_X(x_2) \geq F_X(x_1)$, if $x_2 > x_1$.

If $F_X(x)$ is continuously differentiable, the probability density function (PDF) can be further defined as:

$$p_X(x) = \frac{\mathrm{d}F_X(x)}{\mathrm{d}x} \tag{1.8}$$

namely

$$F_X(x) = \int_{-\infty}^{x} p_X(x')\mathrm{d}x' \tag{1.9}$$

and

$$\Pr\{x_1 < X \leq x_2\} = \int_{x_1}^{x_2} p_X(x)\mathrm{d}x = F_X(x_2) - F_X(x_1) \tag{1.10}$$

It is worth noting that the distribution function defined by Eq. (1.7) is not only applicable to continuous random variables but also to discrete random variables. The difference lies in the fact that the distribution function of discrete random variables is a non-continuous, non-differentiable function that rises in a stepwise manner, as illustrated in Fig. 1.1.

The probability distribution of discrete random variables can be defined as follows:

$$p_k = \Pr\{X = x_k\} = F_X(x_k) - F_X(x_k^-) \tag{1.11}$$

The unit step function (also known as Heaviside function) is introduced to characterize the stepwise variation of the distribution function:

$$\text{if } x < x_1:\ \Pr\{X \leqslant x\} = 0;$$
$$\text{if } x \geqslant x_1:$$
$$\Pr\{X = x_1\} = p;$$
$$\Pr\{X = x_2\} = 1 - p$$

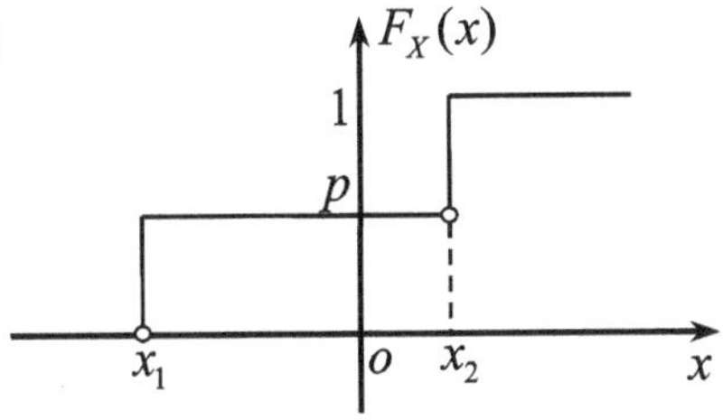

Fig. 1.1 Distribution function of discrete random variables

$$u(x) = \begin{cases} 1, & x \geq 0 \\ 0, & x < 0 \end{cases} \tag{1.12}$$

Thus, the distribution function, as defined by Eq. (1.7), can be represented as follows:

$$F_X(x) = \sum_{k,\, x_k \leq x} p_k = \sum_k p_k u(x - x_k) \tag{1.13}$$

Although the aforementioned distribution function is not differentiable at point $x = x_k (k = 1, 2, \cdots)$ in the classical sense, its derivative may be regarded as infinitely large at these points. Hence, we can introduce a generalized function that fulfills the following two conditions:

$$\delta(x) = \begin{cases} \infty, & x = 0 \\ 0, & x \neq 0 \end{cases} \quad \text{and} \quad \int_{-\infty}^{\infty} \delta(x) \mathrm{d}x = 1 \tag{1.14}$$

Then, differentiating Eq. (1.13) yields the PDF:

$$p_X(x) = \frac{\mathrm{d}F_X(x)}{\mathrm{d}x} = \sum_k p_k \delta(x - x_k) \tag{1.15}$$

The generalized function $\delta(x)$, defined by Eq. (1.14), is commonly referred to as the Dirac Delta function. The derivative of the unit step function is the Dirac Delta function.

Figure 1.2 presents the probability densities of typical continuous and discrete random variables, where the distribution function $F_X(x)$ and probability density $p_X(x)$ are abbreviated as $F(x)$ and $p(x)$, respectively. The basic conditions that the PDF needs to satisfy are:

1. $p_X(x) \geq 0$, for all $-\infty < x < \infty$;
2. $\int_{-\infty}^{\infty} p_X(x) \mathrm{d}x = 1$, this is the so-called compatibility condition.

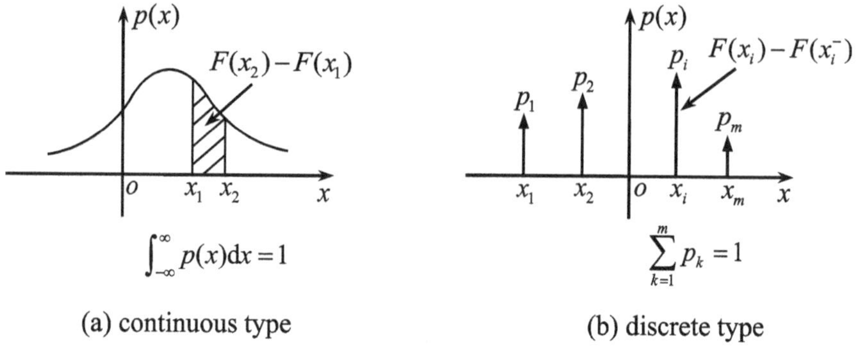

Fig. 1.2 Probability density function

The distribution function or PDF provides a comprehensive description of the statistical regularity of random variables. Commonly used continuous probability distributions in science and engineering include normal, lognormal, exponential, Weibull, and extreme value distributions, among others.

1.1.3 Numerical Characteristics of Random Variables

In many cases, obtaining the probability density of a random variable is quite challenging. Thus, it is often more practical to consider the numerical characteristics of the random variable. The main characteristics are:

Mean Value (Mathematical Expectation)

The mean value of a random variable $X(\varpi)$ is defined as:

$$\mu_X = E[X(\varpi)] = \int_{-\infty}^{\infty} x p_X(x) \mathrm{d}x = \int_{\Omega} X(\varpi) \ell(\mathrm{d}\varpi) \tag{1.16}$$

where $\ell(\mathrm{d}\varpi)$ denotes the probability measure of the micro-subset $\mathrm{d}\varpi \in \Omega$ in the sample space Ω, say $\ell(\mathrm{d}\varpi) = \Pr\{\mathrm{d}\varpi\}$, while $X(\varpi)$ is the random variable defined in Eq. (1.3).

Second-Order Moment (Variance and Standard Deviation)

Random variables can also be described by statistical moments. $\mu_n = E[X^n(\varpi)]$ is called the nth-order origin moment of the random variable $X(\varpi)$, while $\alpha_n = E\{[X(\varpi) - \mu_X]^n\}$ is called the nth-order central moment. It can be observed that the mean value is the first-order origin moment, namely $\mu_X = \mu_1$. The second-order origin moment is:

$$E\left[X^2(\varpi)\right] = \int_{-\infty}^{\infty} x^2 p_X(x)\mathrm{d}x = \int_{\Omega} X^2(\varpi)\ell(\mathrm{d}\varpi) \tag{1.17}$$

which is called mean square value. The second-order central moment is:

$$D_X = E\left\{[X(\varpi) - \mu_X]^2\right\} = \int_{-\infty}^{\infty} (x - \mu_X)^2 p_X(x)\mathrm{d}x = \int_{\Omega} [X(\varpi) - \mu_X]^2 \ell(\mathrm{d}\varpi) \tag{1.18}$$

which is called variance, and its square root is:

$$\sigma_X = \sqrt{D_X} = \sqrt{E\left\{[X(\varpi) - \mu_X]^2\right\}} \tag{1.19}$$

which is called standard deviation.

Skewness

The ratio of the third-order central moment to the cube of the standard deviation is defined as skewness of a random variable, given by:

$$K_s = \frac{\alpha_3}{\sigma_X^3} = \frac{E\left[(X - \mu_X)^3\right]}{\left\{E\left[(X - \mu_X)^2\right]\right\}^{3/2}} \tag{1.20}$$

Kurtosis

The definition of kurtosis is as follows:

$$K_k = \frac{\alpha_4}{\sigma_X^4} - 3 = \frac{E\left[(X - \mu_X)^4\right]}{\left\{E\left[(X - \mu_X)^2\right]\right\}^2} - 3 \tag{1.21}$$

The mean value represents the center of gravity (or centroid) of the PDF. The mean square value corresponds to the moment of inertia of the PDF about the vertical axis $x = 0$. Similarly, the variance can also be interpreted as a moment of inertia of the PDF about the vertical axis, but with the axis of rotation of $x = \mu_X$. A larger variance indicates greater degree of dispersion in the sample, while a smaller variance indicates less the dispersion. In the special case where $\sigma_X^2 = 0$, the dispersion vanishes entirely, and the random variable degenerates into a deterministic value.

Skewness and kurtosis characterize the geometric shape of the PDF, serving as indicators to identify the non-Gaussian nature of the probability distribution. It can be demonstrated that a random variable conforming to normal distribution $N(\mu, \sigma^2)$ has a mean value of μ, a standard deviation of σ, a skewness of $K_s = 0$, and a kurtosis of $K_k = 0$. A skewness of zero signifies that the PDF curve is symmetrical around the mean. A negative skewness implies that the distribution is skewed to the left (longer tail on the left), while a positive skewness implies that it is skewed to the right (longer tail on the right). A kurtosis of zero indicates that the steepness of the PDF curve at its peak is the same as that of the normal distribution. A kurtosis less than zero indicates a relatively flatter distribution, while a kurtosis greater than zero indicates a more peaked distribution (steeper curve near the peak).

For a positive-valued random variable X, the coefficient of variation is also commonly utilized in engineering to represent the degree of variability of the random variable. The coefficient of variation is defined as the ratio of the standard deviation to the mean value, such that:

$$\delta_X = \frac{\sigma_X}{\mu_X} \tag{1.22}$$

1.1.4 Random Vector, Independence

In science and engineering, it is common to encounter multiple random variables. In practice, the primary concern is their correlation and independence.

Let X and Y be two continuous random variables, their joint distribution function is defined as:

$$\begin{aligned} F_{XY}(x, y) &= \Pr\{X(\varpi) \leq x, Y(\varpi) \leq y\} \\ &= \int_{X(\varpi) \leq x, Y(\varpi) \leq y} \ell(\mathrm{d}\varpi) \end{aligned} \tag{1.23}$$

Similarly, here $\ell(\mathrm{d}\varpi)$ denotes the probability measure of subset $\mathrm{d}\varpi$ within the sample space Ω. As a result, the joint probability density of (X, Y) is:

$$p_{XY}(x, y) = \frac{\partial^2 F_{XY}(x, y)}{\partial x \partial y} = \frac{\partial^2 \int_{X(\varpi) \leq x, Y(\varpi) \leq y} \ell(\mathrm{d}\varpi)}{\partial x \partial y} \tag{1.24a}$$

As a matter of fact, Eq. (1.24a) can also be expressed as:

$$p_{XY}(x,y) = \frac{1}{\mathrm{d}x\mathrm{d}y}\int_{x<X(\varpi)\leq x+\mathrm{d}x,\ y<Y(\varpi)\leq y+\mathrm{d}y} \ell(\mathrm{d}\varpi) \tag{1.24b}$$

From this expression, it becomes evident that the significance of "density" in probability density is very clear. Conversely, the probability of the following events can be given as follows:

$$\begin{aligned} &\Pr\{x<X(\varpi)\leq x+\mathrm{d}x, y<Y(\varpi)\leq y+\mathrm{d}y\} \\ &= p_{XY}(x,y)\mathrm{d}x\mathrm{d}y \\ &= \int_{x<X(\varpi)\leq x+\mathrm{d}x,\, y<Y(\varpi)\leq y+\mathrm{d}y} \ell(\mathrm{d}\varpi) \end{aligned} \tag{1.25}$$

Covariance, Correlation Coefficient

The covariance of two random variables is defined as:

$$\begin{aligned} \mathrm{Cov}(X,Y) &= E[(X-\mu_X)(Y-\mu_Y)] \\ &= \int_{-\infty}^{\infty}\int_{-\infty}^{\infty}(x-\mu_X)(y-\mu_Y)p_{XY}(x,y)\mathrm{d}x\mathrm{d}y \\ &= \int_{\Omega}[X(\varpi)-\mu_X][Y(\varpi)-\mu_Y]\ell(\mathrm{d}\varpi) \end{aligned} \tag{1.26}$$

Furthermore, the correlation coefficient of two random variables can be defined as:

$$\rho_{XY} = \frac{\mathrm{Cov}(X,Y)}{\sigma_X\sigma_Y} = \frac{E[(X-\mu_X)(Y-\mu_Y)]}{\sqrt{E\left[(X-\mu_X)^2\right]}\sqrt{E\left[(Y-\mu_Y)^2\right]}} \tag{1.27}$$

It can be shown that $-1 \leq \rho_{XY} \leq 1$. When $|\rho_{XY}| = 1$, there must be $Y = kX + b$, where k, b are constants. When $\rho_{XY} = 1$, we have $k > 0$, and Y and X are positively correlated. In contrast, when $\rho_{XY} = -1$, we have $k < 0$, indicating that Y and X are negatively correlated.

When $\rho_{XY} = 0$, it means X and Y are not related, that is, there is no linear relationship between X and Y.

Conditional Distribution, Independence

Assume the probabilities of random events A and B are non-zero. Then, the event $A \cap B$ (a compound event, abbreviated as AB) can be formed in two steps: "A occurs first, followed by B". The precise definition of this event is "A occurs and B occurs under the condition that A occurs", and its expression is

$$P(AB)=P(A)\cdot P(B|A) \tag{1.28}$$

Hence, given that event A has occurred, the probability of event B occurring is:

$$P(B|A)=\frac{P(AB)}{P(A)} \tag{1.29}$$

If events A and B are independent, then $P(B|A)=P(B)$. Therefore, it follows that:

$$P(AB)=P(A)\cdot P(B) \tag{1.30}$$

This is the condition for random events A and B to be independent.

For two continuous random variables X and Y, if $A=\{x<X\leq x+\mathrm{d}x\}$ and $B=\{y<Y\leq y+\mathrm{d}y\}$, then:

$$P(AB)=\Pr\{x<X\leq x+\mathrm{d}x, y<Y\leq y+\mathrm{d}y\}=p_{XY}(x,y)\mathrm{d}x\mathrm{d}y$$

Since $P(A)=p_X(x)\mathrm{d}x$, $P(B)=p_Y(y)\mathrm{d}y$, if A and B are independent, it follows from Eq. (1.30) that:

$$p_{XY}(x,y)=p_X(x)p_Y(y) \tag{1.31}$$

This is the condition for the independence of random variables X and Y.

According to Eq. (1.29), it can be derived that:

$$\begin{aligned}P(B|A)&=\Pr\{y<Y\leq y+\mathrm{d}y|x<X\leq x+\mathrm{d}x\}=\frac{P(AB)}{P(A)}\\&=\frac{p_{XY}(x,y)\mathrm{d}x\mathrm{d}y}{p_X(x)\mathrm{d}x}=\frac{p_{XY}(x,y)}{p_X(x)}\mathrm{d}y\end{aligned}$$

If $p_{Y|X}(y|x)$ is denoted as the conditional probability density of Y under the condition $X=x$, then $\lim\limits_{\mathrm{d}x\to 0}P(B|A)=p_{Y|X}(y|x)\mathrm{d}y$, and thus we have:

$$p_{Y|X}(y|x)=\frac{p_{XY}(x,y)}{p_X(x)} \tag{1.32}$$

Obviously, thus:

$$p_Y(y)=\int_{-\infty}^{\infty}p_{Y|X}(y|x)p_X(x)\mathrm{d}x=E\left[p_{Y|X}(y|X)\right]=\int_{-\infty}^{\infty}p_{XY}(x,y)\mathrm{d}x \tag{1.33}$$

where $p_Y(y)$ denotes marginal probability density.

The above case of two random variables can be extended to the cases of any number of random variables. For an n-dimensional random variable $(X_1, X_2, \cdots, X_n)$, the joint distribution function is defined as:

$$F_{X_1X_2\cdots X_n}(x_1, x_2, \cdots, x_n) = \Pr\{X_1 \le x_1, X_2 \le x_2, \cdots, X_n \le x_n\} \tag{1.34}$$

and the joint probability density is:

$$p_{X_1X_2\cdots X_n}(x_1, x_2, \cdots, x_n) = \frac{\partial^n F_{X_1X_2\cdots X_n}(x_1, x_2, \cdots, x_n)}{\partial x_1 \partial x_2 \cdots \partial x_n} \tag{1.35}$$

Obviously, the joint probability density and the joint distribution function satisfy the following conditions:

(a) Compatibility:

$$F_{X_1X_2\cdots X_n}(\infty, \infty, \cdots, \infty) = 1, \quad \int_{-\infty}^{\infty} \cdots \int_{-\infty}^{\infty} p_{X_1X_2\cdots X_n}(x_1, x_2, \cdots, x_n) \mathrm{d}x_1 \mathrm{d}x_2 \cdots \mathrm{d}x_n = 1$$

(b) Non-negativity:

$$F_{X_1X_2\cdots X_n}(x_1, x_2, \cdots, x_n) \ge 0, \; p_{X_1X_2\cdots X_n}(x_1, x_2, \cdots, x_n) \ge 0$$

(c) Monotonicity:

$$F_{X_1X_2\cdots X_n}(x_1, x_2, \cdots, x_k, \cdots, x_n) \le F_{X_1X_2\cdots X_n}(x_1, x_2, \cdots, y_k, \cdots, x_n), \forall x_k < y_k$$

If the joint distribution function or joint probability density of n-dimensional random variables is known, the joint distribution of any k ($k \le n$) of these random variables is referred to as k-dimensional marginal distribution. Obviously, to obtain the k-dimensional marginal distribution function, one simply sets the corresponding x values of the random variables beyond the selected k to ∞. Meanwhile, the marginal probability density is defined as the integral of the remaining components within the range of $(-\infty, \infty)$, such that:

$$F_{X_1X_2\cdots X_k}(x_1, x_2, \cdots, x_k) = F_{X_1X_2\cdots X_n}(x_1, x_2, \cdots, x_k, \infty, \cdots, \infty) \tag{1.36}$$

$$p_{X_1X_2\cdots X_k}(x_1, x_2, \cdots, x_k) = \int_{-\infty}^{\infty} \cdots \int_{-\infty}^{\infty} p_{X_1X_2\cdots X_n}(x_1, x_2, \cdots, x_n) \mathrm{d}x_{k+1} \mathrm{d}x_{k+2} \cdots \mathrm{d}x_n \tag{1.37}$$

Note that, although the first k random variables are used here for notational convenience, the relationship actually holds true for any subset of k random variables out of n.

Hence, the joint distribution function or joint probability density encompasses the comprehensive probability information of multi-dimensional random variables, from which the marginal distribution information of any-dimensional random variables can be derived. However, in practice, acquiring joint distribution information is quite challenging, whereas obtaining marginal distribution information is comparatively straightforward. In the case of independent random variables, these two are equivalent. However, for non-independent random variables, how to obtain joint distribution information from marginal distribution information is a problem of great significance.

Decomposition of Related Structures

Consider n-dimensional random variables or an n-dimensional random vector $\boldsymbol{X} = (X_1, X_2, \cdots, X_n)^{\mathrm{T}}$, whose covariance matrix is defined as follows:

$$\boldsymbol{C} = \begin{bmatrix} c_{11} & c_{12} & \cdots & c_{1n} \\ c_{21} & c_{22} & \cdots & c_{2n} \\ \vdots & \vdots & \ddots & \vdots \\ c_{n1} & c_{n2} & \cdots & c_{nn} \end{bmatrix} \tag{1.38}$$

where $c_{ij} = \mathrm{Cov}(X_i, X_j)$. This is a real symmetric matrix. According to the eigen decomposition theorem, there exists a similarity transformation such that

$$\boldsymbol{\Phi}^{\mathrm{T}} \boldsymbol{C} \boldsymbol{\Phi} = \boldsymbol{\Lambda} \tag{1.39}$$

where $\boldsymbol{\Lambda} = \mathrm{diag}\,[\lambda_1, \lambda_2, \cdots, \lambda_n]$ is a diagonal matrix. λ_j $(j = 1, 2, \cdots, n)$ represents the n real eigenvalues of the covariance matrix $\boldsymbol{C}$. $\boldsymbol{\Phi} = [\boldsymbol{\phi}_1, \boldsymbol{\phi}_2, \cdots, \boldsymbol{\phi}_n]$ is the eigenvector matrix of covariance matrix $\boldsymbol{C}$, with its column vectors $\boldsymbol{\phi}_j$ $(j = 1, 2, \cdots, n)$ being the eigenvectors of the covariance matrix. $\boldsymbol{\Phi}$ is an orthogonal matrix, say $\boldsymbol{\Phi}^{\mathrm{T}} \boldsymbol{\Phi} = \boldsymbol{I}$.

Then, define

$$\boldsymbol{Y} = \boldsymbol{\Phi}^{\mathrm{T}} \boldsymbol{X} \tag{1.40}$$

whose inverse transformation is:

$$\boldsymbol{X} = \boldsymbol{\Phi} \boldsymbol{Y} = \sum_{j=1}^{n} \boldsymbol{\phi}_j Y_j \tag{1.41}$$

Then the covariance matrix of $\boldsymbol{Y} = (Y_1, Y_2, \cdots, Y_n)^{\mathrm{T}}$ is defined as follows:

$$\begin{aligned}C_Y &= E\left[(Y-\mu_Y)(Y-\mu_Y)^{\mathrm{T}}\right] \\ &= E\left[\boldsymbol{\Phi}^{\mathrm{T}}(X-\mu_X)(X-\mu_X)^{\mathrm{T}}\boldsymbol{\Phi}\right] \\ &= \boldsymbol{\Phi}^{\mathrm{T}}C\boldsymbol{\Phi} = \Lambda\end{aligned}$$

It can be observed that the standard deviation of Y_j is $\sigma_{Y_j} = \sqrt{\lambda_j}$, that is, the eigenvalues of the covariance matrix of the original random vector are the variances of the corresponding components of the new random vector, while the mean vector is $\boldsymbol{\mu}_Y = E[Y] = E[\boldsymbol{\Phi}^{\mathrm{T}}X] = \boldsymbol{\Phi}^{\mathrm{T}}\boldsymbol{\mu}_X$. Furthermore, if we define $Z_j = \frac{Y_j - \mu_{Y_j}}{\sqrt{\lambda_j}}$, Eq. (1.41) becomes (Li and Chen 2009):

$$X = \boldsymbol{\mu}_X + \sum_{j=1}^{n} \sqrt{\lambda_j} Z_j \boldsymbol{\phi}_j \tag{1.42}$$

where Z_j ($j = 1, 2, \cdots, n$) are the uncorrelated standardized random variables, say $E[Z_j] = 0$, $E[Z_i Z_j] = \delta_{ij}$.

It can be observed that, following the linear transformation outlined in Eq. (1.40), the new random vector $\boldsymbol{Y}$ (or $\boldsymbol{Z}$) becomes pairwise uncorrelated, thereby transforming the originally correlated random vector into an uncorrelated one. Meanwhile, since their eigenvalues often vary greatly, the components with smaller corresponding eigenvalues (i.e., smaller standard deviations) in the new random vector can be omitted, thereby reducing the dimension of the random vector. This will provide immense convenience for the decoupling and dimension-reduction of related random variables, which are frequently encountered in engineering practice.

1.2 Functions of Random Variables

1.2.1 Functions of One-Dimensional Random Variable

Consider the function Y of the random variable $X(\varpi)$, say

$$Y = h[X(\varpi)] = h(X) \tag{1.43}$$

The current central objective is to determine the probability density of $Y = h(X)$, given that the probability density $p_X(x)$ of the random variable $X(\varpi)$ is known. This objective can be accomplished through the equivalent transformation of random events and the invariance of fundamental probability measures. In fact, $X(\varpi)$ establishes a mapping relationship from sample space Ω to the real numbers R, denoted as $X(\varpi) : \Omega \rightarrow R$, whereas its function $Y = h(X) = h[X(\varpi)]$ can be considered as establishing a composite mapping from Ω to R. Essentially, the function of a random variable is a composite function defined on sample points. Based on this, the meaning of $p_X(x)$ is defined by:

$$\begin{aligned} p_X(x)\mathrm{d}x &= \Pr\{X(\varpi) \in (x, x + \mathrm{d}x]\} \\ &= \int_{X(\varpi)\in(x,x+\mathrm{d}x]} \ell(\mathrm{d}\varpi) \end{aligned} \tag{1.44}$$

Regarding the probability density $Y(\varpi) = h[X(\varpi)]$, it is given by:

$$\begin{aligned} p_Y(y)\mathrm{d}y &= \Pr\{Y(\varpi) \in (y, y + \mathrm{d}y]\} \\ &= \Pr\{h[X(\varpi)] \in (y, y + \mathrm{d}y]\} \end{aligned} \tag{1.45}$$

For simplicity, let's first consider the case where $h(\cdot)$ is a monotonically increasing function, as illustrated in Fig. 1.3. As such:

$$\Pr\{h[X(\varpi)] \in (y, y + \mathrm{d}y]\} = \Pr\left\{X(\varpi) \in \left(h^{-1}(y), h^{-1}(y) + \frac{\mathrm{d}h^{-1}(y)}{\mathrm{d}y}\mathrm{d}y\right]\right\} \tag{1.46}$$

Hence, we have:

$$p_Y(y)\mathrm{d}y = \frac{\mathrm{d}h^{-1}(y)}{\mathrm{d}y} p_X\left[h^{-1}(y)\right] \cdot \mathrm{d}y \tag{1.47}$$

Then, we can get

$$p_Y(y) = \frac{\mathrm{d}h^{-1}(y)}{\mathrm{d}y} p_X\left[h^{-1}(y)\right] \tag{1.48}$$

It is readily verifiable that when $h(\cdot)$ is monotonically decreasing, one can simply append a negative sign to the right-hand side of Eq. (1.48), thereby obtaining:

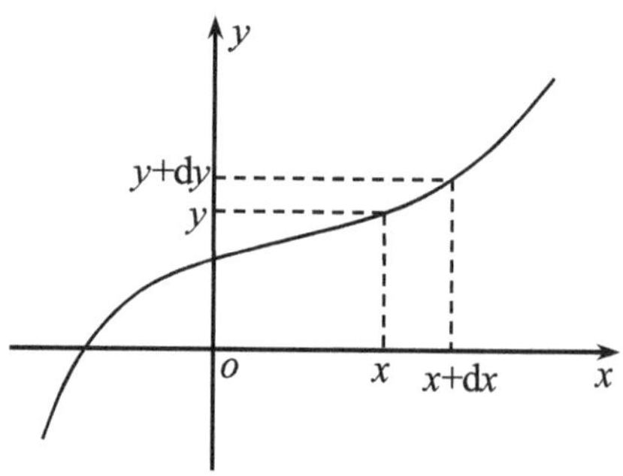

$$\begin{aligned} x &= h^{-1}(y) \\ x + \mathrm{d}x &= h^{-1}(y) + \mathrm{d}h^{-1}(y) \\ &= h^{-1}(y) + \frac{\mathrm{d}h^{-1}(y)}{\mathrm{d}y}\mathrm{d}y \end{aligned}$$

Fig. 1.3 For monotonically increasing functions

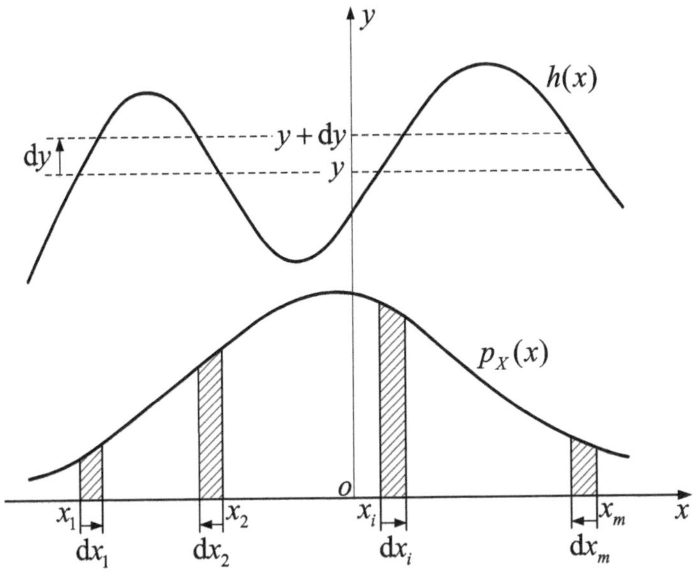

Fig. 1.4 For non-monotonic functions

$$p_Y(y) = \left|\frac{dh^{-1}(y)}{dy}\right| p_X\left[h^{-1}(y)\right] \tag{1.49}$$

Now, consider the case where $h(\cdot)$ is a non-monotonic function, as shown in Fig. 1.4.

Under this circumstance, $\Pr\{h[X(\varpi)] \in (y, y + dy]\}$ is composed of the sum of the measures over multiple segments (for example, m segments) on the x-axis, thus:

$$\begin{aligned}\Pr\{h[X(\varpi)] \in (y, y + dy]\} &= \sum_{k=1}^{m} \Pr\{X(\varpi) \in (x_k, x_k + dx_k]\} \\ &= \sum_{k=1}^{m} \Pr\left\{X(\varpi) \in \left(h_k^{-1}(y), h_k^{-1}(y) + dh_k^{-1}(y)\right]\right\}\end{aligned} \tag{1.50}$$

Hence, we can deduce that:

$$p_Y(y)dy = \sum_{k=1}^{m} \left|\frac{dh_k^{-1}(y)}{dy}\right| p_X\left[h_k^{-1}(y)\right] dy$$

Then, we have:

$$p_Y(y) = \sum_{k=1}^{m} \left| \frac{dh_k^{-1}(y)}{dy} \right| p_X\left[h_k^{-1}(y)\right] \tag{1.51}$$

In fact, the analytical process described above shares an inherent connection with the basic idea of Lebesgue integration, which involves first partitioning the ordinate and subsequently summing the measures of the lengths of the subintervals along the corresponding abscissa.

1.2.2 Functions of Multi-Dimensional Random Variables

Similarly, the basic idea outlined above can be extended to the case involving functions of multi-dimensional random variables. For example, let the function of an n-dimensional random vector $\mathbf{X}(\varpi) = [X_1(\varpi), X_2(\varpi), \cdots, X_n(\varpi)]$ be as follows:

$$Y = h(\boldsymbol{X}) = h(X_1, X_2, \cdots, X_n) \tag{1.52}$$

As a matter of fact, its exact definition is the composite function $Y(\varpi) = h[X_1(\varpi), X_2(\varpi), \cdots, X_n(\varpi)]$. Therefore, similar to Eq. (1.45), there is:

$$\begin{aligned} p_Y(y)\mathrm{d}y &= \Pr\{Y(\varpi) \in (y, y + \mathrm{d}y]\} \\ &= \Pr\{h[X_1(\varpi), X_2(\varpi), \cdots, X_n(\varpi)] \in (y, y + \mathrm{d}y]\} \end{aligned} \tag{1.53}$$

As illustrated in Fig. 1.5, when $y = h(x_1, \cdots, x_n)$ and $y + \mathrm{d}y = h(x_1, \cdots, x_n) + \mathrm{d}h$ enclose an area labeled as $\overline{\Omega}$ on plane $ox_1x_2\cdots x_n$, it is typically represented by the curves $h(x_1, \cdots, x_n) = y$ and $\mathrm{d}y = \sum_{i=1}^{n} \frac{\partial h}{\partial x_i}\mathrm{d}x_i$.

Then, there is (Li and Chen 2009):

$$p_Y(y) = \int \cdots \int \left| \frac{\partial f(y, x_2, \cdots, x_n)}{\partial y} \right| p_X[f(y, x_2, \cdots, x_n), x_2, \cdots, x_n]\mathrm{d}x_2\cdots\mathrm{d}x_n \tag{1.54}$$

where $x_1 = f(y, x_2, \cdots, x_n)$ is derived from $y = h(x_1, x_2, \cdots, x_n)$.

1.2.3 Transformation of Multi-Dimensional Random Variables

Next, we examine the relationship between the joint probability densities of two groups of n-dimensional random variables that are linked in a one-to-one correspondence. Let $\boldsymbol{X} = (X_1, X_2, \cdots, X_n)$ and $\boldsymbol{Y} = (Y_1, Y_2, \cdots, Y_n)$ denote n-dimensional random

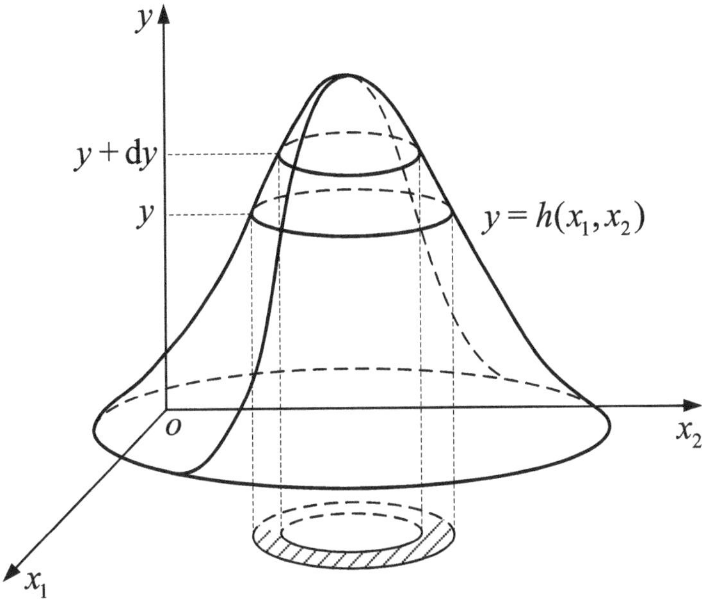

Fig. 1.5 For Functions of multi-dimensional random variables

vectors, with $\boldsymbol{h}$ representing a deterministic transformation from $\boldsymbol{X}$ to $\boldsymbol{Y}$, namely, $\boldsymbol{Y} = \boldsymbol{h}(\boldsymbol{X})$. Now, determine the joint probability density $p_{\boldsymbol{Y}}(\boldsymbol{y})$ of $\boldsymbol{Y}$ based on the joint probability density $p_{\boldsymbol{X}}(\boldsymbol{x})$ of $\boldsymbol{X}$.

Suppose that the transformation $\boldsymbol{h}$ has continuous partial derivatives with respect to each independent variable, and it defines a one-to-one mapping relationship. Therefore, there exists an inverse transformation $\boldsymbol{X} = \boldsymbol{h}^{-1}(\boldsymbol{Y})$, and it possesses continuous partial derivatives as well.

To determine $p_{\boldsymbol{Y}}(\boldsymbol{y})$, consider the mapping of a closed region $\Omega_{\boldsymbol{x}}$ within the sample space of $\boldsymbol{X}$ onto a closed region $\Omega_{\boldsymbol{y}}$ within the sample space of $\boldsymbol{Y}$, under the transformation $\boldsymbol{h}$. According to the law of probability conservation, we can obtain:

$$\int_{\Omega_y} p_{\boldsymbol{Y}}(\boldsymbol{y})\mathrm{d}\boldsymbol{y} = \int_{\Omega_x} p_{\boldsymbol{X}}(\boldsymbol{x})\mathrm{d}\boldsymbol{x} \tag{1.55}$$

By applying the variable transformation rule of multiple integrals, we obtain (Li and Chen 2009):

$$\int_{\Omega_x} p_{\boldsymbol{X}}(\boldsymbol{x})\mathrm{d}\boldsymbol{x} = \int_{\Omega_y} p_{\boldsymbol{X}}\left[\boldsymbol{x} = \boldsymbol{h}^{-1}(\boldsymbol{y})\right] |\boldsymbol{J}_n| \mathrm{d}\boldsymbol{y} \tag{1.56}$$

where $\boldsymbol{J}_n$ represents the n-th order Jacobi determinant of the transformation, and $|\boldsymbol{J}_n|$ denotes its modulus. The determinant $\boldsymbol{J}_n$ is given by the following formula:

$$J_n = \begin{vmatrix} \frac{\partial x_1}{\partial y_1} & \frac{\partial x_1}{\partial y_2} & \cdots & \frac{\partial x_1}{\partial y_n} \\ \frac{\partial x_2}{\partial y_1} & \frac{\partial x_2}{\partial y_2} & \cdots & \frac{\partial x_2}{\partial y_n} \\ \vdots & \vdots & \ddots & \vdots \\ \frac{\partial x_n}{\partial y_1} & \frac{\partial x_n}{\partial y_2} & \cdots & \frac{\partial x_n}{\partial y_n} \end{vmatrix} \tag{1.57}$$

Hence, the joint probability density of $\boldsymbol{Y}$ derived from Eqs. (1.55) and (1.56) is as follows:

$$p_{\boldsymbol{Y}}(\boldsymbol{y}) = p_{\boldsymbol{X}}\left[\boldsymbol{x} = \boldsymbol{h}^{-1}(\boldsymbol{y})\right]|\boldsymbol{J}_n| \tag{1.58}$$

The aforementioned results can also be utilized in the case where the dimension of $\boldsymbol{Y}$ is less than that of $\boldsymbol{X}$. In such instances, let's assume the transformation is as follows:

$$Y_i = h_i(X_1, X_2, \cdots X_n), \quad i = 1, 2, \cdots, m; \quad m < n \tag{1.59}$$

To use the results presented in Eq. (1.58), we expand the m-dimensional random vector $\boldsymbol{Y}$ into an n-dimensional random vector $\mathbf{Z}$. For instance, $\mathbf{Z} = (Z_1, Z_2, \cdots, Z_n)$ can be defined as follows:

$$Z_k = \begin{cases} Y_k, & 1 \le k \le m \\ X_k, & m < k \le n \end{cases} \tag{1.60}$$

In this manner, we can derive the joint probability density of $\mathbf{Z}$. Consequently, the joint probability density of $\boldsymbol{Y}$ is as follows (Liu et al. 2022):

$$\begin{aligned} p_Y(y_1, y_2, \cdots, y_m) &= \int \cdots \int p_Z(z_1, z_2, \cdots, z_n) \mathrm{d}z_{m+1} \cdots \mathrm{d}z_n \\ &= \int \cdots \int p_X\left[\boldsymbol{x}_{1\sim m} = \boldsymbol{f}\left(\boldsymbol{y}, \boldsymbol{x}_{(m+1)\sim n}\right), \boldsymbol{x}_{(m+1)\sim n}\right]|\boldsymbol{J}_n| \mathrm{d}\boldsymbol{x}_{(m+1)\sim n} \end{aligned} \tag{1.61}$$

where $\boldsymbol{x}_{1\sim m} = (x_1, x_2, \cdots, x_m)$, $\boldsymbol{x}_{(m+1)\sim n} = (x_{m+1}, x_{m+2}, \cdots, x_n)$, $\boldsymbol{y} = (y_1, y_2, \cdots, y_m)$; $\boldsymbol{x}_{1\sim m} = \boldsymbol{f}(\boldsymbol{y}, \boldsymbol{x}_{(m+1)\sim n})$ are functions that are inversely solved from $y_i = h_i(x_1, x_2, \cdots, x_n)$ $(i = 1, 2, \cdots, m)$ and expressed in terms of $y_1, y_2, \cdots, y_m, x_{m+1}, x_{m+2}, \cdots, x_n$.

Obviously, when $m = 1$, it pertains to the case of a function of an n-dimensional random variable, as illustrated in Eq. (1.54).

1.2.4 *Numerical Characteristics of Functions of Random Variables*

The numerical characteristics of the function $Y = h(X)$ of the random variable X can be derived from the probability density of Y, which in turn can be obtained via the aforementioned method.

The mean value of the random variable $Y = h(X)$ can be defined as:

$$\mu_Y = E[Y] = \int_{-\infty}^{\infty} y p_Y(y) \mathrm{d}y \tag{1.62}$$

Substituting Eq. (1.45) into Eq. (1.62), we can get:

$$\begin{aligned} \mu_Y &= \int_{-\infty}^{\infty} y \cdot \Pr\{h[X(\varpi)] \in (y, y + \mathrm{d}y]\} \\ &= \int_{-\infty}^{\infty} h(x) \cdot \Pr\{X(\varpi) \in (x, x + \mathrm{d}x]\} \\ &= \int_{-\infty}^{\infty} h(x) p_X(x) \mathrm{d}x \end{aligned} \tag{1.63}$$

Similarly, the variance of Y is as follows:

$$\begin{aligned} \sigma_Y^2 &= E\left[(Y - \mu_Y)^2\right] = \int_{-\infty}^{\infty} (y - \mu_Y)^2 p_Y(y) \mathrm{d}y \\ &= \int_{-\infty}^{\infty} [h(x) - \mu_Y]^2 p_X(x) \mathrm{d}x \end{aligned} \tag{1.64}$$

In practical applications, since it is often difficult to obtain the probability density of Y, the numerical characteristics of Y are usually computed using the above relationship to gain a basic understanding of the probability distribution information of Y. In most cases, when the explicit expression of the function $h(\cdot)$ cannot be obtained, Taylor expansion can be used to derive the approximate mean and variance of Y (Ang and Tang 2007).

1.3 Fundamentals of Stochastic Process

1.3.1 *Basic Concepts of Stochastic Processes*

Given a probability space $(\Omega, \mathcal{F}, P)$ and a parameter set T, if for every pair of $\varpi \in \Omega$ and $t \in T$, there is a definite real number corresponding to them, then this correspondence is termed a stochastic process, denoted as $X(\varpi, t)$.

From a mathematical perspective, the stochastic process $X(\varpi, t)$ is a binary function defined on $\Omega \times T$. Its first argument is a basic random event or basic random variable, hence it is "random"; its second argument is "time", hence it is a process. Therefore, this binary function represents a "stochastic process". For a fixed value of t, $X(\varpi, t)$ serves as a function of the sample point ϖ within the probability space $(\Omega, \mathcal{F}, P)$, and is consequently classified as a random variable, occasionally being figuratively termed a "sectional random variable." For a fixed value of ϖ, $X(\varpi, t)$ is a common function defined on the parameter set T, known as a sample function of the stochastic process. The set of all sample functions is referred to as the sample function space. As such, there are two ways to describe stochastic process.

The first approach is to describe it using a family of random variables. Define stochastic process $X(\varpi, t)$ as a set of random variables indexed by parameter set $t \in T$, with each element t in parameter set T corresponding to a unique random variable. Hence, the number of elements in T is equivalent to the number of random variables in the corresponding random process. In most cases, when T constitutes a continuous interval, a stochastic process can be regarded as being equivalent to an uncountably infinite set of random variables. If T contains only a single element, a finite number of elements, or a countably infinite number of elements, then the corresponding stochastic process is equivalent to either a one-dimensional or multi-dimensional random variable, or a random sequence, respectively.

For a family of random variables, the probabilistic characteristics of stochastic processes are often described using a finite-dimensional distribution family. Specifically, for any given positive integer n, if we consider a selection denoted as $t_1, t_2, \cdots, t_n \in T$, then $(X(t_1), X(t_2), \cdots, X(t_n))$ represents an n-dimensional random vector, and its joint distribution function can be expressed as follows:

$$F_X(x_1, x_2, \cdots, x_n;\ t_1, t_2, \cdots, t_n) = \Pr\{X(t_1) \le x_1, X(t_2) \le x_2,\ \cdots,\ X(t_n) \le x_n\} \quad (1.65)$$

when $t_1,\ t_2,\ \cdots,\ t_n$ takes on all values in T, Eq. (1.65) corresponds to a family of n-dimensional distribution functions, referred to as $\{F_X(x_1, x_2, \cdots, x_n;\ t_1, t_2, \cdots, t_n),\ t_i \in T\}$. When n becomes sufficiently large, the family of n-dimensional distribution functions can be approximately used to describe the probabilistic characteristics of a stochastic process. Indeed, Колмогоров's theorem states that a family of finite-dimensional distribution functions, denoted by $\{F_X(x_1, x_2, \cdots, x_n;\ t_1, t_2, \cdots, t_n),\ n = 1, 2, \cdots,\ t_i \in T\}$, completely determines the probabilistic properties of a stochastic process $X(\varpi, t)$.

The second approach to describe the stochastic process is to use the family of sample functions. Once the probability space $(\Omega, \mathcal{F}, P)$ is determined, each $\varpi \in \Omega$ has a corresponding sample function. Consequently, a stochastic process can also be regarded as a family of sample functions. The number of elements in sample space Ω is equivalent to the number of sample functions that represent random process $X(\varpi, t)$.

Both of the aforementioned descriptions furnish comprehensive probabilistic information regarding the stochastic process and are mathematically equivalent. In specific applications, one may choose different forms of description depending on the requirements of the problem, thereby facilitating the handling of the issue. Typically, theoretical analyses begins with the description in terms of a family of random variables, whereas in practical measurements and data processing, the description based on a family of sample functions is usually employed. These two forms of description are mutually complementary in both theoretical and practical contexts.

For simplicity and to avoid confusion, it is common to abbreviate the random process $X(\varpi, t)$ as $X(t)$.

In practice, another convenient and common method to describe random processes is to use analytical formulas and express them in the form of random functions, namely (Liu et al. 2022):

$$X(t) = g(t; A_1, A_2, \cdots, A_n) \tag{1.66}$$

where the function form of $g(\cdot)$ is known. A_1, A_2, $\cdots$, A_n denotes a set of basic random variables, of which the joint probability distribution is given. Herein, the stochastic process $X(t)$ is a function consisting of random variables, with time t as its parameter.

It should be noted that the sample function of the stochastic process, as defined in Eq. (1.66), is entirely determined by the values of the basic random variables A_1, A_2, $\cdots$, A_n. We designate such a stochastic process $X(t)$ as possessing n degrees of randomness.

1.3.2 *Time-Domain Description of Stochastic Process*

For a random process, its probabilistic characteristics can be completely described by a family of finite-dimensional distribution functions. However, in practical problems, obtaining the entire family of the finite-dimensional distribution functions of the random process is often extremely challenging, or even impossible. In such cases, using second-order statistics is a reasonable and feasible approach.

Mean Function

The mean function of a stochastic process $X(\varpi, t)$ is typically defined as the ensemble average of $X(\varpi, t)$ at time t, such that:

$$\mu_X(t) = E[X(\varpi, t)] \overset{(1)}{=} \int_{-\infty}^{\infty} x p_X(x, t) \mathrm{d}x \overset{(2)}{=} \int_{\Omega} X(\varpi, t) \ell(\mathrm{d}\varpi) \tag{1.67}$$

Correlation Function

The auto-correlation function of a stochastic process $X(\varpi, t)$ is defined as:

$$\begin{aligned} R_X(t_1,t_2) &= E[X(\varpi,t_1)X(\varpi,t_2)] \\ &\overset{(1)}{=} \int_{-\infty}^{\infty}\int_{-\infty}^{\infty} x_1x_2p_X(x_1,x_2;t_1,t_2)\mathrm{d}x_1\mathrm{d}x_2 \\ &\overset{(2)}{=} \int_{\Omega} X(\varpi,t_1)X(\varpi,t_2)\ell(\mathrm{d}\varpi) \end{aligned} \tag{1.68}$$

It is observed that in Eqs. (1.67) and (1.68), the equality sign labeled with (1) employs the first description approach of a stochastic process, whereas the equality sign labeled with (2) utilizes the second description approach.

Centralized second-order moments typically encompass variance functions and covariance functions. For instance, the auto-covariance function of $X(t)$ is

$$\begin{aligned} C_X(t_1,t_2) &= E\{[X(t_1)-\mu_X(t_1)][X(t_2)-\mu_X(t_2)]\} \\ &= R_X(t_1,t_2)-\mu_X(t_1)\mu_X(t_2) \end{aligned} \tag{1.69}$$

The auto-correlation function and the auto-covariance function are numerical characteristics that characterize the statistical dependency between the states of a stochastic process at two different time instants, reflecting the extent of linear correlation between the states of the stochastic process at these two time instants.

If for each parameter $t \in T$, the second-order moment $E[X^2(t)] = R_X(t, t)$ (the variance function) of the random process $X(t)$ exists, then this random process is called a second-order moment process. The auto-correlation function of a second-order moment process always exists.

If the finite-dimensional distribution family of the stochastic process $X(t)$ constitutes a normal distribution family, meaning that for any positive integer n and any set $t_1, t_2, \cdots, t_n \in T$, $(X(t_1), X(t_2), \cdots, X(t_n))$ conforms to an n-dimensional normal distribution, then $X(t)$ is termed a normal process. All statistical properties of a normal process are entirely determined by its mean function and auto-correlation function.

In practical problems, sometimes it is necessary to study two or more random processes simultaneously and their statistical relationships. For such problems, besides analyzing the statistical characteristics of each stochastic process, we must also consider the joint statistical characteristics of several stochastic processes.

The cross-correlation function of two real-valued stochastic processes $X(t)$ and $Y(t)$ is defined as:

$$R_{XY}(t_1,t_2) = E[X(t_1)Y(t_2)] \tag{1.70}$$

Similarly, the cross-covariance function is defined as:

$$\begin{aligned} C_{XY}(t_1, t_2) &= E\{[X(t_1) - \mu_X(t_1)][Y(t_2) - \mu_Y(t_2)]\} \\ &= R_{XY}(t_1, t_2) - \mu_X(t_1)\mu_Y(t_2) \end{aligned} \tag{1.71}$$

If for any $t_1, t_2 \in T$, it always holds that $C_{XY}(t_1, t_2) = 0$, then the stochastic processes $X(t)$ and $Y(t)$ are said to be uncorrelated. For two real-valued stochastic processes $X(t)$ and $Y(t)$ with a mean of zero, uncorrelation is equivalent to orthogonality. Furthermore, if two stochastic processes are mutually independent, they are necessarily uncorrelated; conversely, being uncorrelated does not necessarily imply their mutual independence.

The above definition can be extended to the case of n-dimensional stochastic processes. Accordingly, the correlation function and covariance function of such stochastic processes become the correlation function matrix and covariance function matrix, respectively.

Stationary Characteristics

If the finite-dimensional distribution function family of a stochastic process $X(t)$ does not change over time, that is, for any positive integer n and real number τ, as well as any $t_1, t_2, \cdots, t_n \in T$, when $t_1 + \tau, t_2 + \tau, \cdots, t_n + \tau \in T$, there always exists:

$$F_X(x_1, x_2, \cdots, x_n;\ t_1 + \tau, t_2 + \tau, \cdots, t_n + \tau) = F_X(x_1, x_2, \cdots, x_n;\ t_1, t_2, \cdots, t_n) \tag{1.72}$$

then this stochastic process is stationary (also known as strictly stationary). It can be seen that any finite-dimensional distribution function of a stationary stochastic process is only related to the time difference $t_n - t_1, t_{n-1} - t_1, \cdots$, but not to the starting time t_1.

Strict stationary is not very common in practice. If the mean of a second-order moment process $X(t)$ does not change over time, and the auto-correlation function only depends on the time difference $\tau = t_2 - t_1$, then there exists:

$$\mu_X(t) = \text{constant} \tag{1.73a}$$

$$R_X(t_1, t_2) = R_X(t_2 - t_1) = R_X(\tau) \tag{1.73b}$$

The stochastic process is then called a wide-sense stationary process, and is often directly referred to as a stationary process or a second-order stationary process.

Additionally, when considering two real-valued stationary processes $X(t)$ and $Y(t)$ simultaneously, if their cross-correlation function $R_{XY}(t, t + \tau)$ is also a univariate function of the time difference τ, that is:

$$R_{XY}(t, t + \tau) = E[X(t)Y(t + \tau)] = R_{XY}(\tau) \tag{1.74}$$

then $X(t)$ and $Y(t)$ are said to be stationary correlated, or jointly stationary. According to Eq. (1.74), there exists $R_{XY}(\tau) = R_{YX}(-\tau)$.

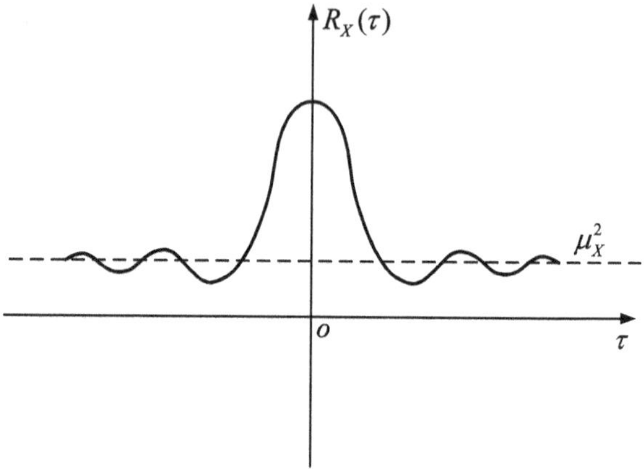

Fig. 1.6 Auto-correlation function

The auto-correlation function of a stationary stochastic process exhibits the following properties:

(a) **Symmetry:** $R_X(\tau) = R_X(-\tau)$,
(b) **Boundedness:** $|R_X(\tau)| \leq R_X(0) = E[X^2(t)]$,
(c) **Amnesia** (or **limited correlation**)**:** If $X(t)$ does not contain periodic components, then $\lim\limits_{\tau \to \infty} R_X(\tau) - \mu_X^2 = 0$.

The image of a typical auto-correlation function is depicted in Fig. 1.6.

The symmetry of the auto-correlation function indicates that the correlation of a stochastic process is non-directional; the "strength of the impact" in the "forward" and "backward" directions is exactly the same. The boundedness indicates that the degree of correlation between any two points does not exceed the correlation between the same point and itself. The amnesia (or limited correlation) indicates that for a general stochastic process, as time increases indefinitely, the influence of the initial conditions will tend to become negligible, meaning that the memory of the initial conditions is more significant within a limited time.

For this purpose, the correlation time can be defined for a second-order moment process with mean value of zero (as shown in Fig. 1.7):

$$\tau_c = \frac{1}{R_X(0)} \int_0^{\infty} R_X(\tau) \mathrm{d}\tau \tag{1.75}$$

When $\tau << \tau_c$, it can be considered that the correlation is strong. However, when $\tau >> \tau_c$, it can be concluded that there is no correlation at all between the two points.

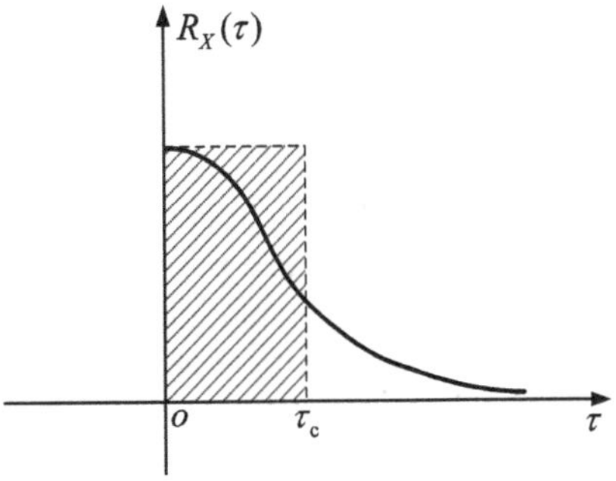

Fig. 1.7 Correlation time

Ergodicity

If any given sample function of the stochastic process $X(\varpi, t)$ encompasses all the probabilistic information of the stochastic process, then this stochastic process possesses ergodicity. Specifically, the mean function and auto-correlation function of a stochastic process can typically be derived from a sample function, and this is generally achieved by performing a time average on the sample function.

The time average of $X(\varpi, t)$ is defined as:

$$\mu_X^{(t)}(\varpi) = \lim_{T \to \infty} \frac{1}{2T} \int_{-T}^{T} X(\varpi, t)\mathrm{d}t \tag{1.76}$$

The time auto-correlation function is defined as:

$$R_X^{(t)}(\varpi, \tau) = \lim_{T \to \infty} \frac{1}{2T} \int_{-T}^{T} X(\varpi, t)X(\varpi, t+\tau)\mathrm{d}t \tag{1.77}$$

For general stochastic processes, the aforementioned average in the time domain relies on the sample function, implying that both $\mu_X^{(t)}(\varpi)$ and $R_X^{(t)}(\varpi, \tau)$ are functions of ϖ, and consequently, they themselves are still random variables. However, if the time average is equivalent to the ensemble average, i.e., $\mu_X^{(t)}(\varpi) = \mu_X(t)$, then the stochastic process is considered to be mean-ergodic. If the condition of $R_X^{(t)}(\varpi, \tau) = R_X(\tau)$ is also satisfied, then the stochastic process remains auto-correlation ergodic. It is evident that any ergodic process is necessarily a stationary stochastic process.

Although the conditions for ergodicity are extremely demanding, in practical research and applications, people frequently find themselves consciously or unconsciously, and even inevitably, resorting to this assumption. For instance, in wind engineering, to obtain the statistical properties of wind speed records, a long-duration wind speed record is typically segmented into multiple parts, with each part considered as an independent record. In fact, it is assumed here that the wind speed process exhibits ergodicity.

1.3.3 Frequency-Domain Description of Stochastic Process

Power Spectrum Density of Stationary Process

The Fourier series and Fourier transform indicate that a time-domain process can be represented as the superposition of a series of harmonic components of different frequencies, and the amplitudes of these harmonic components represent the magnitudes of the corresponding periodic components. It is evident that the Fourier transform of a stochastic process is likewise random. Hence, capturing the probabilistic information of stochastic processes in the frequency domain remains challenging. Utilizing the Fourier transform of the correlation function serves as a reasonable approach of representing such processes in the frequency domain.

Define the Fourier transform of the correlation function $R_X(\tau)$ associated with the stationary stochastic process $X(t)$ as follows:

$$S_X(\omega) = \frac{1}{2\pi}\int_{-\infty}^{\infty} R_X(\tau)\mathrm{e}^{-\mathrm{i}\omega\tau}\mathrm{d}\tau \tag{1.78}$$

Correspondingly, its inverse Fourier transform is:

$$R_X(\tau) = \int_{-\infty}^{\infty} S_X(\omega)\mathrm{e}^{\mathrm{i}\omega\tau}\mathrm{d}\omega \tag{1.79}$$

In Eq. (1.79), if we set $\tau = 0$, the following can be obtained:

$$R_X(0) = E\left[X^2(t)\right] = \int_{-\infty}^{\infty} S_X(\omega)\mathrm{d}\omega \tag{1.80}$$

Here, $E[X^2(t)]$ denotes the instantaneous average power of the stochastic process. For instance, if $X(t)$ stands for voltage, then $X^2(t)$ is directly proportional to the electrical power. Eq. (1.80) demonstrates that $S_X(\omega)\mathrm{d}\omega$ signifies the magnitude of power within the interval $[\omega, \omega + \mathrm{d}\omega]$ for the stochastic process, thus $S_X(\omega)$ is termed the power spectrum density (PSD), often shortened to auto-spectrum density or spectrum density. Given the real symmetric property of $R_X(\tau)$, $S_X(\omega)$ is necessarily a real number, and we also have $S_X(\omega) = S_X(-\omega)$.

Next, let's examine the white noise process. If the stationary process $X(t)$ has zero mean and a correlation function of $R_X(\tau) = 2\pi S_0\delta(\tau)$, according to Eq. (1.75), the correlation time is $\tau_c = 0$, indicating that any two time instants of this stochastic process are uncorrelated. Substituting $R_X(\tau) = 2\pi S_0\delta(\tau)$ into Eq. (1.78), the following can be obtained:

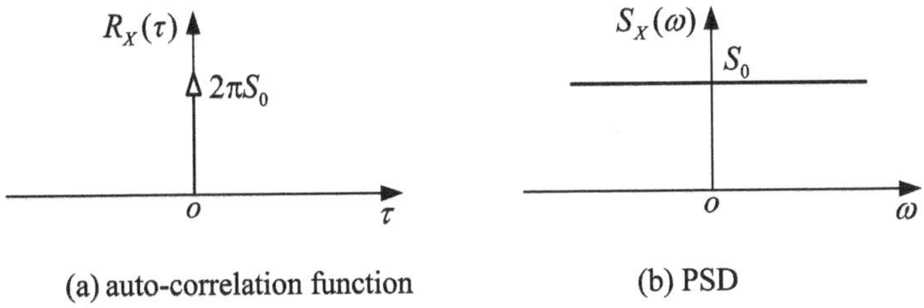

Fig. 1.8 White noise process

$$S_X(\omega) = S_0 \int_{-\infty}^{\infty} \delta(\tau) e^{-i\omega\tau} d\tau = S_0 \tag{1.81}$$

It is evident that the PSD $S_X(\omega)$ remains constant, with all frequency components contributing equally. Such a process is termed a white noise process, with its auto-correlation function and PSD illustrated in Fig. 1.8.

An ideal white noise process is physically unrealizable, as it would require power $E[X^2(t)] = R_X(0) = \infty$ [consider Eq. (1.80)]. Intuitively, the white noise process demands that there be no correlation between any two time instants, implying that the velocity between these two time instants might theoretically be infinite, which is obviously impossible. However, when the correlation time τ_c of the stochastic process is significantly shorter than the characteristic time scale of the system (for instance, the smallest inherent period of interest in dynamical systems), the white noise model serves as a suitable approximation. Furthermore, the finite bandwidth white noise model is also frequently employed in engineering applications. At this point, the PSD is defined as follows:

$$S_X(\omega) = \begin{cases} S_0, & \omega_1 \leq |\omega| \leq \omega_2 \\ 0, & \text{other cases} \end{cases} \tag{1.82}$$

The PSD $S_X(\omega)$ mentioned above is symmetric, that is $S_X(\omega) = S_X(-\omega)$. In practice, negative frequencies do not exist, hence the one-sided PSD function is often introduced:

$$G_X(\omega) = \begin{cases} 2S_X(\omega), & \omega > 0 \\ S_X(0), & \omega = 0 \\ 0, & \text{other cases} \end{cases} \tag{1.83}$$

Cross-Spectrum Density of Joint Stationary Processes

Let $X(t)$ and $Y(t)$ be two stationary processes, and assume that they are jointly stationary (stationarily correlated). The Eq. (1.78) can be generalized to obtain the cross-spectrum density as:

$$S_{XY}(\omega) = \frac{1}{2\pi}\int_{-\infty}^{\infty} R_{XY}(\tau)\mathrm{e}^{-\mathrm{i}\omega\tau}\mathrm{d}\tau \tag{1.84}$$

As previously mentioned, when $X(t)$ is a real stationary process, its PSD $S_X(\omega)$ is a real, non-negative, even function. Now, even when both $X(t)$ and $Y(t)$ are real stationary processes, the cross-spectrum density is typically no longer a real, non-negative, even function, but rather a complex-valued function. The cross-spectrum density has the following properties:

(a) $S_{XY}(\omega)$ and $S_{YX}(\omega)$ are conjugates of each other, say $S_{XY}(\omega) = S^*_{YX}(\omega)$.
(b) The cross-correlation function $R_{XY}(\tau)$ and the cross-spectrum density $S_{XY}(\omega)$ constitute a Fourier transform pair, that is:

$$\begin{cases} S_{XY}(\omega) = \dfrac{1}{2\pi}\displaystyle\int_{-\infty}^{\infty} R_{XY}(\tau)\mathrm{e}^{-\mathrm{i}\omega\tau}\mathrm{d}\tau \\ R_{XY}(\tau) = \displaystyle\int_{-\infty}^{\infty} S_{XY}(\omega)\mathrm{e}^{\mathrm{i}\omega\tau}\mathrm{d}\omega \end{cases} \tag{1.85}$$

(c) The real part of $S_{XY}(\omega)$ is an even function, whereas the imaginary part is an odd function.
(d) The relationship between cross-spectrum density and auto-spectrum density satisfies $|S_{XY}(\omega)|^2 \le S_X(\omega)S_Y(\omega)$.

The mean-square calculus of stochastic processes is highly analogous to ordinary calculus. The mean-square derivatives of all orders of a second-order stationary process $X(t)$ are also stochastic processes. In fact, the cross-correlation function of the n-th order mean-square derivative $X^n(t)$ and the m-th order mean-square derivative $X^m(t)$ is related to the following correlation function of $X(t)$:

$$R_{X^{(n)}X^{(m)}}(\tau) = (-1)^n \frac{\mathrm{d}^{n+m}R_X(\tau)}{\mathrm{d}\tau^{n+m}} \tag{1.86}$$

The cross-spectrum density between $X^n(t)$ and $X^m(t)$ bears the following relationship with the spectrum density of $X(t)$:

$$S_{X^{(n)}X^{(m)}}(\omega) = (-1)^n(\mathrm{i}\omega)^n(\mathrm{i}\omega)^m S_X(\omega) \tag{1.87}$$

When $X(t)$ and $Y(t)$ are jointly stationary processes, the n-th order mean square derivative $X^n(t)$ of $X(t)$ and the m-th order mean square derivative $Y^m(t)$ of $Y(t)$

remain jointly stationary. The relationship between their cross-correlation function and cross-spectrum density is as follows:

$$R_{X^{(n)}Y^{(m)}}(\tau) = (-1)^n \frac{\mathrm{d}^{n+m}R_{XY}(\tau)}{\mathrm{d}\tau^{n+m}} \tag{1.88}$$

$$S_{X^{(n)}Y^{(m)}}(\omega) = (-1)^n (\mathrm{i}\omega)^n (\mathrm{i}\omega)^m S_{XY}(\omega) \tag{1.89}$$

If $X(t)$ is regarded as a displacement process, then $S_X(\omega)$ represents the displacement spectrum density. From Eq. (1.87), we can deduce the relationship between the spectrum density of the velocity process $\dot{X}(t)$ and that of displacement, as well as the relationship between the spectrum density of the acceleration process $\ddot{X}(t)$ and that of displacement, such that:

$$\begin{aligned} S_{\dot{X}}(\omega) &= S_{\dot{X}\dot{X}}(\omega) = \omega^2 S_X(\omega) \\ S_{\ddot{X}}(\omega) &= S_{\ddot{X}\ddot{X}}(\omega) = \omega^4 S_X(\omega) \end{aligned} \tag{1.90}$$

The cross-spectrum density does not possess the same physical significance as the auto-spectrum density. Its introduction is primarily aimed at describing the correlation between two stationary processes in the frequency domain. In earthquake engineering, the Eq. (1.91)

$$\gamma_{XY}(\omega) = \frac{S_{XY}(\omega)}{\sqrt{S_X(\omega)S_Y(\omega)}} \tag{1.91}$$

is typically referred to as the coherence function between $X(t)$ and $Y(t)$, whereas $|\gamma_{XY}(\omega)|$ is designated as the lagged coherence function. Obviously, there exists $0 \leq |\gamma_{XY}(\omega)| \leq 1$. If $|\gamma_{XY}(\omega)| = 0$, these two stationary processes are incoherent, and if $|\gamma_{XY}(\omega)| = 1$, these two stationary processes are completely coherent.

For stationary stochastic processes $X(t)$ and $Y(t)$ that possess zero mean, the cross-covariance function is identical to the cross-correlation function. In this case, if $X(t)$ and $Y(t)$ are uncorrelated, then $R_{XY}(\tau) = 0$, and consequently $S_{XY}(\omega) = 0$, which indicates that $X(t)$ and $Y(t)$ are incoherent. Therefore, uncorrelatedness and incoherence are equivalent.

Spectrum Density of Non-Stationary Processes

As shown in Eqs. (1.68) and (1.70), the auto-correlation function of a non-stationary process depends on two specific time instants, rather than their difference. Therefore, the auto-correlation function of a non-stationary process is a function of two variables, which necessitates the utilization of the theory of the two-dimensional Fourier transform. Next, we briefly present the two-dimensional Fourier transform of the auto-correlation function $R_X(t_1, t_2)$.

Firstly, we denote $\Gamma_X(\omega_1, t_2)$ as the Fourier transform of the auto-correlation function $R_X(t_1, t_2)$ with respect to t_1, that is:

$$\Gamma_X(\omega_1, t_2) = \frac{1}{2\pi}\int_{-\infty}^{\infty} R_X(t_1, t_2)e^{-i\omega_1 t_1}dt_1 \tag{1.92a}$$

By applying the inverse Fourier transform, we obtain:

$$R_X(t_1, t_2) = \int_{-\infty}^{\infty} \Gamma_X(\omega_1, t_2)e^{i\omega_1 t_1}d\omega_1 \tag{1.92b}$$

Now, let $S_X(\omega_1, \omega_2)$ represent the Fourier transform of $\Gamma_X(\omega_1, t_2)$ with respect to t_2. It is noted that it is advantageous to change the sign of the exponent in the definition of $S_X(\omega_1, \omega_2)$, that is:

$$S_X(\omega_1, \omega_2) = \frac{1}{2\pi}\int_{-\infty}^{\infty} \Gamma_X(\omega_1, t_2)e^{+i\omega_2 t_2}dt_2 \tag{1.93a}$$

Accordingly, the inverse Fourier transform is:

$$\Gamma_X(\omega_1, t_2) = \int_{-\infty}^{\infty} S_X(\omega_1, \omega_2)e^{-i\omega_2 t_2}d\omega_2 \tag{1.93b}$$

Thus, substituting Eq. (1.92a) into Eq. (1.93a), and simultaneously substituting Eq. (1.93b) into Eq. (1.92b), we obtain the following double Fourier transform pair (Papoulis and Pillai 2002):

$$\begin{cases} S_X(\omega_1, \omega_2) = \dfrac{1}{(2\pi)^2}\displaystyle\int_{-\infty}^{\infty}\int_{-\infty}^{\infty} R_X(t_1, t_2)e^{-i(\omega_1 t_1 - \omega_2 t_2)}dt_1 dt_2 \\ R_X(t_1, t_2) = \displaystyle\int_{-\infty}^{\infty}\int_{-\infty}^{\infty} S_X(\omega_1, \omega_2)e^{i(\omega_1 t_1 - \omega_2 t_2)}d\omega_1 d\omega_2 \end{cases} \tag{1.94}$$

where $S_X(\omega_1, \omega_2)$ is called the generalized PSD of a non-stationary process, which is simply called the generalized spectrum density.

In mathematics, the generalized spectrum density is related to the PSD of stationary processes. Given that $S_X(\omega_1, \omega_2)$ is positive only in the case $\omega_2 = \omega_1$, while it equals to zero in the case $\omega_2 \neq \omega_1$, denoted as:

$$S_X(\omega_1, \omega_2) = S_X(\omega_1)\delta(\omega_2 - \omega_1)$$

the generalized spectrum density reduces to the PSD of a stationary process in such cases. The generalized spectrum density lacks a definitive physical meaning and exhibits notable deficiencies in its properties. Since $R_X(t_1, t_2)$ is not an even function, the generalized spectrum density is not a real-valued function and is rarely adopted in practical applications.

In fact, the PSD of a non-stationary process $X(t)$ can also be defined as:

$$\begin{cases} S_X(\omega, t) = \dfrac{1}{2\pi}\displaystyle\int_{-\infty}^{\infty} R_X(t, t+\tau)\mathrm{e}^{-\mathrm{i}\omega\tau}\mathrm{d}\tau \\ R_X(t, t+\tau) = \displaystyle\int_{-\infty}^{\infty} S_X(\omega, t)\mathrm{e}^{\mathrm{i}\omega\tau}\mathrm{d}\omega \end{cases} \tag{1.95}$$

where $S_X(\omega, t)$ denotes the evolutionary PSD (EPSD) of a non-stationary process, which is abbreviated as evolutionary spectrum density.

Intuitively, in Eq. (1.95), $S_X(\omega, t)$ has a clearer physical meaning, as it describes the energy distribution over time in the frequency domain. Moreover, it is also easy to apply the PSD of a stationary process to the evolutionary spectrum density of a non-stationary process. Currently, the most widely adopted method is to construct the evolutionary spectrum density using a modulation function. For instance, if the PSD of a stationary process $X(t)$ is $S_X(\omega)$, then the evolutionary spectrum density of a non-stationary process $Y(t) = a(t)X(t)$ is as follows:

$$S_Y(\omega, t) = a^2(t)S_X(\omega) \tag{1.96}$$

1.3.4 Spectral Decomposition of Stochastic Process

Orthogonal Incremental Process

Let $Z(\omega)$ be a second-order moment process with zero mean, where the parameter set is defined in the frequency domain. If for any $\omega_1 < \omega_2 \leq \omega_3 < \omega_4$, it holds that:

$$E\{[Z(\omega_2) - Z(\omega_1)][Z^*(\omega_4) - Z^*(\omega_3)]\} = 0 \tag{1.97}$$

where $Z(\omega)$ is called the orthogonal incremental process. This indicates that the increments over any two non-overlapping intervals are orthogonal (uncorrelated).

In fact, the aforementioned orthogonal incremental process $Z(\omega)$ can also be defined as:

$$E[Z(\omega)] = 0 \tag{1.98a}$$

$$E[\mathrm{d}Z(\omega)\mathrm{d}Z^*(\omega')] = \mathrm{d}\Psi(\omega)\delta_{\omega\omega'} \tag{1.98b}$$

where $\Psi(\omega)$ is a deterministic function. The increments satisfy $\mathrm{d}Z(\omega) = Z(\omega + \mathrm{d}\omega) - Z(\omega)$. $\delta_{\omega\omega'}$ denotes the Kronecker-delta, that is, when $\omega = \omega'$, $\delta_{\omega\omega'} = 1$, while in the case of $\omega \neq \omega'$, $\delta_{\omega\omega'} = 0$.

The orthogonal incremental process can be employed to construct a class of stochastic processes, including both stationary and non-stationary ones.

Spectral Decomposition of Stationary Process

Let $X(t)$ be a mean-square continuous real-valued stationary process with the mean function of $E[X(t)] = 0$. Then $X(t)$ can be expressed as the following Fourier-Stieltjes integral:

$$X(t) = \int_{-\infty}^{\infty} e^{i\omega t} dZ(\omega) \tag{1.99}$$

where $Z(\omega)$denotes the orthogonal incremental process. Eq. (1.99) is referred to as the spectral decomposition of a stationary process, illustrating that a mean-square continuous stationary process can be represented as the superposition of an uncountably infinite number of harmonic waves, each possessing a distinct frequency and mutually uncorrelated random amplitudes.

Since $X(t)$ is a real-valued stationary process, it remains unchanged after conjugation, that is:

$$X(t) = \int_{-\infty}^{\infty} e^{-i\omega' t} dZ^*(\omega') = \int_{-\infty}^{\infty} e^{i\omega t} dZ^*(-\omega) \tag{1.100}$$

Based on Eqs. (1.99) and (1.100), and taking into account that $E[X(t)] = 0$, it is evident that the increment $dZ(\omega)$ must satisfy the following condition:

$$E[dZ(\omega)] = 0, \quad dZ(\omega) = dZ^*(-\omega)$$

Furthermore, using Eq. (1.98b), the auto-correlation function is

$$\begin{aligned} E[X(t)X(t+\tau)] &= E\left[\int_{-\infty}^{\infty} e^{-i\omega' t} dZ^*(\omega') \int_{-\infty}^{\infty} e^{i\omega(t+\tau)} dZ(\omega)\right] \\ &= \int_{-\infty}^{\infty}\int_{-\infty}^{\infty} e^{i(\omega-\omega')t + i\omega\tau} E[dZ(\omega)dZ^*(\omega')] \\ &= \int_{-\infty}^{\infty} e^{i\omega\tau} d\Psi(\omega) \end{aligned} \tag{1.101}$$

Eq. (1.101) indicates that the auto-correlation function of $X(t)$ depends only on the time difference τ. Therefore, $X(t)$ is a wide-sense stationary process.

The deterministic function $\Psi(\omega)$ can be differentiable or non-differentiable. If it is differentiable, we can set $S_X(\omega) = \frac{d\Psi(\omega)}{d\omega}$, and then Eq. (1.101) can be written as:

$$R_X(\tau) = E[X(t)X(t+\tau)] = \int_{-\infty}^{\infty} e^{i\omega\tau} S_X(\omega) d\omega \tag{1.102}$$

This corresponds to Eq. (1.79), where $S_X(\omega)$ represents the PSD. Eq. (1.101) is more general than Eq. (1.102) because the former also includes the case where $\Psi(\omega)$ is non-differentiable. Since the PSD $S_X(\omega)$ is non-negative and $\Psi(\omega)$ is a monotonically non-decreasing function, it is referred to as the spectrum distribution function, denoted by $\Psi_X(\omega)$.

Based on the above analysis, for a zero-mean, real-valued stationary process $X(t)$ with continuous mean square, if the PSD $S_X(\omega)$ is known, the increment $\mathrm{d}Z(\omega)$ should satisfy the following conditions:

$$E[\mathrm{d}Z(\omega)] = 0, \quad \mathrm{d}Z(\omega) = \mathrm{d}Z^*(-\omega), \quad E[\mathrm{d}Z(\omega)\mathrm{d}Z^*(\omega')] = S_X(\omega)\delta_{\omega\omega'}\mathrm{d}\omega \tag{1.103}$$

They are the basic conditions that need to be satisfied for the simulation of stationary process using spectral representation.

Spectral Decomposition of Non-Stationary Process

The orthogonal incremental process can also be employed to construct a type of non-stationary processes, known as evolutionary stochastic processes. A zero-mean real-valued stochastic process $X(t)$ can be expressed as the following Fourier-Stieltjes integral:

$$X(t) = \int_{-\infty}^{\infty} A(t,\omega)\,\mathrm{e}^{\mathrm{i}\omega t}\mathrm{d}Z(\omega) \tag{1.104}$$

where $A(t, \omega)$ is a deterministic function of time t and frequency ω, satisfying $A^*(t, -\omega) = A(t, \omega)$, and is called the time-frequency modulation function. $Z(\omega)$ denotes the orthogonal incremental process.

The auto-correlation function of $X(t)$ is as follows:

$$\begin{aligned} E[X(t_1)X(t_2)] &= \int_{-\infty}^{\infty}\int_{-\infty}^{\infty} A(t_1,\omega)A^*(t_2,\omega')\mathrm{e}^{\mathrm{i}(\omega t_1 - \omega' t_2)}E[\mathrm{d}Z(\omega)\mathrm{d}Z^*(\omega')] \\ &= \int_{-\infty}^{\infty} A(t_1,\omega)A^*(t_2,\omega)\mathrm{e}^{\mathrm{i}\omega(t_1 - t_2)}\mathrm{d}\Psi(\omega) \end{aligned} \tag{1.105}$$

By setting $t_1 = t_2$, we can obtain the mean square function:

$$E\left[X^2(t)\right] = \int_{-\infty}^{\infty} |A(t,\omega)|^2\mathrm{d}\Psi(\omega) \tag{1.106}$$

Equations (1.105) and (1.106) demonstrate that the mean square value is a function of time t, whereas the auto-correlation function relies on two specific time instants, not the time difference. Thus, $X(t)$ is a non-stationary process. If $\Psi(\omega)$ is differentiable, namely $\mathrm{d}\Psi(\omega) = S(\omega)\mathrm{d}\omega$, then we have:

$$E[X(t_1)X(t_2)] = \int_{-\infty}^{\infty} A(t_1, \omega)A^*(t_2, \omega)\mathrm{e}^{\mathrm{i}\omega(t_1 - t_2)}S(\omega)\mathrm{d}\omega \tag{1.107}$$

$$E\left[X^2(t)\right] = \int_{-\infty}^{\infty} |A(t, \omega)|^2 S(\omega)\mathrm{d}\omega \tag{1.108}$$

Therefore, the evolutionary spectrum density can be defined as:

$$S_X(\omega, t) = |A(t, \omega)|^2 S(\omega) \tag{1.109}$$

In the above equations, $\Psi(\omega)$ and $S(\omega)$ denote the spectrum distribution function and spectrum density function of the stationary process $X(t)$, respectively. If $A(t, \omega) = a(t)$, it uniformly alters the spectrum density $S(\omega)$. Under such circumstances, the EPSD $S_X(\omega, t)$ of the non-stationary process is expressed as Eq. (1.96).

Similar to the spectral decomposition of a stationary process, for a zero-mean real non-stationary process $X(t)$ with continuous mean square, if the PSD $S(\omega)$ of its corresponding stationary process is known, and the time-frequency modulation function satisfies $A(t, \omega) = A^*(t, -\omega)$, the basic conditions that the increment $\mathrm{d}Z(\omega)$ in Eq. (1.104) should satisfy are given by Eq. (1.103) [with $S_X(\omega)$ replaced by $S(\omega)$]. They are also the basic conditions that require to be satisfied for the simulating a non-stationary process using spectral representation.

References

Ang AHS, Tang WH (2007) Probability concepts in engineering: emphasis on applications in Civil & Environmental Engineering. Wiley, Hoboken, NJ

Li J, Chen JB (2009) Stochastic dynamics of structures. John Wiley & Sons, Singapore

Liu ZJ, Chen JB, Peng YB (2022) Dynamics of structures. China Architecture & Building Press, Beijing. (in Chinese)

Papoulis A, Pillai SU (2002) Probability, random variables and stochastic processes, 4th edn. McGraw Hill, New York

Chapter 2
Fundamentals of Earthquake Ground Motion

Abstract This chapter introduces fundamental knowledge in earthquake engineering, including the amplitude, duration, and spectral characteristics of ground motions, as well as their non-stationary features and spatial variability. Regarding spectral characteristics, it focuses on the power spectrum, Fourier spectrum, and response spectrum, together with the relationships among these three spectra. Regarding non-stationary features, it presents intensity modulation function and time–frequency modulation function models, and on this basis elaborates on four categories of evolutionary power spectrum models. Regarding spatial variability, it explains the basic concepts of continuous and discrete spatio-temporal stochastic fields of ground motions, and describes coherence function models, including the incoherence effect, wave-passage effect, and local site effect, along with their classical models.

Keywords Earthquake ground motion · Amplitude characteristics · Duration characteristics · Spectral characteristics · Non-stationary characteristics · Spatial variability characteristics

2.1 Amplitude Characteristics of Earthquake Ground Motion

The ground motion amplitude can refer to the maximum value or an equivalent value in a certain sense for any of the seismic acceleration, velocity, or displacement. The seismic peak ground acceleration (PGA), say $a_{\max}$, is the earliest and most intuitive measure proposed to describe the ground motion amplitude, and it has been the most widely accepted and applied in earthquake engineering. However, since the seismic PGA is mainly governed by the high-frequency components of ground motion, which are determined by the local characteristics of the seismic source, it is difficult for this parameter to fully represent the overall characteristics of the seismic source. Consequently, earthquakes may exhibit large PGA without causing significant damages. In response, several related concepts such as static equivalent acceleration, effective peak acceleration (EPA), and root mean square acceleration a_{rms} were

Z. Liu et al., *Dimension-reduction Representation of Stochastic Ground Motion and Engineering Applications*, https://doi.org/10.1007/978-981-92-0168-6_2

proposed to overcome the limitations of seismic PGA. In fact, the values of basis seismic PGA specified in the "Seismic Ground Motion Parameters Zonation Map of China" (GB18306-2015 2015), and the design values of basis seismic PGA in the "Code for Seismic Design of Buildings" (GB50011-2010 2016), are both based on statistical averages.

2.1.1 *Occurrence Probability of Seismic Ground Motion Intensity*

Now, consider the probability distribution of the seismic intensity I that a specific location has experienced. Let $N(I)$ denotes the number of times within t_0 years that earthquakes with intensities greater than I have occurred in this region. Then the average number of occurrences of earthquakes with intensities greater than I in this area within a year can be expressed as $\overline{N}(I)=N(I)/t_0$ (times per year). Furthermore, assume $\log\overline{N}(I)=a-bI$, if only earthquakes with intensities exceeding I_0 are considered, the above expression can be rewritten as $\log\overline{N}(I)=a'-b(I-I_0)$.

It should be noted that since the probability of a region experiencing an earthquake with an intensity greater than I is very small, the average annual number of occurrences of such earthquakes $\overline{N}(I)$ can be regarded as the annual probability of exceedance or the annual average occurrence probability. Furthermore, if the annual average number N_0 of earthquakes with intensities greater than I_0 is known, then the **annual probability of exceedance** can be expressed as follows (Shibata 2020):

$$\overline{N}(I)=N_0 10^{-b(I-I_0)}=N_0 \mathrm{e}^{-\beta(I-I_0)} \tag{2.1}$$

In earthquake engineering, the occurrence of seismic events is generally assumed to follow a Poison process. That is, the annual probability of exceedance $\overline{N}(I)$ of an earthquake at a certain location is considered constant over time, and earthquakes occurring in different years are regarded as independent events. Accordingly, the probability P_n that earthquakes with intensities greater than I occurring n times within t years follows the Poison distribution, such that:

$$P_n=P(n)=\frac{\left[\overline{N}(I)t\right]^n \mathrm{e}^{-\overline{N}(I)t}}{n!}$$

Obviously, when $n=0$, it indicates that no earthquakes with intensities greater than I would occur within t years. In other words, the probability of earthquakes within t years have intensities less than I is $P_0=P(n=0)=\mathrm{e}^{-\overline{N}(I)t}$.

Therefore, the probability that an earthquake with an intensity greater than I occurs at least once within t years (i.e., the probability of exceedance) is:

$$P(n \geq 1) = 1 - P_0 = 1 - \mathrm{e}^{-\overline{N}(I)t} = 1 - \mathrm{e}^{-N_0 t \mathrm{e}^{-\beta(I - I_0)}} \tag{2.2}$$

Meanwhile, the annual probability of exceedance $\overline{N}(I)$ has an inverse relationship with the average return period $\overline{T}$, such that:

$$\overline{T} = \frac{1}{\overline{N}(I)} \tag{2.3}$$

According to Eq. (2.2), for a given probability of exceedance P and time t, the corresponding intensity I can be determined. When t equals the average return period $\overline{T}$, that is, $t = \overline{T}$, it follows from Eqs. (2.2) and (2.3) that the probability of exceedance becomes $P = 1 - \mathrm{e}^{-1} = 0.632$.

If the relationship between the experienced seismic intensity and the annual probability of exceedance at a certain location is known, then, based on the inverse relationship between the annual probability of exceedance and average return period, say Eq. (2.3), the seismic intensity I corresponding to a specific average return period $\overline{T}$ can be determined. This process essentially constitutes seismic hazard analysis.

2.1.2 The Code-Specified Seismic Peak Ground Acceleration

The current "Code for Seismic Design of Buildings" (GB50011-2010 2016) of China adopts a three-level seismic precautionary criterion. The so-called "three-level" refers to frequent ground motion (minor earthquake), basis ground motion (moderate earthquake), and rare ground motion (major earthquake), respectively. For a design period of 50 years, the corresponding probabilities of exceedance are 63.2%, 10%, and 2 ~ 3% (specifically, 2% for regions with seismic intensity of 9, and 3% for regions with intensity of 7 and 6), respectively. The associated return periods are 50 years, 475 years, and 2475–1642 years (specifically, 2475 years for regions with seismic intensity of 9, 1975 years for regions with seismic intensity of 8, and 1642 years for regions with seismic intensity of 7 and 6), respectively. Among these, the seismic precautionary intensity or basis seismic PGA of "moderate earthquake" level is determined from the seismic intensity zonation map or the seismic ground motion parameter zonation map. According to seismic hazard analysis, the ratio of earthquake intensity or PGA among different return periods varies from one location to another. For a given site or a specific location, if frequent and rare ground motions are defined in terms of probability of exceedance or return period, the conventional empirical rule—taking the intensity I of basis ground motion as the baseline, the intensity of frequent ground motion as $(I-1.55)$, and the intensity of rare ground motion as $(I+1.0)$—is generally inaccurate. However, to maintain consistency with previous versions of the code, this convention is still retained in the current version of the code. Meanwhile, to facilitate practical engineering application, the seismic design code provides the correspondence between

Table 2.1 Correspondence between seismic precautionary intensity and basis seismic PGA for soil site II

Seismic precautionary intensity	6-degree	7-degree		8-degree		9-degree
Basis seismic PGA (soil site II)	0.05 g	0.10 g	0.15 g	0.20 g	0.30 g	0.40 g

Table 2.2 Design values of seismic PGA for soil site II ($a_{\mathrm{max,\ II}}$) (unit: cm/s^2)

Basis seismic PGA Return period (years)	Seismic precautionary intensity					
	6-degree	7-degree		8-degree		9-degree
	0.05 g	0.10 g	0.15 g	0.20 g	0.30 g	0.40 g
Frequent ground motion (50)	18	35	55	70	110	140
Basis ground motion (475)	50	100	150	200	300	400
Rare ground motion (1642)	120	220	310	–	–	–
Rare ground motion (1975)	–	–	–	400	510	–
Rare ground motion (2475)	–	–	–	–	–	620

the seismic precautionary intensity and the basis seismic PGA for various regions, as shown in Table 2.1.

Building upon this correspondence relationship, and according to the "Seismic Ground Motion Parameters Zonation Map of China" (GB18306-2015 2015), the seismic PGA for frequent ground motion should not be less than 1/3 of the basis seismic PGA, while the seismic PGA for rare ground motion should be taken as 1.6 ~ 2.3 times of the basis seismic PGA. Meanwhile, in combination with the "Code for Seismic Design of Buildings" (GB50011-2010 2016), the corresponding design values of the seismic PGA for soil site II, say $a_{\mathrm{max,\ II}}$, under different seismic precautionary intensities and return periods are detailed in Table 2.2.

Additionally, according to the "Seismic Ground Motion Parameters Zonation Map of China" (GB18306-2015 2015), a very rare ground motion corresponds to that with an annual probability of exceedance of 10^{-4}, that is, a return period of 10,000 years. The seismic PGA for such a very rare ground motion is recommended to be taken as 2.7 ~ 3.2 times the basis seismic PGA. Typically, the values can be progressively decreased from 3.2 to 2.7 as the seismic precautionary intensity rises from 6 to 9.

To further account for the influences of soil site conditions on seismic PGA, "Seismic Ground Motion Parameters Zonation Map of China" (GB18306-2015 2015) specifies that the design values of seismic PGA a_{max} for soil sites $\mathrm{I_0}$, $\mathrm{I_1}$, III, and IV can be determined based on $a_{\mathrm{max,\ II}}$ and the corresponding site-specific PGA adjustment factor F_a, given by:

$$a_{\mathrm{max}} = F_a \times a_{\mathrm{max,II}} \tag{2.4}$$

where F_a denotes the site-specific PGA adjustment factor, which can be determined according to Table 2.3.

Table 2.3 Values of site-specific PGA adjustment factor F_a

	Seismic precautionary intensity					
	6-degree	7-degree		8-degree		9-degree
Classification of soil sites	0.05 g	0.10 g	0.15 g	0.20 g	0.30 g	0.40 g
I_0	0.72	0.74	0.75	0.76	0.85	0.90
I_1	0.80	0.82	0.83	0.85	0.95	1.00
II	1.00	1.00	1.00	1.00	1.00	1.00
III	1.30	1.25	1.15	1.00	1.00	1.00
IV	1.25	1.20	1.10	1.00	0.95	0.90

2.2 Duration Characteristics of Earthquake Ground Motion

In a strong earthquake, the duration from the onset of ground motion to the point when the motion completely ceases is called the total duration of the strong motion record, also referred to as the recorded duration. Since small-amplitude vibrations at the beginning and end of the record have little engineering significance, earthquake engineering practice typically ignores these portions and focuses instead on the duration of strong shaking that meaningfully affects structures. This period is known as strong motion duration T_d, or sometimes the stationary duration. Similar to the definition of ground motion amplitude parameters, the definition of strong motion duration T_d is not uniform. Common types of definitions include the following (Li and Li 1992):

2.2.1 *Absolute Duration*

The absolute duration is defined based on the absolute value of ground motion acceleration. Specifically, the strong motion duration is taken as the duration between the first- and last-time instances when the absolute acceleration amplitude reaches a predetermined threshold value (e.g., 0.05g).

2.2.2 *Relative Duration*

The relative duration is defined based on the relative value of ground motion acceleration and can be divided into fractional duration and energy duration. The treatment of fractional duration is similar to that of absolute duration, except that the threshold is taken as a proportion of the PGA (ka_{max}), where $k = 0.3 \sim 0.5$ in general. The energy duration is defined as the duration during which the energy accumulated within the strong motion period accounts for a specified proportion of the total energy of the ground motion. For example, the 90% energy duration refers to the time it takes for the cumulative energy to rise from 5% to 95% of the total energy (Xie and Yu 1982; Trifunac and Brady 1975).

2.2.3 Equivalent Duration

The equivalent duration defines the acceleration process of ground motion as an equivalent stationary process with an intensity of a_{rms} and a duration of T_{d} (Xie and Yu 1982; Trifunac and Brady 1975), such that:

$$T_{\mathrm{d}} = \frac{\int_{0}^{T} a^{2}(t)\mathrm{d}t}{a_{\mathrm{rms}}} \tag{2.5}$$

where a_{rms} denotes the standard deviation of the acceleration process $a(t)$ during the stationary strong motion portion of the record. In general, the stationary duration T_{d} should not be less than 10 s.

Figure 2.1 shows the north-south (N-S) component of the El Centro strong motion record in 1940. The total duration of the ground motion is $T = 31.16$ s. The absolute duration corresponding to 0.05g is $T_{\mathrm{d}} = 25.14$ s, and the 90% energy duration is $T_{\mathrm{d}} = 23.84$ s, respectively.

In general, the concept of duration is often used in conjunction with the response spectrum or PGA. Therefore, relative duration is considered superior to absolute duration. In addition, among the various definitions of relative duration, the energy duration is the most widely accepted and commonly applied. The influence of strong-motion duration on structural response is primarily manifested during the nonlinear response stage of a structure. Research findings (Hu 2006) indicate that for inelastic structural systems, a longer strong-motion duration leads to larger structural responses, which generally increase by about 10 ~ 20%. For structures whose

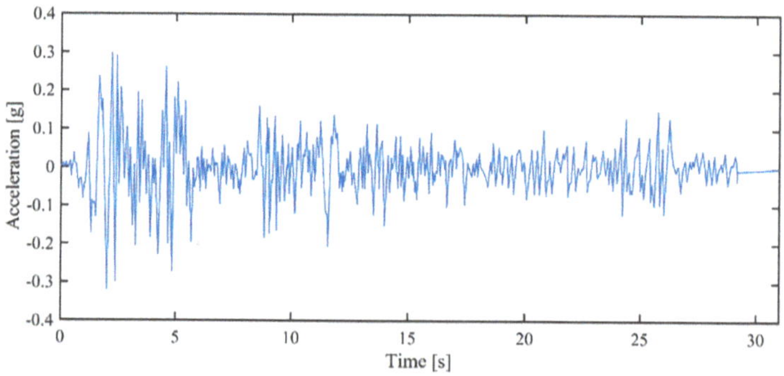

Fig. 2.1 El-Centro strong motion record (N-S)

restoring force curve has a descending branch, the increase in response is even more significant. However, the most critical effect of strong-motion duration on structural safety is the cumulative damage effect, or low-cycle fatigue failure, of which a typical example is soil liquefaction phenomenon.

In the numerical simulation or artificial generation of ground motion acceleration processes, it is necessary to specify the simulated duration T and the time interval Δt. Statistical data on the simulated duration T (sometimes referred to as the total duration or duration) of ground motion time-histories are very limited. As such, T is typically set to $1.5 \sim 2$ times the duration of the strong motions, i.e., $T = (1.5 \sim 2)T_{\text{d}}$. In addition, whether for strong motion records or artificially generated ground motions, T is generally taken as 5 to 10 times the fundamental period of the structure, meaning that the displacement at the top of the structure can undergo approximately 5 ~ 10 full cycles corresponding to its fundamental period. In summary, the duration T of a ground motion time-history should not be less than 5 times the structural fundamental period or 15 s. The time interval Δt can be taken as 0.01 s or 0.02 s commonly.

For simplicity in engineering practice, the "Code for Seismic Design of Buildings" (GB 50011-2010 2016) provides recommended values of the simulated duration T for the five classifications of soil sites. The duration increases gradually from the first to the third design earthquake group within each soil site. Specifically: for soil site I_0, $T = 15 \sim 20$ s; for soil site I_1, $T = 20 \sim 25$ s; for soil site II, $T = 25 \sim 30$ s; for soil site III, $T = 30 \sim 35$ s; and for soil site IV, $T = 35 \sim 40$ s.

2.3 Spectral Characteristics of Earthquake Ground Motion

The spectral composition of seismic ground motion has a significant influence on the dynamic responses of structures. In earthquake engineering, the commonly used spectrum includes power spectrum, Fourier spectrum, and response spectrum. Among these, the response spectrum establishes the relationship between the dynamic characteristics of structures and ground motion properties through the maximum seismic responses of single-degree-of-freedom (SDOF) elastic systems. The Fourier spectrum and power spectrum are fundamental tools used in random vibration analysis to describe the spectral characteristics of stochastic processes, directly reflecting the energy contribution of different frequency components. These three spectra can be theoretically converted into one another under certain assumptions. This section will introduce these three types of spectra and their relationships.

2.3.1 Power Spectrum

Statistical analyses of numerous earthquake records indicate that the mean value of the ground motion acceleration process is approximately zero. Therefore, the second-order statistical characteristics of a stationary ground motion acceleration process can be represented by either the power spectral density (PSD) function or the auto-correlation function. In earthquake engineering, commonly used PSD models primarily include the white noise model, finite bandwidth white noise model, Kanai-Tajimi model, Hu-Zhou model, and Clough-Penzien model. The white noise model assumes that seismic energy is uniformly distributed across the entire frequency range and increases indefinitely over time. Although this assumption does not reflect actual earthquake behavior, it marked the first application of stochastic process theory to describe seismic ground motion. The finite bandwidth white noise model modifies the definition domain of the white noise model but still assumes that the seismic energy is uniformly distributed within a finite frequency range, which still differs significantly from observed earthquake records. To address this, Kanai and Tajimi first proposed the filtered white noise model, which accounts for the influence of soil site properties on the spectral characteristics of ground motion and provides clear physical meaning. This model is now widely known as the Kanai-Tajimi model. Subsequent models such as the Hu-Zhou model and the Clough-Penzien model can be regarded as extensions or improvements of the Kanai-Tajimi model. To date, the Kanai-Tajimi model and its improved versions have been extensively adopted in earthquake engineering, as detailed below:

Kanai-Tajimi Model (K-T Model)

The Kanai-Tajimi model (K-T model) is given as follows (Kanai 1957; Tajimi 1960):

$$S_{\text{K-T}}(\omega) = \frac{\omega_g^4 + 4\zeta_g^2\omega_g^2\omega^2}{\left(\omega^2 - \omega_g^2\right)^2 + 4\zeta_g^2\omega_g^2\omega^2} S_0 \tag{2.6}$$

where ω_g and ζ_g denote the predominant circular frequency and damping ratio of the soil layer, respectively. S_0 represents the PSD of the acceleration white noise process of the bedrock, reflecting the intensity of the ground motion. It is often referred to as the spectrum intensity factor.

In the K-T model, the stationary ground motion acceleration process $\overline{a}(t)$ can be expressed as:

$$\begin{cases} \overline{a}(t) = \ddot{X}_g(t) + w(t) \\ \ddot{X}_g(t) + 2\zeta_g\omega_g\dot{X}_g(t) + \omega_g^2 X_g(t) = -w(t) \end{cases} \tag{2.7}$$

where $\ddot{X}_g(t)$ is the ground acceleration relative to the bedrock. $w(t)$ represents the bedrock acceleration process, which is assumed to be a stationary white noise process with zero-mean and the PSD of S_0.

The K-T model is single-filtered white noise model that effectively represents the filtering effects of a uniform soil layer on seismic waves. This model has clear physical meaning. However, it tends to overestimate the low-frequency content of ground motion. Furthermore, it does not satisfy the condition that ground motion velocity and displacement remain bounded (Kanai 1957; Tajimi 1960). To address these limitations, many researchers in China and abroad have proposed improvements to the K-T model. Two typical double-filtered white noise models are described below.

Hu-Zhou Model (H-Z Model)

The Hu-Zhou model (H-Z model) is presented as below (Hu and Zhou 1962):

$$S_{\text{H-Z}}(\omega) = \frac{\omega_g^4 + 4\zeta_g^2\omega_g^2\omega^2}{\left(\omega^2 - \omega_g^2\right)^2 + 4\zeta_g^2\omega_g^2\omega^2} \times \frac{\omega^{2n}}{\omega^{2n} + \omega_c^{2n}} S_0 \tag{2.8}$$

where ω_c is a parameter that controls the low-frequency content of ground motion. A larger value of ω_c indicates a smaller low-frequency content. The parameter n can be taken as 1, 2, or 3. In practice, it is commonly recommended to take $n = 3$ and $\omega_c = 2.0$ rad/s (Hong et al. 1994).

In fact, the stationary ground motion acceleration process $\overline{a}(t)$ in the H-Z model can be given by (Liu 2007):

$$\begin{cases} \overline{a}(t) = \ddot{X}_g(t) + \ddot{U}(t) \\ \ddot{X}_g(t) + 2\zeta_g\omega_g\dot{X}_g(t) + \omega_g^2 X_g(t) = -\ddot{U}(t) \\ \ddot{U}(t) = \dddot{Y}(t), \quad \dddot{Y}(t) + \omega_c^3 Y(t) = w(t) \end{cases} \tag{2.9}$$

This indicates that the ground motion bedrock acceleration process $\ddot{U}(t)$ is a filtered white noise process. The H-Z model is derived by applying an additional filter to this already filtered white spectrum, thus constituting a double-filtered spectrum.

Clough-Penzien Model (C-P Model)

The Clough-Penzien model (C-P model) is expressed as follows (Clough and Penzien 1975):

$$S_{\text{C-P}}(\omega) = \frac{\omega_g^4 + 4\zeta_g^2\omega_g^2\omega^2}{\left(\omega^2 - \omega_g^2\right)^2 + 4\zeta_g^2\omega_g^2\omega^2} \times \frac{\omega^4}{\left(\omega^2 - \omega_f^2\right)^2 + 4\zeta_f^2\omega_f^2\omega^2} S_0 \tag{2.10}$$

Fig. 2.2 Comparison of the three typical PSD models ($\omega_g = 15.71$rad/s, $\zeta_g = 0.72$, $S_0 = 100\text{cm}^2/\text{s}^3$)

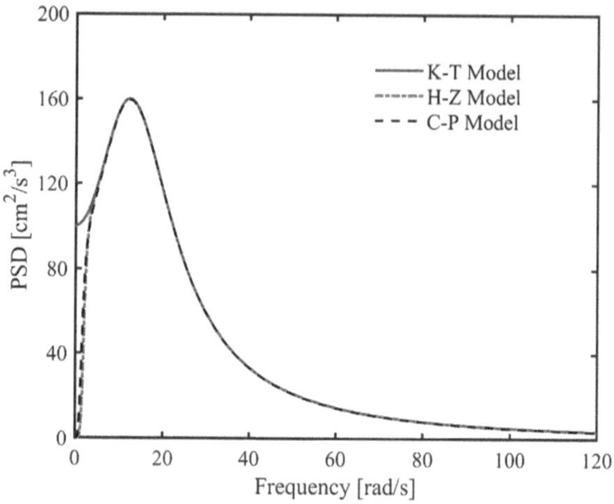

where ω_f and ζ_f denote the predominant circular frequency and damping ratio of the bedrock, respectively. It is recommended to take $\omega_f = 0.1\omega_g$, $\zeta_f = \zeta_g$ (Deodatis 1996). The C-P model assumes both the bedrock and the site soil to behave as second-order linear filters. It represents a stationary double-filtered white noise model, which effectively suppresses low-frequency component of ground motions (Clough and Penzien 1975).

The stationary ground motion acceleration process $\overline{a}(t)$ in the C-P model can be expressed as:

$$\begin{cases} \overline{a}(t) = \ddot{X}_g(t) + \ddot{X}_f(t) \\ \ddot{X}_g(t) + 2\zeta_g\omega_g\dot{X}_g(t) + \omega_g^2 X_g(t) = -\ddot{X}_f(t) \\ \ddot{X}_f(t) + 2\zeta_f\omega_f\dot{X}_f(t) + \omega_f^2 X_f(t) = -w(t) \end{cases} \tag{2.11}$$

In addition, many researchers have proposed other modified models, such as Du model (Du 1998) and so forth. Figure 2.2 presents a comparison among the three typical PSD models mentioned above. It can be seen from the figure that the H-Z and C-P models exhibit nearly identical spectral shapes, both introducing modifications mainly to the low-frequency portion of the K-T model.

The parameter values for the aforementioned PSD models are provided below. To be consistent with the "Code for Seismic Design of Buildings" (GB50011-2010 2016), the predominant circular frequency of the site soil can be calculated by the following expression:

$$\omega_g = \frac{2\pi}{T_g} \tag{2.12}$$

where T_g is the characteristic period, of which the values specified in the "Code for Seismic Design of Buildings" (GB50011-2010 2016) are shown in Table 2.4.

Table 2.4 The specified values of characteristic period T_g (s)

Design earthquake groups	Classification of soil sites				
	I_0	I_1	II	III	IV
The first group	0.20	0.25	0.35	0.45	0.65
The second group	0.25	0.30	0.40	0.55	0.75
The third group	0.30	0.35	0.45	0.65	0.90

In the PSD models given by Eqs. (2.6), (2.8), and (2.10), the formula for calculating the spectral intensity factor is given as follows (Hong et al. 1994):

$$S_0 = \frac{a_{\mathrm{max}}^2}{r^2 \omega_{\mathrm{e}}} \tag{2.13}$$

where a_{max} is the mean value of the PGA, of which the value can be directly taken according to Sect. 2.1. The parameter r is the peak factor. ω_{e} can be obtained from the mean square value $\sigma_{\overline{a}}^2$ of the stationary ground motion acceleration process $\overline{a}(t)$ as follows:

$$\omega_{\mathrm{e}} = \frac{\sigma_{\overline{a}}^2}{S_0} = \frac{R(0)}{S_0} = \frac{1}{S_0} \int_{-\infty}^{\infty} S(\omega)\,\mathrm{d}\omega \tag{2.14}$$

where $S(\omega)$ denotes the two-sided PSD of the stationary ground motion acceleration process described above.

In the K-T model, the spectral intensity factor is given by:

$$S_0 = \frac{a_{\mathrm{max}}^2}{r^2 \pi \omega_{\mathrm{g}} \left(2\zeta_{\mathrm{g}} + \frac{1}{2\zeta_{\mathrm{g}}}\right)} \tag{2.15}$$

For convenience, the expression of the spectral intensity factor given in the K-T model (Eq. (2.15)) is sometimes directly used to calculate the spectral intensity factors of other PSD models. The resulting error is generally small and acceptable.

The calculation formulas for the peak factor r are given as follows:

$$r = \sqrt{2\ln(\nu T_{\mathrm{d}})} + \frac{0.5772}{\sqrt{2\ln(\nu T_{\mathrm{d}})}} \tag{2.16}$$

$$\nu = \frac{1}{\pi}\sqrt{\lambda_2/\lambda_0} \tag{2.17}$$

$$\lambda_k = \int_{-\infty}^{\infty} \omega^k S(\omega)\,\mathrm{d}\omega, \quad k = 0 \text{ or } 2 \tag{2.18}$$

Table 2.5 Parameter values for stationary ground motion PSD models

Design earthquake groups		Classification of soil sites				
		I_0	I_1	II	III	IV
ω_g (rad/s)	The first group	31.42	25.13	17.95	13.96	9.67
	The second group	25.13	20.94	15.71	11.42	8.38
	The third group	20.94	17.95	13.96	9.67	6.98
ζ_g		0.56	0.64	0.72	0.80	0.90

where T_d is the stationary duration of the ground motion, and v is the zero-crossing rate, respectively.

It is worth noting that for certain PSD models of ground motion, the computed second-order spectral moment λ_2 may become infinitely large. Therefore, in most practical cases, it is not feasible to directly compute the peak factor using Eq. (2.16). However, although the peak factor is a function of the mean zero-crossing rate and the ground motion duration, it is relatively insensitive to variations in these parameters. Therefore, it is generally acceptable to adopt a value of $r = 3.0$ or to determine it by fitting the design response spectrum.

For practical use, and based on Eq. (2.12), Table 2.4, and existing studies (Ou et al. 1991; Xue et al. 2003; Liu et al. 2019), recommended parameter values for stationary ground motion PSD models are provided in Table 2.5.

It should be noted that, according to the "Code for Seismic Design of Buildings" (GB50011-2010 2016), the characteristic period T_g must be increased by 0.05 s when evaluating rare earthquake actions. Therefore, the values of predominant circular frequency ω_g given in Table 2.5 only apply to frequent (minor) and design (moderate) earthquakes. For rare earthquakes, the values should be adjusted using Eq. (2.12).

2.3.2 *Response Spectrum*

The earthquake response spectrum establishes a bridge between the characteristics of ground motion and the dynamic responses of structures. Specifically, the response spectrum $S(\omega, \zeta)$ refers to the relationship between the maximum response of a set of SDOF elastic systems (with various natural frequencies ω and the same damping ratio ζ) subjected to a specific ground motion acceleration time-history $a(t)$, and the dynamic characteristics of the systems. Based on the dynamic differential equation of an SDOF elastic system, the relative displacement $x(t)$, relative velocity $\dot{x}(t)$, and absolute acceleration $\ddot{y}(t) = \ddot{x}(t) + a(t)$ of a single-mass system can be derived as follows:

$$x(t) = -\frac{1}{\omega_D}\int_0^t a(\tau)\,\mathrm{e}^{-\zeta\omega(t-\tau)}\sin[\omega_D(t-\tau)]\mathrm{d}\tau \tag{2.19}$$

$$\dot{x}(t) = -\frac{\omega}{\omega_D}\int_0^t a(\tau)\,\mathrm{e}^{-\zeta\omega(t-\tau)}\cos[\omega_D(t-\tau)+\alpha]\mathrm{d}\tau \tag{2.20}$$

$$\ddot{y}(t) = \frac{\omega^2}{\omega_D}\int_0^t a(\tau)\,\mathrm{e}^{-\zeta\omega(t-\tau)}\sin[\omega_D(t-\tau)+2\alpha]\mathrm{d}\tau \tag{2.21}$$

where $\omega_D = \omega\sqrt{1-\zeta^2}$ is the damped natural frequency of the SDOF elastic system, $\tan\alpha = \zeta/\sqrt{1-\zeta^2}$. Since the damping ratio ζ is typically much smaller than 1 (e.g., $\zeta = 0.02$ for steel structures and $\zeta = 0.05$ for reinforced-concrete structures), the approximations $\omega_D \approx \omega$ and $\alpha \approx 0$ are generally acceptable.

First, considering only the maximum seismic responses, Eqs. (2.19)–(2.21) yield:

$$|x(t)|_{\max} = \left|\frac{1}{\omega}\int_0^t a(\tau)\,\mathrm{e}^{-\zeta\omega(t-\tau)}\sin[\omega(t-\tau)]\mathrm{d}\tau\right|_{\max} \tag{2.22}$$

$$|\dot{x}(t)|_{\max} = \left|\int_0^t a(\tau)\,\mathrm{e}^{-\zeta\omega(t-\tau)}\cos[\omega(t-\tau)]\mathrm{d}\tau\right|_{\max} \tag{2.23}$$

$$|\ddot{y}(t)|_{\max} = \omega^2|x(t)|_{\max} \tag{2.24}$$

Next, by allowing the natural frequency ω to vary continuouxsly along the frequency axis, we can obtain the continuous set (envelope) of maximum responses corresponding to different natural frequencies:

$$S_d(\omega,\zeta) = |x(t)|_{\max} \tag{2.25}$$

$$S_v(\omega,\zeta) = |\dot{x}(t)|_{\max} \tag{2.26}$$

$$S_a(\omega,\zeta) = \omega^2 S_d(\omega,\zeta) \tag{2.27}$$

where $S_d(\omega,\zeta)$, $S_v(\omega,\zeta)$ and $S_a(\omega,\zeta)$ denote the response spectrum of the relative displacement, relative velocity, and absolute acceleration, respectively.

When the duration of the ground motion time-history is sufficiently long, the sin $[\omega(t-\tau)]$ and $\cos[\omega(t-\tau)]$ inside the integrals in Eqs. (2.22) and (2.23) differ just by a phase shift of $\pi/2$ over multiple cycles. Further, when only the maximum values are considered, interchanging cos and sin introduces negligible errors. Thus, Eqs. (2.25)–(2.27) can be rewritten as:

$$S_d(\omega,\zeta) = \frac{1}{\omega}\overline{S}_v(\omega,\zeta) \tag{2.28}$$

$$\overline{S}_{v}(\omega,\zeta)=\left|\int_{0}^{t}a(\tau)\mathrm{e}^{-\zeta\omega(t-\tau)}\sin[\omega(t-\tau)]\mathrm{d}\tau\right|_{\max} \tag{2.29}$$

$$S_{a}(\omega,\zeta)=\omega\overline{S}_{v}(\omega,\zeta) \tag{2.30}$$

where $\overline{S}_{v}(\omega,\zeta)$ is the Pseudo-Velocity Spectrum, which generally shows little difference from the true relative velocity response spectrum.

The response spectrum curve is essentially the result of the maximum dynamic responses of a series of SDOF systems with different dynamic characteristics subjected to the same ground motion time-history excitation. Therefore, the response spectrum curve does not represent the characteristics of any specific structure but primarily reflects the spectral characteristics of the ground motion. In other words, the shape of the response spectrum indicates the relative contribution of different frequency components in the ground motion time-history. Furthermore, the earthquake response spectrum describes ground motion characteristics from the unique perspective of the maximum dynamic responses of SDOF elastic systems "sweeping" along the frequency axis according to their dynamic properties. In this way, it establishes a "bridge" between the properties of the ground motion and the structural dynamic responses (Li and Li 1992).

In practical applications, the natural vibration period T_0 of the structure is commonly used to represent the seismic response spectrum. Since $\omega = 2\pi T_0$, the aforementioned response spectrum can also be rewritten as $S_d(T_0,\zeta)$, $S_v(T_0,\zeta)$, and $S_a(T_0,\zeta)$, respectively. Meanwhile, to facilitate comparison and application, response spectrum values are often normalized by dividing by the maximum value of the corresponding physical quantity of the ground motion, resulting in a normalized response spectrum or dynamic amplification factor. For example, the acceleration dynamic amplification factor is defined as:

$$\beta(T_0)=\frac{S_a(T_0,\zeta)}{|a(t)|_{\max}} \tag{2.31}$$

Figure 2.3 shows the response spectrum of absolute acceleration, relative velocity, and relative displacement with respect of the SDOF elastic systems with different damping ratios under the north-south (N-S) component of the El Centro earthquake record.

The factors influencing ground motion are complex and variable, and earthquake records differ significantly from one another, exhibiting strong randomness. Consequently, response spectrum curves also show considerable dispersion. In seismic design for engineering structures, it is desirable to establish a standardized representation of ground motion actions that ensures a certain level of safety-thus giving rise to the design (codes) response spectrum. A simple and straightforward approach is to statistically average the response spectrum curves from a large number of strong motion records, apply mathematical smoothing techniques, and then make adjustments based on both safety and economic considerations, ultimately forming the design response spectrum. It should be noted that the design response spectrum does

Fig. 2.3 Response spectrum of El Centro earthquake record (N-S component)

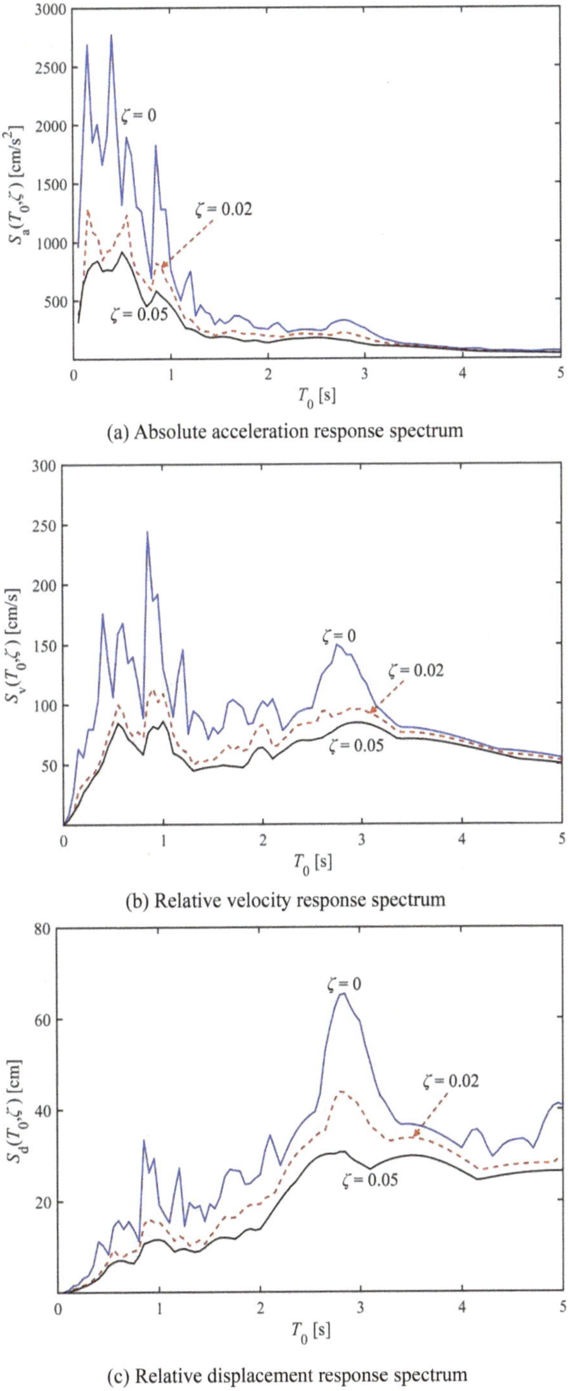

(a) Absolute acceleration response spectrum

(b) Relative velocity response spectrum

(c) Relative displacement response spectrum

not represent the characteristics of any specific measured motion record. Instead, it reflects a comprehensive representation based on the collective behavior of numerous earthquake ground motions, thereby serving as a standardized provision for seismic action in structural design.

In summary, the response spectrum primarily reflects the spectral characteristics of ground motion, while the design response spectrum serves as a prescribed representation of seismic action for structural seismic design. The unifying aspect between them lies in the fact that both describe the relationship between the maximum seismic response and the natural vibration period (or frequency) of SDOF elastic systems. The provisions related to the design response spectrum in "Code for Seismic Design of Buildings" (GB50011-2010 2016) of China will be introduced in detail in Sect. 2.3.4.

2.3.3 *Fourier Spectrum*

For such irregular vibrations as ground motion, it is challenging to directly identify the periodic components and their relative contributions. Fourier analysis decomposes seemingly irregular functions into periodic harmonic components-sine and cosine waves-making it easier to examine their frequency-domain characteristics. The following section will briefly introduce the Fourier spectrum based on the discrete Fourier transform and the Fourier integral (Fourier transform) (Ohsaki 2008).

Discrete Fourier Transform

In practices, the ground motion time-history $a(t)$ is usually represented by N discrete values $a_m = a(t_m)$ measured at specific time instants $t_m = m\Delta t$ ($m = 0, 1, 2, \cdots, N - 1$). In this case, a discrete Fourier transform is needed to conduct Fourier analysis. Here, Δt is the time interval of the ground motion acceleration process, and the duration of the ground motion is $T = N\Delta t$, where N is the number of sampling points (typically an even number).

If the ground motion time-history $a(t)$ is sampled at equally spaced instants, yielding N sample values $a_m (m = 0, 1, 2, \cdots, N - 1)$, then define:

$$\begin{cases} A_k = \dfrac{2}{N} \displaystyle\sum_{m=0}^{N-1} a_m \cos \dfrac{2\pi km}{N}, & k = 0, 1, 2, \cdots, \dfrac{N}{2} - 1, \dfrac{N}{2} \\ B_k = \dfrac{2}{N} \displaystyle\sum_{m=0}^{N-1} a_m \sin \dfrac{2\pi km}{N}, & k = 1, 2, \cdots, \dfrac{N}{2} - 1 \end{cases} \tag{2.32}$$

then a_m can be expressed as a finite Fourier series (in real form) with coefficients A_k and B_k:

$$a_m = \frac{A_0}{2} + \sum_{k=1}^{N/2-1}\left(A_k \cos\frac{2\pi km}{N} + B_k \sin\frac{2\pi km}{N}\right) + \frac{A_{N/2}}{2}\cos\frac{2\pi(N/2)m}{N},\quad m = 0,1,2,\cdots,N-1 \tag{2.33}$$

When the sample point interval is Δt, the time instant of the m-th sample point is $t = m\Delta t$. Thus, the right-hand side of Eq. (2.33) becomes a function of time t. Clearly, this time function will completely pass through the N discrete sample values a_m, but there is no guarantee that it will match the original function $a(t)$ between the sample points. Therefore, it is only an approximation, known as the **finite Fourier approximation** of the original function:

$$\tilde{a}(t) = \frac{A_0}{2} + \sum_{k=1}^{N/2-1}\left(A_k \cos\frac{2\pi kt}{N\Delta t} + B_k \sin\frac{2\pi kt}{N\Delta t}\right) + \frac{A_{N/2}}{2}\cos\frac{2\pi(N/2)t}{N\Delta t} \tag{2.34}$$

Furthermore, by setting $A_k = G_k \cos\varphi_k$, $B_k = G_k \sin\varphi_k$, and considering $\omega_k = k\Delta\omega = 2\pi k/T = 2\pi k/(N\Delta t)$, Eq. (2.34) can be rewritten as:

$$\tilde{a}(t) = \frac{G_0}{2} + \sum_{k=1}^{N/2-1} G_k \cos(\omega_k t - \varphi_k) + \frac{G_{N/2}}{2}\cos\left(\omega_{N/2} t\right) \tag{2.35}$$

where the first term $G_0/2$ represents the mean value of the samples, indicating the offset of the entire curve from the zero line. It does not represent oscillatory motion.

In Eq. (2.35), $G_k = \sqrt{A_k^2 + B_k^2}$ represents the amplitude of the frequency component corresponding to ω_k, and $\varphi_k = \arctan\left(\frac{B_k}{A_k}\right)$ represents the phase angle of ω_k, respectively. Note that the range of φ_k is $(-\pi, \pi)$, which is not restricted to the principal value of the arctangent function. In addition, the minimum frequency (or fundamental frequency) is $\omega_{\min} = \Delta\omega = \frac{2\pi}{T} = \frac{2\pi}{N\Delta t}$, while the maximum frequency (or Nyquist frequency) is $\omega_{\max} = \omega_{N/2} = \frac{N}{2}\Delta\omega = \frac{\pi}{\Delta t}$.

It should be noted that the **Fourier amplitude spectrum** is generally not represented directly by the amplitude G_k ($k = 1, 2, \cdots, N/2 - 1$) and the corresponding frequency ω_k, but rather by the product of G_k with $T/2$ or $N\Delta t/2$. Therefore, the dimension of the Fourier amplitude spectrum corresponds to the product of the dimension upon the time-history amplitude and the dimension upon the time. Similarly, a plot of the phase angle φ_k versus the corresponding frequency ω_k is called the **Fourier phase spectrum**.

In fact, by applying Euler's formula and defining the complex amplitude $F_k = (A_k - \mathrm{i}\,B_k)/2$, the discrete Fourier transform can be written in complex form as:

$$a_m = \sum_{k=0}^{N-1} F_k e^{i(2\pi km/N)}, \quad m = 0, 1, 2, \cdots, N-1 \tag{2.36}$$

$$F_k = \frac{1}{N} \sum_{m=0}^{N-1} a_m e^{-i(2\pi km/N)}, \quad k = 0, 1, 2, \cdots, N-1 \tag{2.37}$$

Equations (2.36) and (2.37) constitute a discrete Fourier transform pair. Clearly, the amplitude satisfies $|F_k| = G_k/2$.

Fourier Integral (Fourier Transform)

For a non-periodic function with finite energy over a finite duration, Fourier analysis can be performed using the Fourier integral or Fourier transform. In this case, the ground motion time-history $a(t)$ is treated as a continuous function of time t, and its Fourier integral form is written as:

$$a(t) = \frac{1}{2\pi} \int_{-\infty}^{\infty} \left\{ \int_{-\infty}^{\infty} a(\tau) e^{-i\omega\tau} d\tau \right\} e^{i\omega t} d\omega \tag{2.38}$$

Then Eq. (2.38) forms a Fourier transform pair, such that:

$$a(t) = \frac{1}{2\pi} \int_{-\infty}^{\infty} F(\omega) e^{i\omega t} d\omega \tag{2.39}$$

$$F(\omega) = \int_{-\infty}^{\infty} a(t) e^{-i\omega t} dt \tag{2.40}$$

From Eq. (2.39), it can be observed that $a(t)$ is a superposition of harmonic functions $e^{i\omega t}$ with complex amplitude of $\frac{1}{2\pi} F(\omega) d\omega = \frac{F(\omega)}{T}$. This indicates that the complex amplitude F_k in Eq. (2.36) is equivalent to the complex amplitude $F(\omega)/T$ in Eq. (2.39). The Fourier transform $F(\omega)$ is typically a continuous spectrum in complex form, commonly referred to as the **Fourier spectrum**. For a real function $a(t)$, $F(\omega)$ is conjugate symmetric, say its real part is an even function of ω, and its imaginary part is an odd function.

Now, the Fourier spectrum $F(\omega)$ can be expressed in terms of its real and imaginary parts, as well as in terms of its amplitude and phase:

$$F(\omega) = A(\omega) - iB(\omega) = G(\omega) e^{-i\varphi(\omega)} \tag{2.41}$$

Taking into account the conjugate symmetry, Eq. (2.39) can be expressed as:

$$a(t) = \frac{1}{\pi}\int_0^{\infty}[A(\omega)\cos\omega t + B(\omega)\sin\omega t]\,d\omega \\ = \frac{1}{\pi}\int_0^{\infty}G(\omega)\cos[\omega t - \varphi(\omega)]\,d\omega \tag{2.42}$$

where $G(\omega) = |F(\omega)| = \sqrt{A^2(\omega) + B^2(\omega)}$ is the **Fourier amplitude spectrum**, $\varphi(\omega) = \arctan\frac{B(\omega)}{A(\omega)}$ is the **Fourier phase spectrum**. $A(\omega)$ denotes the Fourier cosine transform, and $B(\omega)$ is the Fourier sine transform, specifically:

$$A(\omega) = \int_{-\infty}^{\infty} a(\tau)\cos\omega\tau\,d\tau = \mathrm{Re}[F(\omega)] \tag{2.43}$$

$$B(\omega) = \int_{-\infty}^{\infty} a(\tau)\sin\omega\tau\,d\tau = -\mathrm{Im}[F(\omega)] \tag{2.44}$$

It can be seen that the Fourier amplitude spectrum $G_kT/2$ obtained from the discrete Fourier transform is fully equivalent to the amplitude spectrum $G(\omega)$ derived from the continuous Fourier transform. The Fourier amplitude spectrum $G(\omega)$ (or $G_kT/2$) and the Fourier phase spectrum $\varphi(\omega)$ together provide a comprehensive description of the spectral characteristics of a ground motion process from two different perspectives. They not only reflect the distribution of seismic energy in the frequency domain-showing the energy carried by harmonic vibrations at different frequencies-but also include information about the phase distribution of each frequency component. This distinguishes the Fourier spectrum from the response spectrum and the PSD.

2.3.4 *Relationships Among the Three Spectra*

Response Spectrum and Power Spectrum

When generating artificial ground motion acceleration time-histories to match the design response spectrum specified in the "Code for Seismic Design of Buildings" (GB50011-2010 2016) of China, it is necessary to utilize the approximate conversion relationship between the acceleration response spectrum $S_a(\omega, \zeta)$ and the PSD $S(\omega)$ (Kaul 1978):

$$S(\omega) = -\frac{\zeta}{\pi\omega}S_a^2(\omega,\zeta) \times \frac{1}{\ln\left[-\frac{\pi}{\omega T}\ln(1-p)\right]} \tag{2.45}$$

where T is the ground motion duration, and p is the exceeding probability associated with the response spectrum, respectively.

Table 2.6 The maximum value of the horizontal seismic influence coefficient α_{max}

	6-degree	7-degree		8-degree		9-degree
Earthquake action	0.05g	0.10g	0.15g	0.20g	0.30g	0.40g
Frequent ground motion	0.04	0.08	0.12	0.16	0.24	0.32
Basis ground motion	0.12	0.23	0.34	0.45	0.68	0.90
Rare ground motion	0.28	0.50	0.72	0.90	1.20	1.40

In the "Code for Seismic Design of Buildings" (GB50011-2010 2016) of China, the design response spectrum is given by:

$$S_a(\omega,\zeta)=S_a(T_0,\zeta)=\begin{cases}[0.45+(10\eta_2-4.5)T_0]\alpha_{max}g, & 0<T_0\le 0.1\\ \eta_2\alpha_{max}g, & 0.1<T_0\le T_g\\ (T_g/T_0)^{\gamma}\eta_2\alpha_{max}g, & T_g<T_0\le 5T_g\\ [\eta_2 0.2^{\gamma}-\eta_1(T_0-5T_g)]\alpha_{max}g, & 5T_g<T_0\le 6.0\end{cases} \tag{2.46}$$

In this equation:

α_{max}—the maximum value of the horizontal seismic influence coefficient, as listed in Table 2.6,
T_0—Natural vibration period of the structure (unit: s);
g—Acceleration due to gravity, g ≈ 1000 cm/s^2;
T_g—Characteristic period of the site soil, as given in Table 2.4;
γ—Attenuation index of the descending segment of the response spectrum curve;
η_1—Adjustment coefficient for the slope of the linear descending segment, when $\eta_1<0$, it should be taken as 0;
η_2—Damping adjustment factor, when $\eta_2<0.55$, it should be taken as 0.55.

In Eq. (2.46) for the design response spectrum, the parameters γ, η_1 and η_2 can be computed respectively as follows:

$$\gamma=0.9+\frac{0.05-\zeta}{0.3+6\zeta}$$

$$\eta_1=0.02+\frac{0.05-\zeta}{4+32\zeta}\quad \text{and}\quad \eta_1\ge 0$$

$$\eta_2=1+\frac{0.05-\zeta}{0.08+1.6\zeta}\quad \text{and}\quad \eta_2\ge 0.55$$

Figure 2.4 shows the design response spectrum curve specified in the "Code for Seismic Design of Buildings" (GB50011-2010 2016) of China. As illustrated by the curves: the first segment (0 ~ 0.1 s) is a steeply rising straight line, the second segment (0.1 s~T_g) is a horizontal straight line, the third segment (T_g~$5T_g$) is a

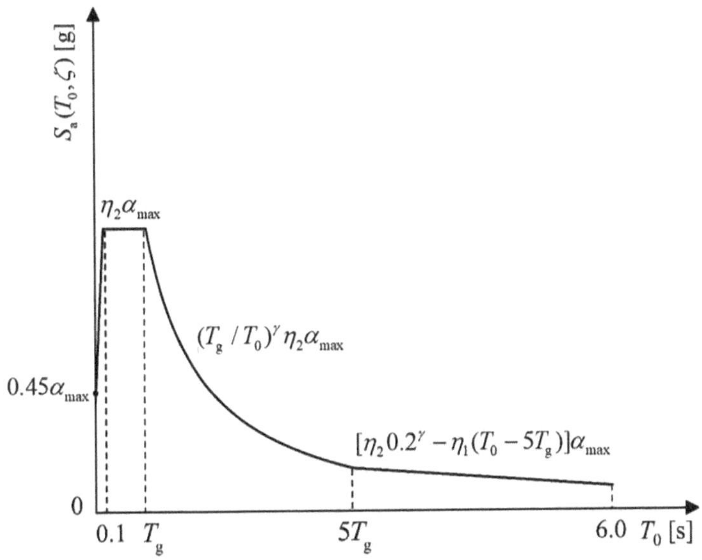

Fig. 2.4 Design response spectrum curve in the "Code for Seismic Design of Buildings" of China

rapidly descending curve, and the fourth segment ($5T_g$~6.0s) is a gently descending straight line.

To adequately capture the characteristics of the design response spectrum, the discrete control points selected along the spectrum should be distributed as uniformly as possible on a logarithmic scale of the period. Therefore, the linear horizontal axis of the design response spectrum can be converted into a logarithmic coordinate, expressed as:

$$\lg T_{0,k} = \lg T_{0,1} + \frac{1 - \lg T_{0,1}}{M - 1}(k - 1), \quad k = 1, 2, \cdots, M \tag{2.47}$$

where lg is the base-10 logarithm, $T_{0,\,k}$ denotes the k-th control point of the period, M is the total number of discrete control points within the 10-second period range. For the acceleration response spectrum values in the period range of T_0=6 s ~ 10 s, the absolute acceleration response spectrum $S_a(T_0, \zeta)$ at $T_0 = 10$ s can be taken as $0.14\alpha_{max}g$ ($\zeta = 0.05$) or $0.18\alpha_{max}g$ ($\zeta = 0.02$). For periods between 6 s $< T_0 <$ 10 s, linear interpolation may be applied.

In the numerical simulation or artificial generation of ground motion acceleration time-histories, the PSD function is typically discretized at equal frequency intervals, with the total number of discrete points N being much bigger than the total number of control points M on the response spectrum curve. As such, when converting the design response spectrum into a PSD, the discrete control points $T_{0,\,k}$ on the design response spectrum curve are first determined using Eq. (2.47). These are then converted to control points expressed in terms of circular frequency via $\omega_k = 2\pi/$

$T_{0,\,k}$. Next, using Eq. (2.45), the PSD values at each control point can be calculated. Finally, the PSD values at equal frequency intervals are obtained through linear interpolation, i.e.,

$$S(\omega_i) = S(\omega_k) + \frac{\omega_i - \omega_k}{\omega_{k+1} - \omega_k}[S(\omega_{k+1}) - S(\omega_k)], \quad i = 1, 2, \cdots, N; \; k = 1, 2, \cdots, M \tag{2.48}$$

where $S(\omega_k)$ and $S(\omega_{k+1})$ denote the PSD values at the k-th and (k+1)-th control points, respectively. $S(\omega_i)$ denotes the PSD value at the i-th point arranged at equal frequency intervals, in which $\omega_i = i\Delta\omega$.

In general, once the PSD of the ground motion acceleration process is obtained using Eq. (2.45), it becomes straightforward to generate an ensemble of ground motion acceleration time-histories. These can then be used to compute the seismic acceleration response spectrum. If the generated response spectrum does not satisfactorily match the target design response spectrum, iterative adjustment methods may be applied until the desired accuracy is achieved.

Fourier Amplitude Spectrum and Response Spectrum

From Eqs. (2.23) and (2.26), the relative velocity response spectrum for an undamped SDOF elastic system can be expressed as:

$$\begin{aligned} S_{\mathrm{v}}(\omega, 0) &= |\dot{x}(t)|_{\max} \\ &= \left| \int_0^t a(\tau) \cos\omega(t-\tau) \mathrm{d}\tau \right|_{\max} \\ &= \left| \cos(\omega t) \int_0^t a(\tau)\cos(\omega\tau)\mathrm{d}\tau + \sin(\omega t) \int_0^t a(\tau)\sin(\omega\tau)\mathrm{d}\tau \right|_{\max} \\ &= \left| \sqrt{\left[\int_0^t a(\tau)\cos(\omega\tau)\mathrm{d}\tau\right]^2 + \left[\int_0^t a(\tau)\sin(\omega\tau)\mathrm{d}\tau\right]^2} \cos[\omega t - \theta(t)] \right|_{\max} \end{aligned} \tag{2.49}$$

where

$$\theta(t) = \arctan \frac{\int_0^t a(\tau)\sin(\omega\tau)\mathrm{d}\tau}{\int_0^t a(\tau)\cos(\omega\tau)\mathrm{d}\tau}$$

Considering Eqs. (2.43) and (2.44), the Fourier amplitude spectrum of a ground motion acceleration time-history $a(\tau)$ with duration T is given by:

$$
\begin{aligned}
G(\omega) &= |F(\omega)| \\
&= \sqrt{A^2(\omega) + B^2(\omega)} \\
&= \sqrt{\left[\int_0^T a(\tau)\cos(\omega\tau)\mathrm{d}\tau\right]^2 + \left[\int_0^T a(\tau)\sin(\omega\tau)\mathrm{d}\tau\right]^2}
\end{aligned}
\tag{2.50}
$$

Since ω represents the circular frequency, $2\pi/\omega$ always denotes a period. It is important to note that in the relative velocity response spectrum given by Eq. (2.49), $2\pi/\omega$ refers to the natural vibration period of the SDOF elastic system under ground motion excitation, whereas in the Fourier amplitude spectrum shown in Eq. (2.50), $2\pi/\omega$ represents the period of the harmonic components of the ground motion. If this distinction is not considered, and both the relative velocity response spectrum of an undamped SDOF system and the Fourier amplitude spectrum of the ground motion acceleration are plotted against the period, it can be observed that for most ground motions, the relative velocity response spectrum is slightly larger than the Fourier amplitude spectrum, i.e., $S_v(\omega, 0) \geq G(\omega)$.

Figure 2.5 compares the undamped relative velocity response spectrum and the Fourier amplitude spectrum of the north-south (N-S) component of the 1940 El Centro earthquake record, showing that the two curves are generally similar in shape.

In fact, after the ground motion ceases, the structure commonly continues to undergo free vibration. If the maximum response of the undamped SDOF system occurs after the ground motion ends, then $S_v(\omega, 0) = G(\omega)$, meaning that the maximum relative velocity of the undamped SDOF system during free vibration after the earthquake equals the Fourier amplitude spectrum $G(\omega)$. If the maximum response occurs before the ground motion ends, it is generally observed that $S_v(\omega, 0) > G(\omega)$, and the difference between the two is usually small.

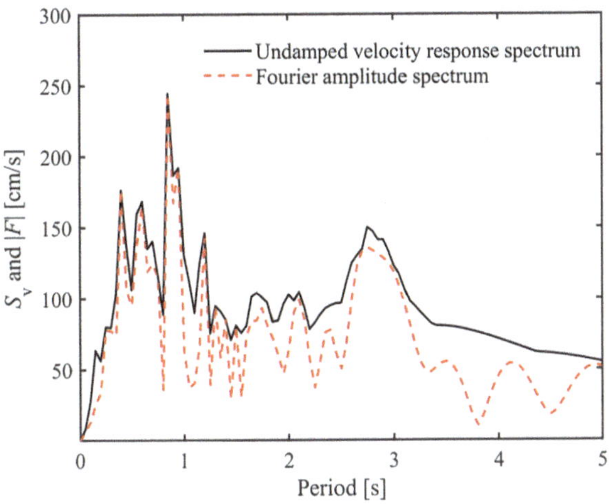

Fig. 2.5 Comparison of the relative velocity response spectrum and the Fourier amplitude spectrum

Fourier Amplitude Spectrum and Power Spectrum

Based on Eqs. (2.39) and (2.40), the total energy of the ground motion signal $a(t)$ is given by:

$$\begin{aligned}\int_{-\infty}^{\infty} a^2(t)\,\mathrm{d}t &= \int_{-\infty}^{\infty} a(t)\left(\frac{1}{2\pi}\int_{-\infty}^{\infty} F(\omega)\mathrm{e}^{\mathrm{i}\omega t}\mathrm{d}\omega\right)\mathrm{d}t \\ &= \frac{1}{2\pi}\int_{-\infty}^{\infty} F(\omega)\left(\int_{-\infty}^{\infty} a(t)\mathrm{e}^{\mathrm{i}\omega t}\mathrm{d}t\right)\mathrm{d}\omega \\ &= \frac{1}{2\pi}\int_{-\infty}^{\infty} F(\omega)F^*(\omega)\mathrm{d}\omega \\ &= \frac{1}{2\pi}\int_{-\infty}^{\infty} |F(\omega)|^2\mathrm{d}\omega\end{aligned} \tag{2.51}$$

Equation (2.51) is known as **Parseval theorem**, which states that the total energy of a signal is conserved in both the time and the frequency domains.

For a ground motion signal $a(t)$—an ordinary function with energy existing only within a finite time interval $[0,\ T]$—the average power (mean square value) over the interval $[0,\ T]$ can be presented as:

$$\frac{1}{T}\int_0^T a^2(t)\mathrm{d}t = \frac{1}{T}\int_{-\infty}^{\infty} a^2(t)\mathrm{d}t = \int_{-\infty}^{\infty}\frac{|F(\omega)|^2}{2\pi T}\mathrm{d}\omega$$

Therefore, the PSD of the signal can be defined as:

$$S_a(\omega) = \frac{|F(\omega)|^2}{2\pi T} \tag{2.52}$$

Thus, the physical meaning of the PSD $S_a(\omega)$ is as follows: the total area under $S_a(\omega)$ equals the mean value of the signal $a(t)$, and $S_a(\omega)\mathrm{d}\omega$ represents the contribution of the frequency components within the interval $(\omega, \omega + \mathrm{d}\omega)$ to this mean square value of the signal.

Essentially, the PSD characterizes the properties of a stochastic process. As a result, the relationship between the PSD and the Fourier amplitude spectrum can only be established through the expected value of the latter, as shown below:

$$S(\omega) = \frac{E\left[|F(\omega)|^2\right]}{2\pi T} \tag{2.53}$$

This relationship indicates that the PSD of a stochastic ground motion process can be obtained as the mean of the squared Fourier amplitude spectra from a set of sample functions.

It should be noted that the Fourier transform pair between the PSD in Eq. (2.53) and the corresponding auto-correlation function is given by:

$$\begin{cases} S(\omega) = \dfrac{1}{2\pi}\displaystyle\int_{-\infty}^{\infty} R(\tau)\mathrm{e}^{-\mathrm{i}\omega\tau}\mathrm{d}\tau \\ R(\tau) = \displaystyle\int_{-\infty}^{\infty} S(\omega)\mathrm{e}^{\mathrm{i}\omega\tau}\mathrm{d}\omega \end{cases} \tag{2.54}$$

For another commonly used form of the Fourier transform pair, where the expression of the auto-correlation function includes a factor of $1/2\pi$, the term $1/2\pi$ on the right-hand side of Eq. (2.53) should accordingly be omitted.

2.4 Non-Stationary Characteristics of Earthquake Ground Motion

The duration of ground motion typically ranges from a few seconds to several tens of seconds. During this period, the ground motion intensity increases rapidly from a small value to its peak, then remains approximately constant on average, and finally decays gradually to nearly zero. Overall, the complete ground motion process consists of three stages: the rising segment, the stationary segment, and the decaying segment, exhibiting non-stationary intensity characteristics (Hu 2006; Yang and Tian 2014). Meanwhile, during the propagation of seismic waves, different frequency components of the ground motion attenuate at varying rates, resulting in non-stationary frequency characteristics as well. In fact, both the intensity and frequency non-stationary characteristics of ground motions significantly influence the dynamic response of structures (Yang and Tian 2014; Clough and Penzien 1975). Therefore, to comprehensively describe the non-stationary characteristics and randomness of ground motions, it is generally necessary to adopt a non-stationary stochastic process model.

2.4.1 *Intensity (Uniform) Modulation Function*

Due to the complexity of non-stationary problems, early studies often simplified the non-stationary ground motion acceleration process $a(t)$ as the product of a stationary process $\overline{a}(t)$ and a deterministic function representing intensity non-stationarity, namely, the intensity modulation function (also called the uniform modulation function) $f(t)$, followed by:

$$a(t) = f(t)\bar{a}(t) \tag{2.55}$$

In this way, the non-stationary ground motion acceleration process can be represented as a non-stationary process with intensity (or uniform) modulation. It should be noted that this intensity-modulated model is only a simplified approximation, as it accounts solely for the intensity non-stationarity but does not consider the frequency non-stationarity of ground motion.

Generally, the intensity modulation function $f(t)$ should satisfy the following conditions:

$$f(t) > 0, \quad \max[f(t)] = 1 \tag{2.56}$$

For a smooth and continuous intensity modulation function, the above conditions can be rewritten as:

$$f'(t^*) = 0, \ f''(t^*) < 0, \quad f(t^*) = 1 \tag{2.57}$$

where t^* corresponds to the arrival time of the PGA.

Below are several commonly used intensity modulation functions.

Bogdanoff-Goldberg-Bernard Model

In 1961, Bogdanoff et al. proposed a smooth and continuous intensity modulation function model (Bogdanoff et al. 1961):

$$f(t) = at\exp(-bt); \quad a > 0, \quad b > 0 \tag{2.58}$$

According to the conditions satisfied by the intensity modulation function $f(t)$ in Eq. (2.57), we obtain:

$$t^* = \frac{1}{b}, \quad a = \mathrm{e} \times b \tag{2.59}$$

where e is the base of the natural logarithm, also known as Euler's number, and can be taken as $\mathrm{e} \approx 2.718281828$. Therefore, the Bogdanoff-Goldberg-Bernard model is a single-parameter, smooth, and continuous model. The parameter b controls both the arrival time of the peak acceleration and the rate of intensity attenuation, with a recommended value of $b = (1/10 \sim 1/3)\mathrm{s}^{-1}$ (Yang and Tian 2014).

Wang Model

In 1965, Wang proposed a smooth, continuous, and two-parameter intensity modulation function (Ou and Wang 1988):

$$f(t)=\left[\frac{t}{c}\exp\left(1-\frac{t}{c}\right)\right]^{d};\quad c>0,\quad d\geq 1 \tag{2.60}$$

where the parameter c controls the arrival time of the PGA, and the parameter d controls the shape of the intensity modulation function $f(t)$. The recommended values are $c = (3 \sim 10)$ s and $d = 1 \sim 3$, respectively. Evidently, the Wang model satisfies the conditions for intensity modulation functions given in Eq. (2.57).

Shinozuka-Sayo Model

In 1967, Shinozuka et al. proposed another smooth and continuous intensity modulation function model (Shinozuka and Sato 1967):

$$f(t)=c\left(\mathrm{e}^{-at}-\mathrm{e}^{-bt}\right);\quad a,b,c>0,\quad b>a \tag{2.61}$$

Similarly, based on the conditions satisfied by the intensity modulation function $f(t)$ defined in Eq. (2.57), we can get:

$$t^{*}=\frac{\ln b-\ln a}{b-a},\quad c=\frac{1}{\mathrm{e}^{-at^{*}}-\mathrm{e}^{-bt^{*}}} \tag{2.62}$$

Thus, the Shinozuka-Sayo model is usually written in a two-parameter function form:

$$f(t)=\frac{\exp(-at)-\exp(-bt)}{\exp(-at^{*})-\exp(-bt^{*})};\quad b>a>0 \tag{2.63}$$

Research indicates that as the values of a and b increase, the peak acceleration arrives earlier and attenuates more rapidly. The recommended values are $a = 0.2 \sim 0.4$ and $b = 0.4 \sim 0.8$, respectively, with the parameters having units of s^{-1}.

It should be noted that the advantage of the Shinozuka-Sayo model lies in its ability to be extended into a time-frequency modulation function model, enabling the simulation of a fully non-stationary ground motion process with both intensity and frequency non-stationarities. For further details, refer to Sect. 2.4.2.

Amin-Ang Model

In 1968, Amin and Ang proposed a segmented intensity modulation function (Amin and Ang 1968):

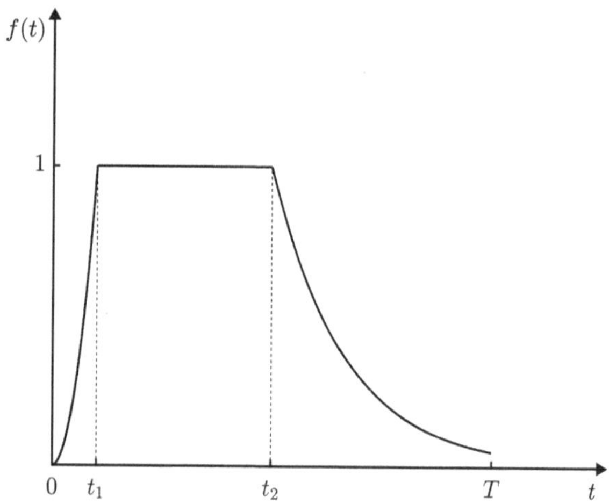

Fig. 2.6 Schematic diagram of the Amin-Ang model

$$f(t)=\begin{cases}(t/t_1)^2, & 0\le t\le t_1\\ 1, & t_1<t<t_2\\ e^{-c(t-t_2)}, & t_2\le t\le T\end{cases} \tag{2.64}$$

where the parameters t_1 and t_2 are the start and end times of the strong motion stationary segment, respectively. The parameter c controls the attenuate rate of the descending segment, as illustrated in Fig. 2.6.

The Amin-Ang model has the advantage of effectively capturing the three stages of a ground motion process: the rising, stationary, and attenuating segments. However, its drawbacks include the use of a segmented formulation, which makes the expression relatively complex. Additionally, it involves many parameters that are often difficult to determine. Numerous researchers have investigated the parameter values of the Amin-Ang model, yet their results show considerable variation (Amin and Ang 1968; Liu et al. 1982; Huo et al. 1991; Xue et al. 2003). For the Amin-Ang model, the recommended values are: t_1 = (0.08~0.16)T, t_2 = (0.45~0.60)T and $c=-\frac{\ln 0.05}{T-t_2}$ (determined based on the 5% of the maximum value), respectively.

Overall, both the intensity modulation function and the duration can describe the time span of a strong motion excitation, thus they are closely related (Hu 2006). Taking the Amin-Ang model as an example, and following the definition of relative duration, the duration T_d corresponding to $a = ka_{max}$ is the time interval between the intersection points of the horizontal line at level k (parallel to the time axis t) with the rising segment $f(t) = (t/t_1)^2$ and the attenuating segment $f(t)=e^{-c(t-t_2)}$, namely:

$$T_d=t_2-\frac{\ln k}{c}-\sqrt{k}t_1 \tag{2.65}$$

It can be observed that when $k = 1$, the relative duration is $T_d = t_2 - t_1$, which is the stationary duration. When $k < 1$, the relative duration is $T_d > t_2 - t_1$. When $k = 0$, the relative duration is $T_d = T$, which is the total duration. Note that when $k = 0$, the horizontal line t does not intersect the attenuating segment $f(t) = e^{-c(t-t_2)}$. In this case, T_d should be taken as the endpoint of the attenuating branch, that is, $T_d = T$.

2.4.2 *Time-Frequency Modulation Function*

As mentioned earlier, the non-stationarity of ground motion includes both intensity non-stationarity and frequency non-stationarity. When both aspects are considered simultaneously, the motion is referred to as time-frequency non-stationary or fully non-stationary. In this case, the modulation function applied to a stationary ground motion process is called a time-frequency modulation function.

Exponential Decay Model

In 1997, Lin et al. proposed an exponential-decay non-uniform (time-frequency) modulation function (Lin and Zhang 2004):

$$A(t,\omega) = \beta(t,\omega)f(t) = \exp\left(-\eta_0 \frac{\omega t}{\omega_\alpha t_\alpha}\right)f(t), \quad \eta_0 > 0 \tag{2.66}$$

where $\beta(t, \omega)$ is the frequency modulation function, $f(t)$ is the amplitude modulation function, which can be any intensity modulation function from Sect. 2.4.1. ω_α and t_α are introduced to relate frequency ω and time t in a dimensionless manner. They can be chosen arbitrarily in principle, but it is recommended to take $\omega_\alpha = \omega_g$ and $t_\alpha = T_d$ or T. In this way, by selecting an appropriate modulation factor η_0, the high-frequency components can be made to attenuate more rapidly than the low-frequency components, better representing the actual decay characteristics of ground motion.

Deodatis-Shinozuka Model

In 1989, Deodatis and Shinozuka proposed a type of time-frequency modulation function (Deodatis and Shinozuka 1989):

$$A(t,\omega) = \frac{\exp(-a \times t) - \exp[-(c \times \omega + b) \times t]}{\exp(-a \times t^*) - \exp[-(c \times \omega + b) \times t^*]}; \quad \omega > 0, \quad t > 0 \tag{2.67}$$

where

Table 2.7 The values of the parameter *a*

		Classification of soil sites				
Parameter	Design earthquake groups	I_0	I_1	II	III	IV
a (s^{-1})	The first group	0.40	0.30	0.25	0.20	0.15
	The second group	0.30	0.25	0.20	0.15	0.12
	The third group	0.25	0.20	0.15	0.12	0.10

$$t^* = \frac{\ln(c \times \omega + b) - \ln(a)}{c \times \omega + (b - a)}, \quad \omega > 0 \tag{2.68}$$

According to "Code for Seismic Design of Buildings" (GB50011-2010 2016) of China, Liu et al. (2017a) recommended the values of parameters a, b, and c as $b = a + 0.001$, $c = 0.005$. Additionally, the independent parameter a can be assigned according to Table 2.7 (Liu et al. 2017a).

In fact, the time-frequency modulation function expressed in Eq. (2.67) is an extension of the Shinozuka-Sayo intensity modulation function given in Eq. (2.63). When $c = 0$, Eq. (2.67) degenerates into the intensity modulation function form of Eq. (2.63). Furthermore, Eq. (2.67) exhibits several favorable characteristics. For example, as time increases, the high-frequency components of the seismic waves gradually diminish, while the low-frequency components become more dominant. In addition, by appropriately modifying the time-frequency modulation function in Eq. (2.67), other new forms can also be derived.

Improved Deodatis-Shinozuka Model

In 2017, Liu et al. further improved the Deodatis-Shinozuka model and proposed the following time-frequency modulation function (Liu et al. 2017a):

$$A(t, \omega) = \frac{\exp(-a \times t) - \exp\left[-\left(c\left|\omega - \omega_g\right| + b\right) \times t\right]}{\exp(-a \times t^*) - \exp\left[-\left(c\left|\omega - \omega_g\right| + b\right) \times t^*\right]}; \quad \omega > 0 \quad t > 0 \tag{2.69}$$

in which

$$t^* = \frac{\ln\left(c \times \left|\omega - \omega_g\right| + b\right) - \ln(a)}{c \times \left|\omega - \omega_g\right| + (b - a)}; \quad \omega > 0 \tag{2.70}$$

where the values of parameters a, b, and c are exactly the same to those used in the afore-mentioned Deodatis-Shinozuka model.

2.4.3 Evolutionary Power Spectral Density Model

The PSD function $S(\omega)$ of stationary ground motion processes $\overline{a}(t)$ has been introduced in Sect. 2.3.1. The following section presents the evolutionary PSD (EPSD) $S_a(\omega, t)$ for non-stationary ground motion processes $a(t)$.

Type I Model of EPSD

$$S_a^{(1)}(\omega, t) = f^2(t)S(\omega) \tag{2.71}$$

where $S_a^{(1)}(\omega, t)$ represents the Type I model of EPSD (two-sided spectrum) of the non-stationary ground motion acceleration process $a(t)$. $f(t)$ is the intensity modulation function, which can be selected from the models introduced in Sect. 2.4.1. $S(\omega)$ is the two-sided PSD of the stationary ground motion process $\overline{a}(t)$, which can be chosen from the models described in Sect. 2.3.1.

Type II Model of EPSD

$$S_a^{(2)}(\omega, t) = |A(t, \omega)|^2 S(\omega) \tag{2.72}$$

where $S_a^{(2)}(\omega, t)$ represents the Type II model of EPSD (two-sided spectrum) of the non-stationary ground motion acceleration process $a(t)$. $A(t, \omega)$ is the time-frequency modulation function, which can be obtained from the models introduced in Sect. 2.4.2.

Indeed, the models defined by Eqs. (2.71) and (2.72) are two classical EPSD models in earthquake engineering. Furthermore, Eq. (2.71) can be regarded as a special case of Eq. (2.72).

Type III Model of EPSD

Another category of EPSD model for fully non-stationary ground motion processes was proposed in previous studies (Solari and Spinelli 1984; Liu et al. 2015):

$$S_a^{(3)}(\omega, t) = f^2(t)S(\omega, t) \tag{2.73}$$

where $S_a^{(3)}(\omega, t)$ denotes the Type III model of EPSD (two-sided spectrum) of the non-stationary ground motion acceleration process $a(t)$. Typically, $S(\omega, t)$ can be regarded as an extension of the PSD $S(\omega)$ corresponding to the stationary stochastic

Table 2.8 Parameter values for the generalized Clough-Penzien model

Parameters	Classification of soil sites				
	I_0	I_1	II	III	IV
d(rad/s)	8	7	6	4	2
e	0.2	0.2	0.2	0.2	0.2

process, referred to as the generalized PSD or time-varying PSD. For example, for the Clough-Penzien model of a stationary ground motion process, as shown in Eq. (2.10), its generalized PSD (time-varying PSD) can be defined as follows:

$$S(\omega,t)=\frac{\omega_{\mathrm{g}}^{4}(t)+4\zeta_{\mathrm{g}}^{2}(t)\omega_{\mathrm{g}}^{2}(t)\omega^{2}}{\left[\omega^{2}-\omega_{\mathrm{g}}^{2}(t)\right]^{2}+4\zeta_{\mathrm{g}}^{2}(t)\omega_{\mathrm{g}}^{2}(t)\omega^{2}}\times\frac{\omega^{4}}{\left[\omega^{2}-\omega_{\mathrm{f}}^{2}(t)\right]^{2}+4\zeta_{\mathrm{f}}^{2}(t)\omega_{\mathrm{f}}^{2}(t)\omega^{2}}\times S_{0}(t) \tag{2.74}$$

in which

$$\omega_{\mathrm{g}}(t)=\omega_{0}-d\times\left(\frac{t}{T}\right)^{n},\quad \zeta_{\mathrm{g}}(t)=\zeta_{0}+e\times\left(\frac{t}{T}\right)^{n} \tag{2.75}$$

$$\omega_{0}=\overline{\omega}_{\mathrm{g}}+\frac{d}{2},\quad \zeta_{0}=\overline{\zeta}_{\mathrm{g}}-\frac{e}{2} \tag{2.76}$$

$$\omega_{\mathrm{f}}(t)=0.1\omega_{\mathrm{g}}(t),\quad \zeta_{\mathrm{f}}(t)=\zeta_{\mathrm{g}}(t) \tag{2.77}$$

$$S_{0}(t)=\frac{a_{\max}^{2}}{r^{2}\pi\omega_{\mathrm{g}}(t)\left(2\zeta_{\mathrm{g}}(t)+\frac{1}{2\zeta_{\mathrm{g}}(t)}\right)} \tag{2.78}$$

where the values of $\overline{\omega}_{\mathrm{g}}$ and $\overline{\zeta}_{\mathrm{g}}$ can be assigned according to Table 2.5. The parameter n, which reflects the time-dependency of site parameters, may be taken as $n = 1$ or 2. The parameters d and e characterize the frequency non-stationarity of the ground motion. Based on the "Code for Seismic Design of Buildings" (GB50011-2010 2016) of China, Table 2.8 provides the recommended values for these parameters.

Type IV Model of EPSD

By integrating the characteristics of Type II and Type III models of EPSD, the Type IV model can be proposed (Liu & Liu 2017b; Liu et al. 2019):

$$S_{a}^{(4)}(\omega,t)=|A(t,\omega)|^{2}S(\omega,t) \tag{2.79}$$

where $S_{a}^{(4)}(\omega,t)$ denotes the Type IV model of EPSD (two-sided spectrum) of the non-stationary ground motion acceleration process $a(t)$.

In fact, Eq. (2.73) can be regarded as a special case of Eq. (2.72), where the time-frequency modulation function in Eq. (2.72) is $|f(t)|\sqrt{S(\omega, t)}$, and the corresponding stationary white noise process has a PSD of 1. Similarly, Eq. (2.79) is also a special case of Eq. (2.72), in which the time-frequency modulation function is $|A(t, \omega)|\sqrt{S(\omega, t)}$, and the corresponding PSD function of the stationary white noise process is 1.

In earthquake engineering, the Type I model of EPSD is the most widely used due to its simple formulation. However, this model only accounts for the intensity non-stationarity of ground motion and fails to consider the frequency non-stationarity, which plays a crucial role in the nonlinear response analysis of structures. In contrast, Type II, Type III, and Type IV models of EPSD are capable of more comprehensively capturing the dual non-stationary characteristics of ground motions in both intensity and frequency, and therefore have attracted increasing attention in recent years. These models have been extensively applied in the simulation of non-stationary ground motion processes. It is important to acknowledge that the key to applying the EPSD models lies in accurately determining the time-frequency modulation function $A(t, \omega)$ and the functional form of the time-varying PSD $S(\omega, t)$.

2.5 Spatial Variability Characteristics of Earthquake Ground Motion

In addition to exhibiting full non-stationarity in both intensity and frequency, ground motions also demonstrate significant spatial variability characteristics. This means that during the same seismic event, ground motions differ across various locations on the surface. Even sites separated by just a few tens of meters, the amplitude, phase, and spectral characteristics of the vibrations can vary considerably, and the spatial variability characteristics are highly complex. In fact, large span or elongated structures, such as bridges, pipelines, dams, and high-speed railways, are highly sensitive to the spatial variability of ground motion. As such, in engineering seismic analysis, it is generally necessary to consider the spatial variability of the ground motion field.

2.5.1 Description of the Seismic Ground Motion Stochastic Field

The ground motion stochastic field, denoted as $f_0(\boldsymbol{u}, t) = f_0(x, y, z; t)$, takes coordinates (x, y, z) as spatial parameters and t as the temporal parameter. It constitutes a four-dimensional univariate (4D - 1V) spatio-temporal stochastic field, where $\boldsymbol{u} = (x, y, z)$ represents the coordinates of any point in space. Essentially, $f_0(\boldsymbol{u}, t)$ is a continuous spatio-temporal stochastic field. However, in practical applications, it may sometimes be simplified to a discrete spatio-temporal stochastic field.

Continuous Spatio-Temporal Stochastic Field

A continuous spatio-temporal stochastic field treats both the spatial parameter $\boldsymbol{u}$ and the temporal parameter t as continuous functions, and is analyzed using random wave theory by introducing the wavenumber. For the continuous spatio-temporal stochastic field $f_0(\boldsymbol{u}, t)$, assuming it has zero-mean, its auto-correlation function is $R_{f_0}(\boldsymbol{u}, t; \boldsymbol{u}+\boldsymbol{\xi}, t+\tau)$, and its evolutionary frequency-wavenumber spectrum is $S_{f_0}(\boldsymbol{u}, t; \boldsymbol{\kappa}, \omega)$, where $\boldsymbol{\xi} = (\xi_x, \xi_y, \xi_z)^{\mathrm{T}}$ is the increment of the spatial variable $\boldsymbol{u}$. $\boldsymbol{\kappa} = (\kappa_x, \kappa_y, \kappa_z)^{\mathrm{T}}$ is the wavenumber vector corresponding to the spatial variable $\boldsymbol{u}$.

If $f_0(\boldsymbol{u}, t)$ is assumed to be a temporally non-stationary and spatially non-homogeneous continuous spatio-temporal stochastic field, its evolutionary frequency-wavenumber spectrum can be defined as:

$$S_{f_0}(\boldsymbol{u}, t; \boldsymbol{\kappa}, \omega) = [A(\boldsymbol{u}, t; \boldsymbol{\kappa}, \omega)]^2 S(\boldsymbol{\kappa}, \omega) \tag{2.80}$$

where $A(\boldsymbol{u}, t; \boldsymbol{\kappa}, \omega)$ is the deterministic modulation function, and $S(\boldsymbol{\kappa}, \omega)$ is the frequency-wavenumber spectrum, respectively.

Then the auto-correlation function can be written as:

$$\begin{aligned} R_{f_0}(\boldsymbol{u}, t; \boldsymbol{u}+\boldsymbol{\xi}, t+\tau) &= E\left[f_0(\boldsymbol{u}+\boldsymbol{\xi}, t+\tau) f_0^*(\boldsymbol{u}, t)\right] \\ &= \int_{-\infty}^{\infty}\int_{-\infty}^{\infty} A(\boldsymbol{u}, t; \boldsymbol{\kappa}, \omega) A(\boldsymbol{u}+\boldsymbol{\xi}, t+\tau; \boldsymbol{\kappa}, \omega) S(\boldsymbol{\kappa}, \omega)\, \mathrm{e}^{\mathrm{i}(\boldsymbol{\kappa}\cdot\boldsymbol{\xi}+\omega\tau)} \mathrm{d}\boldsymbol{\kappa}\mathrm{d}\omega \end{aligned} \tag{2.81}$$

where the superscript "*" denotes the complex conjugate, and the vector inner product is given by $\boldsymbol{\kappa} \cdot \boldsymbol{\xi} = \kappa_x\xi_x + \kappa_y\xi_y + \kappa_z\xi_z$.

If $f_0(\boldsymbol{u}, t)$ is assumed to be a temporally stationary and spatially non-homogeneous continuous spatio-temporal stochastic field, its evolutionary frequency-wavenumber spectrum can be given by:

$$S_{f_0}(\boldsymbol{u}, \boldsymbol{\kappa}, \omega) = [A(\boldsymbol{u}, \boldsymbol{\kappa})]^2 S(\boldsymbol{\kappa}, \omega) \tag{2.82}$$

The corresponding auto-correlation function is:

$$R_{f_0}(\boldsymbol{u}, \boldsymbol{u}+\boldsymbol{\xi}, \tau) = \int_{-\infty}^{\infty}\int_{-\infty}^{\infty} A(\boldsymbol{u}, \boldsymbol{\kappa}) A(\boldsymbol{u}+\boldsymbol{\xi}, \boldsymbol{\kappa}) S(\boldsymbol{\kappa}, \omega)\, \mathrm{e}^{\mathrm{i}(\boldsymbol{\kappa}\cdot\boldsymbol{\xi}+\omega\tau)} \mathrm{d}\boldsymbol{\kappa}\mathrm{d}\omega \tag{2.83}$$

If $f_0(\boldsymbol{u}, t)$ is assumed to be a temporally non-stationary and spatially homogeneous continuous spatio-temporal stochastic field, its evolutionary frequency-wavenumber spectrum can be written as:

$$S_{f_0}(t, \boldsymbol{\kappa}, \omega) = [A(t, \omega)]^2 S(\boldsymbol{\kappa}, \omega) \tag{2.84}$$

The corresponding auto-correlation function is:

$$R_{f_0}(t, t + \tau, \boldsymbol{\xi}) = \int_{-\infty}^{\infty} \int_{-\infty}^{\infty} A(t, \omega) A(t + \tau, \omega) S(\boldsymbol{\kappa}, \omega) \, \mathrm{e}^{\mathrm{i}(\boldsymbol{\kappa} \cdot \boldsymbol{\xi} + \omega \tau)} \mathrm{d}\boldsymbol{\kappa} \mathrm{d}\omega \tag{2.85}$$

Obviously, the **frequency-wavenumber spectrum** $S(\boldsymbol{\kappa}, \omega)$ is the evolutionary frequency-wavenumber spectrum of a temporally stationary and spatially homogeneous continuous spatio-temporal stochastic field $f_0(\boldsymbol{u}, t)$:

$$S(\boldsymbol{\kappa}, \omega) = \frac{1}{2\pi} \int_{-\infty}^{\infty} S(\omega) \gamma(\boldsymbol{\xi}, \omega) \mathrm{e}^{-\mathrm{i}\boldsymbol{\kappa} \cdot \boldsymbol{\xi}} \mathrm{d}\boldsymbol{\xi} \tag{2.86}$$

where $\gamma(\boldsymbol{\xi}, \omega)$ is the **spatial coherence function**, and $S(\omega)$ is the stationary PSD of ground motion, assuming that all the points acting on the ground surface share the same stationary PSD.

In summary, for a continuous spatio-temporal stochastic field $f_0(\boldsymbol{u}, t)$, its complete probabilistic information is determined by the evolutionary frequency-wavenumber spectrum $S_{f_0}(\boldsymbol{u}, t; \boldsymbol{\kappa}, \omega)$, where the key lies in defining the modulation function $A(\boldsymbol{u}, t; \boldsymbol{\kappa}, \omega)$ and the frequency-wavenumber spectrum $S(\boldsymbol{\kappa}, \omega)$.

Discrete Spatio-Temporal Stochastic Field

Let P_i and P_j be two locations on the ground surface, with spatial coordinates $\boldsymbol{u}_i = (x_i, y_i, z_i)$ and $\boldsymbol{u}_j = (x_j, y_j, z_j)$. Let $f_0(\boldsymbol{u}_i, t)$ and $f_0(\boldsymbol{u}_j, t)$ denote the ground motion processes at these two points, respectively. Clearly, $f_0(\boldsymbol{u}_i, t)$ and $f_0(\boldsymbol{u}_j, t)$ are two stochastic processes.

Assuming that $f_0(\boldsymbol{u}, t)$ is a zero-mean stationary and homogeneous spatio-temporal stochastic field, the complete probabilistic information of the stochastic processes at any two points P_i and P_j can be determined by their cross-correlation function:

$$R_{f_0}(\boldsymbol{u}_i, \boldsymbol{u}_j; t_1, t_2) = E[f_0(\boldsymbol{u}_i, t_1) f_0(\boldsymbol{u}_j, t_2)] \tag{2.87}$$

Since the stochastic field $f_0(\boldsymbol{u}, t)$ is stationary in time and homogeneous in space, we have:

$$R_{f_0}(\boldsymbol{u}_i, \boldsymbol{u}_j; t_1, t_2) = R_{f_0}(\boldsymbol{u}_j - \boldsymbol{u}_i, t_2 - t_1) = R_{f_0}(\boldsymbol{\xi}_{ij}, \tau) \tag{2.88}$$

Herein, $\boldsymbol{\xi}_{ij} = \boldsymbol{u}_j - \boldsymbol{u}_i$ is the position vector between the two points, $\tau = t_2 - t_1$ denotes the time lag. For a stationary and homogeneous spatio-temporal stochastic field, the cross-correlation function has the following symmetries:

$$R_{f_0}(\boldsymbol{\xi}_{ij}, \tau) = R_{f_0}(\boldsymbol{\xi}_{ji}, \tau), \quad R_{f_0}(\boldsymbol{\xi}_{ij}, \tau) = R_{f_0}(\boldsymbol{\xi}_{ij}, -\tau) \tag{2.89}$$

Furthermore, from Eq. (2.89) it follows that:

$$R_{f_0}(\boldsymbol{\xi}_{ji}, \tau) = R_{f_0}(\boldsymbol{\xi}_{ij}, -\tau) \tag{2.90}$$

In Eq. (2.88), if the cross-correlation function depends only on the distance between P_i and P_j, i.e., on $|\boldsymbol{\xi}_{ij}|$, then:

$$R_{f_0}(\boldsymbol{u}_i, \boldsymbol{u}_j; t_1, t_2) = R_{f_0}(\boldsymbol{\xi}_{ij}, \tau) = R_{f_0}(|\boldsymbol{\xi}_{ij}|, \tau) \tag{2.91}$$

and the stochastic field is said to be spatially isotropic.

Under the assumptions of stationarity, homogeneity, and isotropy, the cross-correlation function can be further simplified to:

$$R_{f_0}(\xi_{ij}, \tau) = R_{ij}(\tau) \tag{2.92}$$

where $\xi_{ij} = |\boldsymbol{\xi}_{ij}|$ is the distance between the two points.

According to the Wiener-Khinchin theorem, the cross-PSD function (two-sided spectrum) between two stochastic processes at points P_i and P_j is given by:

$$S_{ij}(\omega) = \frac{1}{2\pi}\int_{-\infty}^{\infty} R_{ij}(\tau)\exp(-\mathrm{i}\omega\tau)\mathrm{d}\tau \tag{2.93}$$

Now, the coherence function $\gamma_{ij}(\omega)$ is defined as:

$$\gamma_{ij}(\omega) = \frac{S_{ij}(\omega)}{\sqrt{S_i(\omega)S_j(\omega)}}, \quad -\infty < \omega < \infty \tag{2.94}$$

where $S_i(\omega)$ and $S_j(\omega)$ are the auto-PSD functions of the stochastic processes at points P_i and P_j, respectively. The auto-PSD function $S_i(\omega)$ is a real, non-negative, and even function.

Thus, the cross-PSD function can be calculated using the coherence function:

$$S_{ij}(\omega) = \gamma_{ij}(\omega)\sqrt{S_i(\omega)S_j(\omega)} \tag{2.95}$$

Generally, the cross-PSD function has the following properties:

$$S_{ij}(\omega) = S_{ji}(-\omega) = S_{ji}^{*}(\omega) \tag{2.96}$$

This indicates that the cross-PSD function $S_{ij}(\omega)$ is a conjugate symmetric complex function.

Now, we can construct a non-stationary multivariate stochastic process. In engineering practice, it is often sufficient to consider a finite number of spatial locations. In such cases, the four-dimensional univariate (4D - 1V) continuous spatio-temporal stochastic field $f_0(\boldsymbol{u}, t)$ can be discretized into a one-dimensional multivariate (1D - nV) stochastic vector process $\boldsymbol{X}(t)$. Specifically, by selecting n points of interest P_1, P_2, $\cdots$, P_n, the ground motion process at each point P_i is denoted as $f_0(\boldsymbol{u}_i, t) = X_i(t)$ $(i = 1, 2, \cdots, n)$, and all are assumed to be zero-mean non-stationary stochastic processes. These n non-stationary processes $X_i(t)$ $(i = 1, 2, \cdots, n)$ can then be assembled into a stochastic vector process $\boldsymbol{X}(t)$, i.e., $\boldsymbol{X}(t) = [X_1(t), X_2(t), \cdots, X_n(t)]^{\mathrm{T}}$, thereby discretizing the continuous spatio-temporal stochastic field $f_0(\boldsymbol{u}, t)$ into a zero-mean non-stationary stochastic vector process $\boldsymbol{X}(t)$. For this non-stationary vector process $\boldsymbol{X}(t)$, its two-sided EPSD function matrix can be defined as:

$$\boldsymbol{S}_X(\omega, t) = \begin{bmatrix} S_{11}(\omega, t) & S_{12}(\omega, t) & \cdots & S_{1n}(\omega, t) \\ S_{21}(\omega, t) & S_{22}(\omega, t) & \cdots & S_{2n}(\omega, t) \\ \vdots & \vdots & \ddots & \vdots \\ S_{n1}(\omega, t) & S_{n2}(\omega, t) & \cdots & S_{nn}(\omega, t) \end{bmatrix} \tag{2.97}$$

Among them, the elements in the EPSD function matrix $\boldsymbol{S}_X(\omega, t)$ can be expressed in the following form:

$$S_{ii}(\omega, t) = S_i(\omega, t) = |A_i(t, \omega)|^2 S_i(\omega), \quad i = 1, 2, \cdots, n \tag{2.98}$$

$$S_{ij}(\omega, t) = A_i(t, \omega) A_j(t, \omega) \sqrt{S_i(\omega) S_j(\omega)} \gamma_{ij}(\omega), \quad i, j = 1, 2, \cdots, n \quad j \neq i \tag{2.99}$$

where $A_i(t, \omega)$ is the time-frequency modulation function, $S_i(\omega)$ $(i = 1, 2, \cdots, n)$ denotes the stationary PSD (two-sided) associated with the non-stationary ground motion process $X_i(t)$, and $\gamma_{ij}(\omega)$ is the coherence function, defined in Eq. (2.94), respectively.

In Eqs. (2.98) and (2.99), $S_i(\omega, t)$ is referred to as the evolutionary auto-PSD of the non-stationary ground motion process $X_i(t)$, and $S_{ij}(\omega, t)$ is the evolutionary cross-PSD between the non-stationary ground motion processes $X_i(t)$ and $X_j(t)$ $(j \neq i)$, respectively.

Furthermore, the auto-correlation function $R_i(t, t + \tau)$ of the non-stationary ground motion process $X_i(t)$, and the cross-correlation function $R_{ij}(t, t + \tau)$ between $X_i(t)$ and $X_j(t)$ $(j \neq i)$ can be written as:

$$R_{ii}(t,t+\tau)=R_i(t,t+\tau)=\int_{-\infty}^{\infty}A_i(t,\omega)A_i(t+\tau,\omega)S_i(\omega)\mathrm{e}^{\mathrm{i}\omega\tau}\mathrm{d}\omega,\quad i=1,2,\cdots,n \tag{2.100}$$

$$R_{ij}(t,t+\tau)=\int_{-\infty}^{\infty}A_i(t,\omega)A_j(t+\tau,\omega)\sqrt{S_i(\omega)S_j(\omega)}\,\gamma_{ij}(\omega)\mathrm{e}^{\mathrm{i}\omega\tau}\mathrm{d}\omega,\quad i,j=1,2,\cdots,n,\quad j\neq i \tag{2.101}$$

In summary, the complete probabilistic information of the discrete spatio-temporal stochastic field, or the non-stationary stochastic vector process $\boldsymbol{X}(t)$, is determined by the EPSD function matrix $\boldsymbol{S_X}(\omega,t)$. The key to its definition lies in specifying the stationary PSD $S_i(\omega)$, the time-frequency modulation function $A_i(t,\omega)$, and the coherence function $\gamma_{ij}(\omega)$.

2.5.2 Coherence Function Model

As mentioned earlier, the spatial variability characteristics of the ground motion field are described by the coherence function (a complex function of frequency ω), which is defined as:

$$\gamma_{ij}(\omega)=\frac{S_{ij}(\omega)}{\sqrt{S_i(\omega)S_j(\omega)}}=\left|\gamma_{ij}(\omega)\right|\exp\left[\mathrm{i}\theta_{ij}(\omega)\right] \tag{2.102}$$

where $S_{ij}(\omega)$ represents the cross-PSD of the stationary ground motion processes at surface points i and j, $S_i(\omega)$ and $S_j(\omega)$ are the auto-PSD functions of the stationary ground motion processes at surface points i and j, respectively. $|\gamma_{ij}(\omega)|$ is commonly referred to as the lagged coherence function, which measures the degree of correlation between the ground motion processes at the two points and satisfies the condition $0\le|\gamma_{ij}(\omega)|\le 1$. $\theta_{ij}(\omega)$ is the phase angle:

$$\theta_{ij}(\omega)=\arctan\frac{\mathrm{Im}\left[\gamma_{ij}(\omega)\right]}{\mathrm{Re}\left[\gamma_{ij}(\omega)\right]} \tag{2.103}$$

where Im and Re are the imaginary and real parts of the complex function, respectively.

The factors influencing the coherence function of ground motions between two points include the propagation medium, propagation path, and local site conditions. In 1996, Der Kiureghian conducted a theoretical study on ground motion coherence functions and classified the main influencing factors into three categories: incoherence effect (also referred to as lagged coherence effect), wave-passage effect, and site-response effect (Der Kiureghian 1996). A theoretical model for the ground motion coherence function was also proposed (Der Kiureghian 1996; Dumanogluid A and Soyluk 2003):

$$\gamma_{ij}(\omega)=\gamma_{ij}^{\text{incoh}}(\omega)\cdot\gamma_{ij}^{\text{wave}}(\omega)\cdot\gamma_{ij}^{\text{site}}(\omega)=\gamma_{ij}^{\text{incoh}}(\omega)\cdot\exp\left[\mathrm{i}\left(\theta_{ij}^{\text{wave}}(\omega)+\theta_{ij}^{\text{site}}(\omega)\right)\right] \tag{2.104}$$

where $\gamma_{ij}^{\text{incoh}}(\omega)$ represents the real-valued incoherence effect, caused by the reflection and refraction of seismic waves in the site soil. $\gamma_{ij}^{\text{wave}}(\omega)$ means the complex-valued wave-passage effect, determined by the propagation velocity of seismic waves in the medium and the distance and relative position between the two points. $\theta_{jk}^{\text{site}}(\omega)$ is the complex-valued site-response effect, resulting from differences in local soil site conditions.

Clearly, from Eqs. (2.102) and (2.104), we can obtain:

$$\left|\gamma_{ij}(\omega)\right|=\gamma_{ij}^{\text{incoh}}(\omega),\quad \theta_{ij}(\omega)=\theta_{ij}^{\text{wave}}(\omega)+\theta_{ij}^{\text{site}}(\omega) \tag{2.105}$$

In the following sections, the incoherence effect, wave-passage effect, and site-response effect will be introduced separately.

Incoherence Effect

Based on observational records from dense arrays such as SMART-1 and LSST, several coherence function models for ground motion acceleration have been proposed. Some commonly used models are presented below.

Harichandran-Vanmarcke Model

In 1986, based on recordings of four earthquakes from the dense array SMART-1, Harichandran and Vanmarcke proposed the following expression for the lagged coherence function model (Harichandran and Vanmarcke 1986):

$$\gamma_{ij}^{\text{incoh}}(\omega)=A\exp\left[-2\frac{\xi_{ij}(1-A+\alpha A)}{\alpha\cdot\vartheta(\omega)}\right]+(1-A)\exp\left[-2\frac{\xi_{ij}(1-A+\alpha A)}{\vartheta(\omega)}\right] \tag{2.106}$$

in which

$$\vartheta(\omega)=k\left[1+(\omega/\omega_0)^b\right]^{-1/2},\quad \omega>0 \tag{2.107}$$

Table 2.9 Parameter values for the Harichandran-Vanmarcke model

References	Model parameters				
	A	α	k (m)	ω_0 (rad/s)	b
Harichandran and Vanmarcke (1986)	0.736	0.147	5210	6.85	2.78
Harichandran et al. (1996)	0.636	0.0186	31,200	9.49	2.95
Deodatis (1996)	0.626	0.022	19,700	12.692	3.47

where ξ_{ij} is the distance (scalar) between points i and j; A, α, k, ω_0, b are the model parameters. Table 2.9 provides the values of these parameters as reported in selected references (Harichandran and Vanmarcke 1986; Harichandran et al. 1996; Deodatis 1996).

Luco-Wong Model

In 1986, Luco and Wong proposed another expression for the lagged coherence function model as follows (Luco and Wong 1986):

$$\gamma_{ij}^{\text{incoh}}(\omega) = \exp\left[-\left(\eta\frac{\omega\xi_{ij}}{v_s}\right)^2\right] \tag{2.108}$$

where ξ_{ij} is the distance (scalar) between points i and j. v_s is the shear wave velocity (scalar) of seismic wave propagation in the medium, and η is the incoherence coefficient, respectively. Luco and Wong (1986) recommended a value of $\frac{\eta}{v_s} \approx (2 \sim 3) \times 10^{-4}\text{m}^{-1}\cdot\text{s}$, while Tubino et al. (2003) suggested $\eta \in [0,\ 0.5]$.

Loh-Yeh Model

In 1988, Loh and Yeh proposed the following lagged coherence function model (Loh and Yeh 1988):

$$\gamma_{ij}^{\text{incoh}}(\omega) = \exp\left[-\alpha\frac{\omega\xi_{ij}}{2\pi v_{\text{app}}}\right], \quad \omega > 0 \tag{2.109}$$

where v_{app} is the apparent wave velocity (scalar) of the seismic ground wave, representing the propagation velocity along the straight line between points i and j. α is the wave number of the ground motion. Based on the 40 acceleration records from the dense array SMART-1, Loh and Yeh (1988) recommended a value of $\alpha = 0.125$. In 1995, Loh and Ku replaced the apparent wave velocity v_{app} in the model with the shear wave velocity v_s.

Notes on the lagged coherence function model (Oliveira et al. 1991; Loh and Ku 1995):

1. The lagged coherence function decreases with increasing separation distance and frequency. It is general a function of the scalar distance between the two relevant points, and the relative position between the points and the apparent wave velocity does not need to be considered.
2. The apparent wave velocity tends to increase with frequency, but it exhibits significant variability across different frequencies. Since many factors affect the apparent wave velocity, in engineering applications it can be reasonably assumed as a constant value depending on the site characteristics. Typically, a representative constant value can be adopted. It is commonly approximated as $v_{\text{app}} = v_s$.
3. Mathematically, the lagged coherence function is a real and even function with respect to frequency $\omega \in (-\infty, \infty)$. However, in engineering practice, just the positive frequency range $\omega \in (0, \infty)$ is considered, resulting in the above lagged coherence functions are generally defined over positive frequencies $\omega \in (0, \infty)$.

Wave-Passage Effect

Due to the finite propagation velocity of seismic waves in the medium, when the distance between ground observation points is large, the difference in arrival times of seismic waves at points i and j leads to a phase difference between the ground motions at the two points. This effect is referred to as the wave-passage effect. The coherence function caused by the wave-passage effect can be expressed as (Vanmarcke 1983; Zerva and Zervas 2002):

$$\gamma_{ij}^{\text{wave}}(\omega) = \exp\left[\mathrm{i}\theta_{ij}^{\text{wave}}(\omega)\right] \tag{2.110}$$

in which

$$\theta_{ij}^{\text{wave}}(\omega) = -\frac{j-i}{|j-i|} \times \frac{\omega \xi_{ij}}{v_{\text{app}}} \tag{2.111}$$

where ξ_{ij} is the distance (scalar) between points i and j, the apparent wave velocity v_{app} represents the propagation velocity along the straight line between points i and j.

Site-Response Effect

The difference in the coherence function caused by variations in local soil sites is referred to as the site-response effect. The coherence function induced by the site-response effect can be expressed as (Der Kiureghian 1996):

$$\gamma_{ij}^{\text{site}}(\omega) = \exp\left[\mathrm{i}\theta_{ij}^{\text{site}}(\omega)\right] \tag{2.112}$$

in which

$$\theta_{ij}^{\mathrm{site}}(\omega) = \arctan\frac{\mathrm{Im}\left[H_i(\omega)H_j^*(\omega)\right]}{\mathrm{Re}\left[H_i(\omega)H_j^*(\omega)\right]} = \arctan\frac{\mathrm{Im}\left[H_i(\omega)H_j(-\omega)\right]}{\mathrm{Re}\left[H_i(\omega)H_j(-\omega)\right]} \tag{2.113}$$

where $H_i(\omega)$ and $H_j(\omega)$ represent the transfer functions of absolute acceleration at ground observation points i and j, respectively, given by:

$$H_i(\omega) = \frac{\omega_{g,i}^2 + 2\mathrm{i}\zeta_{g,i}\omega_{g,i}\omega}{\left(\omega_{g,i}^2 - \omega^2\right) + 2\mathrm{i}\zeta_{g,i}\omega_{g,i}\omega}, \quad H_j(\omega) = \frac{\omega_{g,j}^2 + 2\mathrm{i}\zeta_{g,j}\omega_{g,j}\omega}{\left(\omega_{g,j}^2 - \omega^2\right) + 2\mathrm{i}\zeta_{g,j}\omega_{g,j}\omega} \tag{2.114}$$

where $\omega_{g,\ i}$ and $\zeta_{g,\ i}$ represent the natural circular frequency and damping ratio of the SDOF soil layer at point i, respectively.

Actually, according to the formula arctan(A) − arctan(B) = arctan[(A − B)/(1 + AB)], Eq. (2.113) can also be rewritten as:

$$\theta_{ij}^{\mathrm{site}}(\omega) = \theta_i^{\mathrm{site}}(\omega) - \theta_j^{\mathrm{site}}(\omega) \tag{2.115}$$

where

$$\theta_i^{\mathrm{site}}(\omega) = \arctan\frac{\mathrm{Im}[H_i(\omega)]}{\mathrm{Re}[H_i(\omega)]}, \quad \theta_j^{\mathrm{site}}(\omega) = \arctan\frac{\mathrm{Im}\left[H_j(\omega)\right]}{\mathrm{Re}\left[H_j(\omega)\right]} \tag{2.116}$$

References

Amin M, Ang AHS (1968) Nonstationary stochastic models of earthquake motions. J Eng Mech Div 94(2):559–584

Bogdanoff JL, Goldberg JE, Bernard MC (1961) Response of a simple structures to a random earthquake like disturbance. Bull Seismol Soc Am 51(2):293–310

Clough RW, Penzien J (1975) Dynamics of structures. McGraw Hill, New York

Deodatis G (1996) Non-stationary stochastic vector processes: seismic ground motion applications. Prob Eng Mech 11:149–168

Deodatis G, Shinozuka M (1989) Simulation of seismic ground motion using stochastic waves. J Eng Mech 115(12):2723–2736

Der Kiureghian A (1996) A coherency model for spatially varying ground motions. Earthq Eng Struct Dynam 25(1):99–111

Du XL (1998) A random seismic input model used in seismic reliability design and analysis of hydraulic buildings. Earthq Eng Eng Vib 18(4):76–81. (in Chinese)

Dumanogluid A A, Soyluk K (2003) A stochastic analysis of long span structures subjected to spatially varying ground motions including the site-response effect. Eng Struct 25(10): 1301–1310

GB18306–2015 (2015) Seismic ground motion parameters zonation map of China. Standards Press of China, Beijing. (in Chinese)

GB50011-2010 (2016) Code for seismic design of buildings. China Architecture & Building Press, Beijing. (in Chinese)
Harichandran RS, Vanmarcke EH (1986) Stochastic variation of earthquake ground motion in space and time. J Eng Mech 112(2):154–174
Harichandran RS, Hawwari A, Sweidan BN (1996) Response of long-span bridges to spatially varying ground motion. J Struct Eng 122(5):476–484
Hong F, Jiang JR, Li YT (1994) Power spectral models of earthquake ground motions and evaluation of its parameters. Earthq Eng Eng Vib 14(2):46–52. (in Chinese)
Hu YX (2006) Earthquake engineering, 2nd edn. Seismological Press, Beijing. (in Chinese)
Hu YX, Zhou XY (1962) The response of elastic systems under stationary and non-stationary ground motions. In: Collection of earthquake engineering research reports, vol 1. Science Press, Beijing, pp 33–50. (in Chinese)
Huo JR, Hu YX, Hong QM (1991) Study on envelope function of acceleration time history. Earthq Eng Eng Vib (1):1–12. (in Chinese)
Kanai K (1957) Semi-empirical formula for the seismic characteristics of the ground motion. Bull Earthq Res Inst 35(2):309–325
Kaul MK (1978) Stochastic characterization of earthquakes through their response spectrum. Earthq Eng Struct Dynam 6(5):497–509
Li J, Li GQ (1992) Introduction to earthquake engineering. Seismological Press, Beijing. (in Chinese)
Lin JH, Zhang YH (2004) Pseudo excitation method of random vibration. Science Press, Beijing. (in Chinese)
Liu ZJ (2007) Orthogonal expansion method of engineering stochastic dynamic loads and its application. Tongji University, Shanghai. (in Chinese)
Liu ZJ, Liu ZX (2017b) Simulation of fully non-stationary ground motion based on seismic design response spectrum. J Vibr Eng 30(3):457–465. (in Chinese)
Liu XH, Liu JW, Chen YQ (1982) Influence of damping ratio on response spectrum. J Seismol Res (1):124–132. (in Chinese)
Liu ZJ, Wang L, Dan QW et al (2015) Generalized evolutionary spectrum of non-stationary ground motion and its applications in seismic design of hydraulic structures. J Hydraul Eng 46(9): 1028–1036. (in Chinese)
Liu ZJ, Liu ZH, Liu W (2017a) Probability model of fully non-stationary ground motion with the target response spectrum compatible. J Vib Shock 36(2):32–38. (in Chinese)
Liu ZJ, Liu ZX, Ruan XX et al (2019) POD-based dimension reduction representation of stochastic ground motion fields. Sci Sin Technol 49(5):589–601. (in Chinese)
Liu ZX, Liu ZJ, Xu YZ (2019) Probability density evolution analysis of a shear-wall structure under stochastic ground motions by shaking table test. Soil Dyn Earthq Eng 122:53–66
Loh CH, Ku BD (1995) An efficient analysis of structural response for multiple-support seismic excitations. Eng Struct 17(1):15–26
Loh CH, Yeh YT (1988) Spatial variation and stochastic modelling of seismic differential ground movement. Earthq Eng Struct Dynam 16(4):583–596
Luco JE, Wong HL (1986) Response of a rigid foundation to a spatially random ground motion. Earthq Eng Struct Dynam 14(6):891–908
Ohsaki Y (2008) Introduction to spectral analysis of ground motions (trans: Tian Q), 2nd edn. Seismological Press, Beijing. (in Chinese)
Oliveira CS, Hao H, Penzien J (1991) Ground motion modeling for multiple-input structural analysis. Struct Saf 10(1–3):79–93
Ou JP, Wang GY (1988) Random vibration of structures. Higher Education Press, Beijing. (in Chinese)
Ou JP, Niu DT, Du XL (1991) Random earthquake ground motion model and its parameter determination used in aseismic design. Earthq Eng Eng Vib 11(3):45–54. (in Chinese)
Shibata A (2020) Structural seismic analysis (trans: Qu Z), 3rd edn. China Architecture & Building Press, Beijing. (in Chinese)

Shinozuka M, Sato Y (1967) Simulation of nonstationary random process. J Eng Mech Div 93(1): 11–40

Solari G, Spinelli P (1984) Time-domain analysis of tall buildings response to wind action. In: Proceedings of the 3rd International Conference on Tall Buildings, Hong Kong & Guangzhou, pp 278–284

Tajimi H (1960) A statistical method of determining the maximum responses of a building structure during an earthquake. In: Proceeding of the 2nd world conference on earthquake engineering, Japan, pp 781–797

Trifunac MD, Brady AG (1975) A study on the duration of strong earthquake ground motion. Bull Seismol Soc Am 65(3):581–626

Tubino F, Carassale L, Solari G (2003) Seismic response of multi-supported structures by proper orthogonal decomposition. Earthq Eng Struct Dynam 32:1639–1654

Vanmarcke EH (1983) Random fields: Analysis and synthesis. MIT Press, Cambridge, MA

Xie LL, Yu SJ (1982) Principles of strong motion observation and analysis. Seismological Press, Beijing. (in Chinese)

Xue SD, Wang XS, Cao Z (2003) Parameters study on seismic random model based on the new seismic code. Chin Civil Eng J 36(5):5–10. (in Chinese)

Yang QS, Tian YJ (2014) Earthquake ground motions and artificial generation. Science Press, Beijing. (in Chinese)

Zerva A, Zervas V (2002) Spatial variation of seismic ground motions: an overview. Appl Mech Rev 55(3):271–297

Chapter 3
Dimension-Reduction Representation of Univariate Stochastic Ground Motions: Frequency-Domain Analysis Method

Abstract In seismic analysis of engineering structures, one-dimensional one-variate (1D-1V) non-stationary stochastic processes are commonly employed to simulate uniform seismic actions for point-like structures with relatively small spans, such as towering structures and high-rise buildings. Among available approaches, the spectral representation method (SRM) for non-stationary stochastic processes, commonly considered as frequency-domain analysis method, is one of the most widely adopted strategies for simulating stochastic earthquake ground motions. Therefore, this chapter examines the fundamental theory of SRM for simulating non-stationary stochastic processes, including the original SRM, the conventional Monte Carlo method and the novel dimension-reduction simulation methods. Numerical examples of non-stationary stochastic ground motion are presented to verify the effectiveness and advantages of the proposed approach. Finally, the necessary and sufficient conditions for the dimension-reduction simulation method are clarified, along with its physical interpretation, full probabilistic nature, and its main strengths and limitations.

Keywords Non-stationary stochastic processes · Original spectral representation · Dimension-reduction simulation methods · Conventional Monte Carlo methods · Random orthogonal functions · Stochastic ground motion

Earthquake ground motions exhibit strong randomness. As far back as the 1940s, Housner (1947) suggested employing stochastic processes to characterize seismic ground motions. For several decades thereafter, the simulation of stochastic ground motions has consistently been a focal point of research in earthquake engineering. In seismic analysis of engineering structures, one-dimensional one-variable (1D-1V) non-stationary stochastic processes are commonly employed to simulate the uniform seismic actions experienced by point-like structures with relatively small spans, such as towering structures and high-rise buildings. Among them, the spectral representation method (SRM) of non-stationary stochastic processes, which is also commonly considered as frequency-domain analysis method, is one of the most widely used approaches for simulating stochastic earthquake ground motions. Therefore, this chapter will delve into the elementary theory of SRM for simulating

Z. Liu et al., *Dimension-reduction Representation of Stochastic Ground Motion and Engineering Applications*, https://doi.org/10.1007/978-981-92-0168-6_3

non-stationary stochastic processes, as well as the advancements in the SRM-based dimension-reduction simulation methods.

3.1 Original Spectral Representation of Non-Stationary Stochastic Processes

Based on the theory of spectral decomposition associated with stochastic processes, a real-valued non-stationary process $X(t)$ with zero-mean can be formulated in Fourier-Stieltjes integral form, such that (Priestley 1965):

$$X(t) = \int_{-\infty}^{\infty} A(t,\omega)\mathrm{e}^{\mathrm{i}\omega t}\mathrm{d}Z(\omega) \tag{3.1}$$

where $A(t, \omega)$ signifies the deterministic real-valued time-frequency modulation function, and there exists $A(t, -\omega) = A(t, \omega)$. $Z(\omega)$ denotes a complex orthogonal incremental process, and its increment $\mathrm{d}Z(\omega)$ satisfies the following basic conditions:

$$\mathrm{d}Z(-\omega) = \mathrm{d}Z^*(\omega), \quad E[\mathrm{d}Z(\omega)] = 0, \quad E[\mathrm{d}Z(\omega)\mathrm{d}Z^*(\omega')] = S(\omega)\delta_{\omega\omega'}\mathrm{d}\omega \tag{3.2}$$

where $E[\cdot]$ denotes the mathematical expectation. $\delta_{\omega\omega'}$ is the Kronecker-delta, that is, when $\omega = \omega'$, there exists $\delta_{\omega\omega'} = 1$, and in case of $\omega \neq \omega'$, there exists $\delta_{\omega\omega'} = 0$. $S(\omega)$ is the two-sided PSD function of the corresponding stationary process, satisfying $S(-\omega) = S(\omega)$ and $S(\omega_0) = S(0) = 0$. Commonly, the stationary PSD function $S(\omega)$ is a non-negative real-valued one.

Now, let's define the complex increment $\mathrm{d}Z(\omega)$as follows:

$$\mathrm{d}Z(\omega) = \mathrm{d}U(\omega) - \mathrm{i}\mathrm{d}V(\omega) \tag{3.3}$$

where the real increments $\mathrm{d}U(\omega)$ and $\mathrm{d}V(\omega)$ hold the basic conditions as below (Liu et al. 2025):

$$\mathrm{d}U(-\omega) = \mathrm{d}U(\omega), \quad \mathrm{d}V(-\omega) = -\mathrm{d}V(\omega) \tag{3.4a}$$

$$E[\mathrm{d}U(\omega)] = E[\mathrm{d}V(\omega)] = 0 \tag{3.4b}$$

$$E[\mathrm{d}U(\omega)\mathrm{d}U(\omega)] = E[\mathrm{d}U(\omega)\mathrm{d}U(-\omega)] = \frac{1}{2}S(\omega)\mathrm{d}\omega \tag{3.4c}$$

$$E[\mathrm{d}V(\omega)\mathrm{d}V(\omega)] = -E[\mathrm{d}V(\omega)\mathrm{d}V(-\omega)] = \frac{1}{2}S(\omega)\mathrm{d}\omega \tag{3.4d}$$

$$E[\mathrm{d}U(\omega)\mathrm{d}U(\omega')] = E[\mathrm{d}V(\omega)\mathrm{d}V(\omega')] = 0, \quad \omega' \neq \pm\omega \tag{3.4e}$$

$$E[\mathrm{d}U(\omega)\mathrm{d}V(\omega')] = 0 \tag{3.4f}$$

Obviously, Eq. (3.3) and Eq. (3.4) meet all the basic conditions defined in Eq. (3.2).

Introducing Eq. (3.3) and Eq. (3.4a) into Eq. (3.1), and taking into account the Euler's formula $e^{\pm i\omega t} = \cos(\omega t) \pm i\sin(\omega t)$, we can achieve the real-valued finite series form of the non-stationary process $X(t)$, followed by:

$$
\begin{aligned}
X(t) &= \int_{-\infty}^{\infty} A(t,\omega)e^{i\omega t}[\mathrm{d}U(\omega) - i\mathrm{d}V(\omega)] \\
&= \int_{-\infty}^{0} A(t,\omega)e^{i\omega t}[\mathrm{d}U(\omega) - i\mathrm{d}V(\omega)] + \int_{0}^{\infty} A(t,\omega)e^{i\omega t}[\mathrm{d}U(\omega) - i\mathrm{d}V(\omega)] \\
&= \int_{0}^{\infty} A(t,\omega)e^{-i\omega t}[\mathrm{d}U(\omega) + i\mathrm{d}V(\omega)] + \int_{0}^{\infty} A(t,\omega)e^{i\omega t}[\mathrm{d}U(\omega) - i\mathrm{d}V(\omega)] \\
&= 2\int_{0}^{\infty} A(t,\omega)[\cos(\omega t)\mathrm{d}U(\omega) + \sin(\omega t)\mathrm{d}V(\omega)] \\
&\approx 2\sum_{k=1}^{N} A(t,\omega_k)[\cos(\omega_k t)\Delta U(\omega_k) + \sin(\omega_k t)\Delta V(\omega_k)]
\end{aligned}
\tag{3.5}
$$

where $\omega_k = (k-0.5)\Delta\omega$, $\Delta\omega$ denotes the frequency interval. N is the number of discrete frequency terms. Typically, to guarantee the necessary accuracy, N should be large sufficiently, whereas $\Delta\omega$ should be small enough.

Moreover, considering the conditions included in Eqs. 3.4b–f, the orthogonal increments $\Delta U(\omega_k)$ and $\Delta V(\omega_k)$ can be expressed as the following:

$$
\Delta U(\omega_k) = \sqrt{\frac{1}{2}S(\omega_k)\Delta\omega}\, R_k, \quad \Delta V(\omega_k) = \sqrt{\frac{1}{2}S(\omega_k)\Delta\omega}\, I_k; \quad k = 1, 2, \cdots, N \tag{3.6}
$$

where $\{R_k, I_k\}_{k=1}^{N}$ denotes a set of real-valued standard orthogonal random variables, holding the basic conditions as follows:

$$
E[R_k] = E[I_k] = 0, \quad E[R_k R_m] = E[I_k I_m] = \delta_{km}, \quad E[R_k I_m] = 0 \tag{3.7}
$$

Substituting Eq. (3.6) into Eq. (3.5) leads to the following finite series form for the non-stationary process $X(t)$, given by (Liu et al. 2025):

$$
X(t) \approx \sum_{k=1}^{N} \sqrt{2S_X(\omega_k, t)\Delta\omega}[R_k\cos(\omega_k t) + I_k\sin(\omega_k t)] \tag{3.8}
$$

where $S_X(\omega, t) = A^2(t,\omega)S(\omega)$ is the two-sided EPSD function of the non-stationary stochastic process $X(t)$.

Here, the mean square relative error of the non-stationary process $X(t)$ is provided, such that (Liu et al. 2016):

$$\varepsilon(N) = 1 - \frac{\int_0^{\omega_\mathrm{u}} \int_0^T S_X(\omega, t)\mathrm{d}t\mathrm{d}\omega}{\int_0^{\infty} \int_0^T S_X(\omega, t)\mathrm{d}t\mathrm{d}\omega} \tag{3.9}$$

where $\omega_\mathrm{u} = (N-0.5)\Delta\omega$ denotes the upper cut-off frequency, and T denotes the duration of the non-stationary stochastic process, receptively. In practical engineering, the error is frequently considered to be $\varepsilon(N) \leqslant 5\%$.

As evident from Eq. (3.7), due to the absence of the specified probability distribution for the set of standard orthogonal random variables $\{R_k, I_k\}_{k=1}^N$, Eq. (3.8) cannot be utilized directly for simulating non-stationary stochastic processes. Consequently, Eq. (3.8) can be termed the original SRM (O-SRM) of non-stationary stochastic processes.

It should be noted that when employing the SRM to simulate 1D-1V non-stationary stochastic processes, the generated sample functions display periodicity, with the period denoted as $T' = 2\pi/\Delta\omega$. This suggests that, with ω_u held constant, as $\Delta\omega$ decreases and N increases, T' becomes larger, leading to the simulated sample function exhibiting longer period. Therefore, to avoid distortion, according to the sampling theorem, the time interval of the SRM within the time domain should satisfy the condition $\Delta t \leq \pi/\omega_\mathrm{u}$.

3.2 Conventional Monte Carlo Methods

To simulate non-stationary stochastic processes, it is necessary to specify the probability distribution of the set of standard orthogonal random variables $\{R_k, I_k\}_{k=1}^N$ in Eq. (3.8). Traditionally, there are usually two approaches. One is to directly define $\{R_k, I_k\}_{k=1}^N$ as mutually independent standard Gaussian random variables, which leads to the conventional random amplitude simulation formula. The other is to define $\{R_k, I_k\}_{k=1}^N$ as trigonometric function with mutually independent random phase angles, which leads to the conventional random phase simulation formula. Regardless of the form in which $\{R_k, I_k\}_{k=1}^N$ is defined, it must satisfy the basic conditions (also recognized as necessary conditions) involved in Eq. (3.7). Essentially, both the conventional random amplitude method and the conventional random phase method fall under the category of conventional Monte Carlo method. This section will delve into the conventional Monte Carlo approach for simulating non-stationary stochastic processes, elucidating the relationship and origin of the conventional random amplitude method and the conventional random phase method.

3.2.1 Conventional Random Amplitude Method (Non-Deterministic Spectral Amplitude Method)

In the O-SRM of non-stationary processes, say Eq. (3.8), if the set of the orthogonal random variables $\{R_k, I_k\}_{k=1}^{N}$ is expressed as a set of mutually independent standard Gaussian random variables, the conventional random amplitude simulation formula, also recognized as the non-deterministic spectral amplitude method, can be written as follows:

$$\begin{aligned} X_1(t) &= \sum_{k=1}^{N} \sqrt{2S_X(\omega_k, t)\Delta\omega}[R_k \cos(\omega_k t) + I_k \sin(\omega_k t)] \\ &= \sum_{k=1}^{N} \sqrt{2S_X(\omega_k, t)\Delta\omega}\, \mathcal{A}_k \cos(\omega_k t - \alpha_k) \end{aligned} \tag{3.10}$$

in which

$$R_k = \mathcal{A}_k \cos \alpha_k, \quad I_k = \mathcal{A}_k \sin \alpha_k \tag{3.11}$$

where $X_1(t)$ denotes the simulated non-stationary process applying the conventional random amplitude method. Note that $\mathcal{A}_k$ denotes the random amplitude and α_k denotes the random phase, respectively. Hence, the conventional random amplitude method essentially considers the randomness of both amplitude and phase.

It should be clarified that, since $\{R_k, I_k\}_{k=1}^{N}$ comprises a set of mutually independent standard Gaussian random variables, according to Eq. (3.11), it can be proved that $\{\mathcal{A}_k, \alpha_k\}_{k=1}^{N}$ is also a set of mutually independent random variables, with $\mathcal{A}_k$ obeying Rayleigh distribution, say $p_{\mathcal{A}_k}(x) = x\exp(-x^2/2)$, $x \geqslant 0$ and α_k satisfying the uniform distribution over the interval (0, 2π], respectively. Conversely, if $\{\mathcal{A}_k, \alpha_k\}_{k=1}^{N}$ comprises a set of mutually independent random variables, where $\mathcal{A}_k$ conforms to Rayleigh distribution, say $p_{\mathcal{A}_k}(x) = x\exp(-x^2/2)$, $x \geqslant 0$ and α_k holds the uniform distribution over the interval (0, 2π], based on Eq. (3.11), it can also be demonstrated that $\{R_k, I_k\}_{k=1}^{N}$ is a set of mutually independent standard Gaussian random variables. It is evident that the form of the random variables $\{R_k, I_k\}_{k=1}^{N}$ defined in Eq. (3.11) completely fulfills the basic conditions of Eq. (3.7).

Furthermore, we can also define $\mathcal{A}_k = \sqrt{-2\ln\beta_k}$, where β_k obeys the uniform distribution over the interval (0, 1]. As a result, an alternative form of the random amplitude method can then be acquired:

$$X_1(t) = 2\sum_{k=1}^{N} \sqrt{-S_X(\omega_k, t)\Delta\omega \ln\beta_k}\, \cos(\omega_k t - \alpha_k) \tag{3.12}$$

where $\{\beta_k, \alpha_k\}_{k=1}^{N}$ is a set of mutually independent random variables, and α_k is distributed uniformly over the interval (0, 2π].

3.2.2 *Conventional Random Phase Method (Deterministic Spectral Amplitude Method)*

In the O-SRM of non-stationary processes, say Eq. (3.8), if the set of standard orthogonal random variables $\{R_k, I_k\}_{k=1}^{N}$ is defined as a set of trigonometric functions with mutually independent random phase, the conventional random phase formula, also recognized as the deterministic spectral amplitude method, can be obtained as follows (Shinozuka and Deodatis 1991):

$$X_2(t) = 2\sum_{k=1}^{N}\sqrt{S_X(\omega_k, t)\Delta\omega}\cos(\omega_k t - \alpha_k) \tag{3.13}$$

in which

$$R_k = \sqrt{2}\cos\alpha_k, \quad I_k = \sqrt{2}\sin\alpha_k \tag{3.14}$$

where $X_2(t)$ denotes the simulated non-stationary process applying the conventional random phase method. $\{\alpha_k\}_{k=1}^{N}$ is a set of mutually independent random phase with uniform distribution over the interval (0, 2π]. Similarly, the form of the random variables $\{R_k, I_k\}_{k=1}^{N}$ defined in Eq. (3.14) can satisfy the basic conditions of Eq. (3.7).

It should be noted that the sets of the independent random variables defined in Eq. (3.11) and Eq. (3.14) both completely satisfy the basic conditions provided in Eq. (3.7). As can be deduced from the above, the random amplitude method (Eqs. (3.10) and (3.12)) and the random phase method (Eq. (3.13)) are both derived from the O-RSM of non-stationary stochastic processes (Eq. (3.8)). Therefore, the simulation formulas, say Eqs. (3.10), (3.12) and (3.13), can be considered as the special cases of Eq. (3.8).

Obviously, in the case of the random amplitude method (Eqs. (3.10) and (3.12)), the simulated non-stationary stochastic process $X_1(t)$ possesses a randomness degree of $2N$. Conversely, in the case of the random phase method (Eq. (3.13)), the simulated non-stationary stochastic process $X_2(t)$ exhibits a randomness degree of N, which is merely half of that in $X_1(t)$. As a result, it is evident that by means of defining diverse random orthogonal function forms for the set of standard orthogonal random variables within the O-SRM, the simulated non-stationary stochastic process can possess different levels of randomness degree.

3.3 Novel Dimension-Reduction Simulation Methods

Although the conventional Monte Carlo method for simulating non-stationary stochastic processes has yielded fruitful research outcomes and is widely applied in recent years, it still faces two major challenges. Firstly, the large number of random variables and samples poses a significant challenge to the computational efficiency of nonlinear stochastic dynamic analysis for complex engineering structures. Secondly, owing to its inherent nature of random sampling, the conventional Monte Carlo method is unable to acquire comprehensive probabilistic information from the sample set, causing it difficult to conduct stochastic dynamic response analysis and reliability assessment of complex engineering structures within the framework of full probability. Through the research on the conventional Monte Carlo method, we found that if the set of standard orthogonal random variables in the O-SRM is defined as different forms of random orthogonal functions, the simulation formula can possess different levels of randomness degree. Inspired by this, a novel theory of dimension-reduction representation of non-stationary stochastic processes based on random orthogonal functions is proposed. In this theory, the set of standard orthogonal random variables in the O-SRM is defined as a cluster of orthogonal functions with a minimal number of elementary random variables. By introducing the constraint form of random orthogonal functions, dimension-reduction simulation of non-stationary stochastic processes can be effectively achieved.

3.3.1 *Random Orthogonal Functions*

For the 1D-1V non-stationary stochastic processes, the set of standard orthogonal random variables $\{R_k, I_k\}_{k=1}^{N}$ in the O-SRM, say Eq. (3.8), can be defined as the following form of random orthogonal function (Liu et al. 2016):

$$R_k = g_k(\boldsymbol{\Theta}), \quad I_k = h_k(\boldsymbol{\Theta}); \quad k = 1, 2, \cdots, N \tag{3.15}$$

where $g_k(\cdot)$ and $h_k(\cdot)$ signify the deterministic orthogonal function forms. $\boldsymbol{\Theta} = (\Theta_1, \Theta_2, \cdots, \Theta_q)$ denotes the q-dimensional elementary random vector, of which the joint PDF $p_{\boldsymbol{\Theta}}(\boldsymbol{\theta})$ is known. If the elements in the q-dimensional random vector $\boldsymbol{\Theta} = (\Theta_1, \Theta_2, \cdots, \Theta_q)$ are independent of each other, the joint PDF is calculated as $p_{\boldsymbol{\Theta}}(\boldsymbol{\theta}) = \prod_{l=1}^{q} p_{\Theta_l}(\theta_l)$, where $p_{\Theta_l}(\theta_l)$ denotes the PDF of the elementary random variable $\Theta_l (l = 1, 2, \cdots, q)$.

Obviously, the defined form of Eq. (3.15) should satisfy the basic conditions included in Eq. (3.7), which can be recognized as a necessary condition for simulation. Thus, Eq. (3.7) can be reformulated as follows (Liu et al. 2016):

$$E[R_k] = E[g_k(\boldsymbol{\Theta})] = \int_{\Omega} g_k(\boldsymbol{\theta}) p_{\boldsymbol{\Theta}}(\boldsymbol{\theta}) \mathrm{d}\boldsymbol{\theta} = 0 \tag{3.16a}$$

$$E[I_k] = E[h_k(\boldsymbol{\Theta})] = \int_{\Omega} h_k(\boldsymbol{\theta}) p_{\boldsymbol{\Theta}}(\boldsymbol{\theta}) \mathrm{d}\boldsymbol{\theta} = 0 \tag{3.16b}$$

$$E[R_k R_m] = E[g_k(\boldsymbol{\Theta}) g_m(\boldsymbol{\Theta})] = \int_{\Omega} g_k(\boldsymbol{\theta}) g_m(\boldsymbol{\theta}) p_{\boldsymbol{\Theta}}(\boldsymbol{\theta}) \mathrm{d}\boldsymbol{\theta} = \delta_{km} \tag{3.16c}$$

$$E[I_k I_m] = E[h_k(\boldsymbol{\Theta}) h_m(\boldsymbol{\Theta})] = \int_{\Omega} h_k(\boldsymbol{\theta}) h_m(\boldsymbol{\theta}) p_{\boldsymbol{\Theta}}(\boldsymbol{\theta}) \mathrm{d}\boldsymbol{\theta} = \delta_{km} \tag{3.16d}$$

$$E[R_k I_m] = E[g_k(\boldsymbol{\Theta}) h_m(\boldsymbol{\Theta})] = \int_{\Omega} g_k(\boldsymbol{\theta}) h_m(\boldsymbol{\theta}) p_{\boldsymbol{\Theta}}(\boldsymbol{\theta}) \mathrm{d}\boldsymbol{\theta} = 0 \tag{3.16e}$$

where Ω denotes the distribution interval of the elementary random vector $\boldsymbol{\Theta}$.

In the O-SRM (Eq. (3.8)) of the 1D-1V non-stationary stochastic processes, the forms of random orthogonal function (Eq. (3.15)) that satisfy the basic conditions involved in Eq. (3.16) are diverse, encompassing trigonometric functions and orthogonal polynomials. Moreover, the constructed set of standard orthogonal random variables $\{R_k, I_k\}_{k=1}^{N}$ can be either non-Gaussian or Gaussian. This book will introduce them individually in the following different chapters. Here, we will focus solely on introducing random orthogonal functions that incorporate trigonometric function forms.

For the sake of convenience, consider the following random orthogonal functions based on trigonometric function from, given by:

$$X_n = \sqrt{2}\cos(n\Theta + \varphi), \quad Y_n = \sqrt{2}\sin(n\Theta + \varphi); \quad n = 1, 2, \cdots, N \tag{3.17}$$

where the elementary random variable Θ uniformly distributes over the interval $(0, 2\pi]$, and φ is a deterministic value in $(0, 2\pi]$ generally valued by $\varphi = \pi/4$. It is easy to verify that $\{X_n, Y_n\}_{n=1}^{N}$ defined in Eq. (3.17) fully satisfies the basic conditions involved in Eq. (3.16). Therefore, it is a set of standard orthogonal random variables.

In fact, the set of standard orthogonal random variables $\{X_n, Y_n\}_{n=1}^{N}$ defined by Eq. (3.17) is identically distributed, possessing the same PDF as follows:

$$p_{X_n}(\theta) = p_{Y_n}(\theta) = \begin{cases} \dfrac{1}{\pi\sqrt{2-\theta^2}}, & -\sqrt{2} < \theta < \sqrt{2} \\ 0, & \text{other} \end{cases}; \quad n = 1, 2, \cdots, N \tag{3.18}$$

And the same cumulative distribution function (CDF) is:

$$F_{X_n}(\theta) = F_{Y_n}(\theta) = \begin{cases} 0, & \theta < -\sqrt{2} \\ \dfrac{1}{2} + \dfrac{1}{\pi}\arcsin\dfrac{\theta}{\sqrt{2}}, & -\sqrt{2} \leq \theta \leq \sqrt{2}; \quad n = 1, 2, \cdots, N \\ 1, & \theta > \sqrt{2} \end{cases} \tag{3.19}$$

Obviously, the probability distribution of the set of standard orthogonal random variables $\{X_n, Y_n\}_{n=1}^N$ defined by Eq. (3.17) is the same non-Gaussian one. Then, we utilize the CDF with the same distribution (Eq. (3.19)) of $\{X_n, Y_n\}_{n=1}^N$ to construct two sets of standard random variables with the same uniform distribution and the same Gaussian distribution, respectively.

Firstly, suppose the CDF of the random variable X is $F_X(x)$. It can be proved that if the random variable is defined as $Z = F_X(X)$, then Z is a random variable uniformly distributed over the interval (0, 1] (Liu et al. 2022). According to this, it is straightforward to construct a set of random variables with the same uniform distribution over the interval (0, 1], such that:

$$Z_n^{(1)} = F_{X_n}(X_n) = \frac{1}{2} + \frac{1}{\pi}\arcsin\frac{X_n}{\sqrt{2}} = \frac{1}{2} + \frac{1}{\pi}\arcsin[\cos(n\Theta + \varphi)], \quad n = 1, 2, \cdots, N \tag{3.20a}$$

or

$$Z_n^{(2)} = F_{Y_n}(Y_n) = \frac{1}{2} + \frac{1}{\pi}\arcsin\frac{Y_n}{\sqrt{2}} = \frac{1}{2} + \frac{1}{\pi}\arcsin[\sin(n\Theta + \varphi)], \quad n = 1, 2, \cdots, N \tag{3.20b}$$

Apparently, the mean value and the standard derivation of the set of random variables $\{Z_n^{(1)}, Z_n^{(2)}\}_{n=1}^N$ are $\frac{1}{2}$ and $\frac{1}{2\sqrt{3}}$, respectively. Further, through the following standardized processing:

$$\begin{cases} X_n^{(1)} = \dfrac{Z_n^{(1)} - 1/2}{1/2\sqrt{3}} = \dfrac{2\sqrt{3}}{\pi}\arcsin[\cos(n\Theta + \varphi)] \\ Y_n^{(1)} = \dfrac{Z_n^{(2)} - 1/2}{1/2\sqrt{3}} = \dfrac{2\sqrt{3}}{\pi}\arcsin[\sin(n\Theta + \varphi)] \end{cases}, \quad n = 1, 2, \cdots, N \tag{3.21}$$

the obtained $\{X_n^{(1)}, Y_n^{(1)}\}_{n=1}^N$ is thus a set of standard random variables with the same uniform distribution.

Secondly, based on the equal-probability-inverse-transformation method for random variables (Liu et al. 2022), we can construct a set of standard random variables with the same Gaussian distribution:

$$\begin{cases} X_n^{(2)} = \Phi^{-1}\left(Z_n^{(1)}\right) = \Phi^{-1}\left(\frac{1}{2} + \frac{1}{\pi} \arcsin[\cos(n\Theta + \varphi)]\right) \\ Y_n^{(2)} = \Phi^{-1}\left(Z_n^{(2)}\right) = \Phi^{-1}\left(\frac{1}{2} + \frac{1}{\pi} \arcsin[\sin(n\Theta + \varphi)]\right) \end{cases}, \quad n = 1, 2, \cdots, N \tag{3.22}$$

where Φ^{-1} represents the inverse function of the CDF Φ with the standard Gaussian distribution.

It should be noted that the orthogonality of the standard random variable set defined in Eqs. (3.21) and (3.22) cannot be strictly satisfied, with a few exceptions where orthogonality cannot be completely fulfilled. Strictly speaking, Eqs. (3.21) and (3.22) are not the sets of standard orthogonal random variables. Therefore, when utilizing Eqs. (3.21) and (3.22), it is necessary to implement orthogonalization processing to transform them into the sets of standard orthogonal random variables.

It should be noted that the set of standard orthogonal random variables $\{X_n, Y_n\}_{n=1}^N$ with the same non-Gaussian distribution defined by Eq. (3.17), the set of orthogonalized standard random variables $\{X_n^{(1)}, Y_n^{(1)}\}_{n=1}^N$ with the same uniform distribution defined by Eq. (3.21), and the set of orthogonalized standard random variables $\{X_n^{(2)}, Y_n^{(2)}\}_{n=1}^N$ with the same Gaussian distribution defined by Eq. (3.22) all required to be converted into the set of standard orthogonal random variables $\{R_k, I_k\}_{k=1}^N$ in the O-SRM (Eq. (3.8)) through a deterministic one-to-one mapping relationship. In fact, this deterministic transformation relationship can be realized using the built-in functions of the MATLAB toolbox, namely rand ('state', 0) and temp = randperm (N). For instance, the deterministic one-to-one mapping relationship of $X_n \to R_k$ and $Y_n \to I_k$ is n = temp(k), which constitutes a sufficient condition for the dimension-reduction simulation approach of 1D-1V non-stationary stochastic processes. Thus, the target set of standard orthogonal random variables $\{R_k, I_k\}_{k=1}^N$ can be uniquely determined.

Finally, in response to the afore-mentioned two conventional simulation methods, say the random amplitude method and the random phase method, we can also propose two corresponding dimension-reduction simulation methods, as outlined below.

3.3.2 Dimension-Reduction Simulation Method Via Random Amplitude

For the novel dimension-reduction simulation method considering the randomness of both amplitude and phase, the set of standard orthogonal random variables $\{R_k, I_k\}_{k=1}^N$ can be expressed as below:

$$\begin{cases} R_k = \sqrt{-2\ln\left\{\frac{1}{2} + \frac{1}{\pi}\arcsin\left[\sin\left(\overline{k}\Theta_1 + \varphi\right)\right]\right\}}\cos\left(\overline{k}\Theta_2\right) \\ I_k = \sqrt{-2\ln\left\{\frac{1}{2} + \frac{1}{\pi}\arcsin\left[\sin\left(\overline{k}\Theta_1 + \varphi\right)\right]\right\}}\sin\left(\overline{k}\Theta_2\right) \end{cases}, \quad k, \overline{k} = 1, 2, \cdots, N \tag{3.23}$$

where Θ_1 and Θ_2 are mutually independent elementary random variables, both uniformly distributing over the interval $(0, 2\pi]$. φ is a constant in the range $(0, 2\pi]$, which is taken as $\varphi = \pi/4$ herein.

It is easy to verify that the set of standard orthogonal random variables $\{R_k, I_k\}_{k=1}^{N}$ defined in Eq. (3.23) fully satisfies the basic conditions included in Eq. (3.16), say the necessary conditions for simulation have been fulfilled. As a matter of fact, the set of standard orthogonal random variables $\{R_k, I_k\}_{k=1}^{N}$ defined in Eq. (3.23) has Gaussian distribution.

In Eq. (3.23), the deterministic one-to-one mapping relationship between $\overline{k}\left(\overline{k} = 1, 2, \cdots, N\right)$ and k ($k = 1, 2, \cdots, N$) is $\overline{k} = \text{temp}(k)$, serving as the sufficient condition required for simulation. The implementation method of this correspondence is the same as described in the previous section, and will not be elaborated here.

Hence, through substituting Eq. (3.23) into the O-SRM, i.e., Eq. (3.8), the dimension-reduction representation on the basis of random amplitude for simulating 1D-1V non-stationary stochastic processes can be achieved, such that (Liu et al. 2025):

$$\widehat{X}_1(t) = \sum_{k=1}^{N} \sqrt{2S_X(\omega_k, t)\Delta\omega}\,\mathcal{A}_{\overline{k}}(\Theta_1)\cos\left(\omega_k t - \overline{k}\Theta_2\right) \tag{3.24}$$

in which

$$\mathcal{A}_{\overline{k}}(\Theta_1) = \sqrt{-2\ln\left\{\frac{1}{2} + \frac{1}{\pi}\arcsin\left[\sin\left(\overline{k}\Theta_1 + \varphi\right)\right]\right\}} \tag{3.25}$$

where the random variables Θ_1 and Θ_2 respectively describe the randomness of amplitude and phase.

3.3.3 Dimension-Reduction Simulation Method Via Random Phase

For the novel dimension-reduction simulation method taking into account just the randomness of phase, the standard orthogonal random variable set $\{R_k, I_k\}_{k=1}^{N}$ can be expressed as follows:

$$R_k = \sqrt{2}\cos\left(\bar{k}\Theta + \varphi\right), \quad I_k = \sqrt{2}\sin\left(\bar{k}\Theta + \varphi\right); \quad k, \bar{k} = 1, 2, \cdots, N \tag{3.26}$$

where the elementary random variable Θ holds uniform distribution over the interval $(0, 2\pi]$. φ is a constant in the range $(0,\ 2\pi]$, which is taken as $\varphi = \pi/4$ herein. Apparently, the set of standard orthogonal random variable $\{R_k, I_k\}_{k=1}^{N}$ defined in Eq. (3.26) also sufficiently satisfies the basic conditions in Eq. (3.16).

Furthermore, substituting Eq. (3.26) into the O-SRM, i.e., Eq. (3.8), the dimension-reduction representation on the basis of random phase for simulating 1D-1V non-stationary stochastic processes can be achieved, such that (Liu et al. 2025):

$$\widehat{X}_2(t) = 2\sum_{k=1}^{N}\sqrt{S_X(\omega_k, t)\Delta\omega}\cos\left(\omega_k t - \bar{k}\Theta - \varphi\right) \tag{3.27}$$

where the elementary random variable Θ depicts the randomness of phase. In the same way, the deterministic one-to-one mapping relationship between $\bar{k}\left(\bar{k} = 1, 2, \cdots, N\right)$ and k $(k = 1, 2, \cdots, N)$ is $\bar{k} = \text{temp}(k)$.

3.4 Numerical Examples

In this section, the simulation of non-stationary ground motion stochastic processes is taken as numerical examples. Firstly, the full probability selection method of representative points for elementary random variables is introduced. Then, the dimension-reduction simulation methods based on random amplitude and random phase are both employed respectively to generate representative time-history set of non-stationary seismic ground motion processes. Through comparing with the conventional Monte Carlo methods, the accuracy and efficiency of the dimension-reduction simulation methods are demonstrated. Finally, by adopting the dimension-reduction simulation approaches and integrating them with the direct iteration method for EPSD, the representative time-history set of non-stationary ground motion processes that consistent with the design response spectrum specified in current seismic code is generated.

3.4.1 Method of Selecting Representative Points

To accomplish dimension-reduction simulation of non-stationary ground motion stochastic processes, it is first crucial to select the representative point set of the elementary random vector $\boldsymbol{\Theta} = (\Theta_1, \Theta_2, \cdots, \Theta_q)$. For cases involving just one elementary random variable, the equal probability point selection method can be directly applied, where each representative point has the same assigned probability. In this section, concerning Eqs. (3.17) and (3.27), since the elementary random variables follow a uniform distribution over the interval $(0, 2\pi]$, selecting points with equal probability is equivalent to selecting points with equal interval. Suppose that the representative point set of the elementary random variable Θ is $\{\theta_i\}_{i=1}^{n_{\rm sel}}$, where $n_{\rm sel}$ denotes the number of representative points, such that:

$$\theta_i = (i - 0.5) \times 2\pi / n_{\rm sel}; \quad i = 1, 2, \cdots, n_{\rm sel} \tag{3.28}$$

For the cases with two elementary random variables, say Eq. (3.24), the number theoretical method (NTM) (Li and Chen 2007) is applied herein. Firstly, it is necessary to establish a set of good lattice points (GLP) within the 2-dimensional unit space of $[0, 1]^2$. This set of uniformly distributed points can be constructed by utilizing the Fibonacci sequence. For the generated vector $\boldsymbol{h}$, there exists:

$$n_{\rm sel} = F_l, \quad h_1 = 1, \quad h_2 = F_{l-1} \tag{3.29}$$

where F_l represents an element in the Fibonacci sequence, which satisfies the following recurrence relation:

$$F_l = F_{l-1} + F_{l-2}, \quad F_0 = F_1 = 1; \quad l = 2, 3, 4, \cdots \tag{3.30}$$

Thus, the initial representative point sets $\{\vartheta_{1,i}, \vartheta_{2,i}\}_{i=1}^{n_{\rm sel}}$ can be obtained as follows (Hua and Wang 1981):

$$\begin{cases} \vartheta_{1,i} = \dfrac{i}{F_l} \\ \vartheta_{2,i} = \dfrac{F_{l-1} i}{F_l} - \operatorname{int}\left(\dfrac{F_{l-1} i}{F_l}\right) \end{cases}; \quad i = 1, 2, \cdots, F_l \tag{3.31}$$

where $0 < \vartheta_{l,\,i} \leq 1$ are uniformly scattered in the 2-dimensional unit space $[0, 1]^2$, say the GLP sets. int(·) denotes the operation of taking the largest integer that is not greater than the value inside the parentheses.

Then, the target representative point sets of the elementary random vector $\boldsymbol{\Theta} = (\Theta_1, \Theta_2)$, say $\{\theta_{1,i}, \theta_{2,i}\}_{i=1}^{n_{\rm sel}}$, where $0 < \theta_{l,\,i} \leq 2\pi$, can be obtained from the GLP sets $\{\vartheta_{1,i}, \vartheta_{2,i}\}_{i=1}^{n_{\rm sel}}$ by the linear transformation as $\theta_{l,\,i} = 2\pi\vartheta_{l,i}$. Figure 3.1 shows the initial representative point sets in the 2-dimensional space for the case that $n_{\rm sel} = 233$ using the NTM (Fig. 3.1a) and the Monte Carlo sampling (Fig. 3.1b), respectively. Clearly, the initial representative points generated using the NTM are uniformly distributed in the space $[0, 1]^2$. However, the initial representative points generated using the Monte Carlo method are scattered disorderly within the same space. In fact, the NTM is a deterministic sampling method that requires just one time of sampling to obtain a deterministic representative point set, eliminating

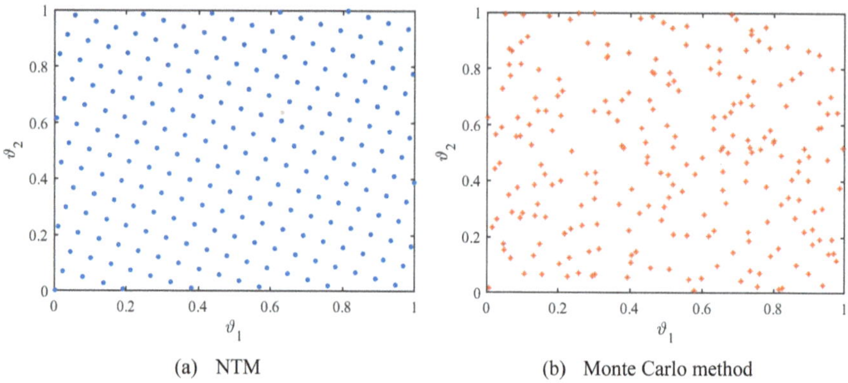

Fig. 3.1 Comparison of the initial representative point sets selected by NTM and Monte Carlo method in case of $n_{\text{sel}} = 233$

repeated sampling. This is significantly different from the random sampling nature of the Monte Carlo method.

In fact, the NTM is a spatial subdivision method based on equal probability, where the assigned probability of each representative point represents the proportion of the distribution space represented by the point in the entire distribution space. Since the representative points selected through the NTM are low discrepancy sequences spreading uniformly over the unit space, the assigned probability of each representative point is $P_i = 1/n_{\text{sel}}$, and all representative points constitute a complete probability set, say $\sum_{i=1}^{n_{\text{sel}}} P_i = 1$. Consequently, the NTM achieves a complete subdivision of the random vector distribution space. Meanwhile, each representative point can generate a corresponding representative sample, meaning that n_{sel} representative points can generate n_{sel} corresponding representative samples. Therefore, the generated set of representative samples can constitute a complete probability set. Benefiting from this, the representative ground motion time-history set generated by the proposed dimension-reduction simulation methods processes the complete probabilistic information of the simulated random process, which is naturally consistent with the probability density evolution theory. Specifically, within a conservative stochastic dynamic system, the transmission of probabilistic information from random sources such as external excitations and system parameters to the probabilistic output of structural dynamic responses is clear and comprehensive. Hence, the dimension-reduction simulation methods establish a solid foundation for integrating with the third-generation structural design theories, such as the probability density evolution theory, to conduct complete probabilistic stochastic dynamic response analysis and dynamic reliability evaluation of complex engineering structures.

3.4.2 Non-Stationary Stochastic Ground Motion Processes

Here, for the purpose of generating time-frequency fully non-stationary stochastic ground motions, the second type of EPSD model introduced in Chap. 2 (Eq. (2.72)) is employed herein, say $S_X(\omega, t) = |A(t, \omega)|^2 S(\omega)$. The improved Deodatis & Shinozuka model (Eqs. (2.69) and (2.70)) is adopted as the time-frequency modulation function $A(t, \omega)$. The PSD of the corresponding stationary process is characterized by the Clough-Penzien model (Eq. (2.10)). In the numerical examples, suppose that the seismic fortification intensity is 8 degree, the soil site is II and the design earthquake group is the second group. Then the specific parameter values for simulation are listed in Table 3.1.

The dimension-reduction simulation method based on random amplitude with two elementary random variables and the dimension-reduction simulation method based on random phase with one elementary random variable are respectively employed to generate the representative time-history sets of 1D-1V non-stationary stochastic ground motion processes. Specifically, for the dimension-reduction simulation method based on random amplitude, the target representative point set $\{\theta_{1,i}, \theta_{2,i}\}_{i=1}^{n_{\text{sel}}}$ of the basic random vector $\boldsymbol{\Theta} = (\Theta_1, \Theta_2)$ is substituted into Eq. (3.24), and implement the deterministic one-to-one mapping between $\overline{k}(\overline{k} = 1, 2, \cdots, N)$ and k $(k = 1, 2, \cdots, N)$, then the representative time-history set of non-stationary stochastic ground motion acceleration processes can be generated. Similarly, for the dimension-reduction simulation method based on random phase, by substituting the target representative point set $\{\theta_i\}_{i=1}^{n_{\text{sel}}}$ of the basic random variable Θ into Eq. (3.27), and implement the deterministic one-to-one mapping between $\overline{k}(\overline{k} = 1, 2, \cdots, N)$ and k $(k = 1, 2, \cdots, N)$, the representative time-history set of non-stationary stochastic seismic acceleration processes can be produced. Obviously, the number of the representative time-histories generated by the dimension-reduction simulation methods is also $n_{\text{sel}} = 233$.

Table 3.1 Parameter values for simulating 1D-1V non-stationary stochastic ground motion processes

Parameter	Value	Parameter	Value
PGA $a_{\max}$ (cm/s^2)	200	Parameter of modulation function a (s^{-1})	0.2
Dominant frequency of soil ω_g (rad/s)	15.71	Duration time T (s)	25
Critical damping of soil ζ_g	0.72	Upper cut-off frequency ω_u (rad/s)	240
Dominant frequency of rock ω_f (rad/s)	1.571	Frequency interval $\Delta\omega$ (rad/s)	0.15
Critical damping of rock ζ_f	0.72	Number of frequency intervals N	1600
Peak factor r	2.83	Time interval Δt (s)	0.01
Number of representative points n_{sel}	233	The mean square relative error $\varepsilon(N)$ (%)	2.0

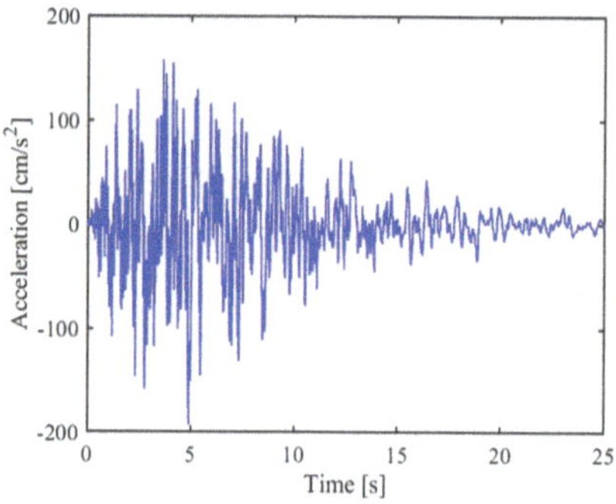

(a) DR-RA

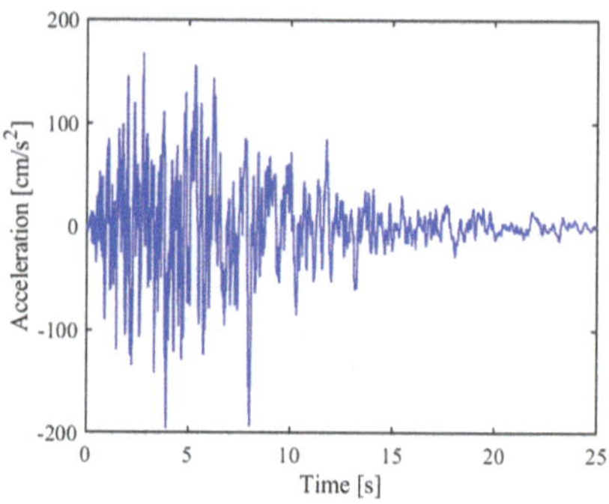

(b) DR-RP

Fig. 3.2 Acceleration representative time-histories applying the dimension-reduction methods

Figure 3.2 shows the 50-th representative time-history curves of non-stationary seismic acceleration processes generated by the dimension-reduction simulation methods based on random amplitude (DR-RA) and random phase (DR-RP), respectively. As can be seen from the figures, the acceleration time-histories exhibit obvious non-stationary characteristics.

In order to quantitatively assess the difference between the target non-stationary stochastic process and the simulated process, two families of average relative errors are defined as follows.

The average relative error upon the mean (Mean-ARE) of non-stationary stochastic process $X(t)$ is defined as

$$\varepsilon_{\mathrm{m},1} = \frac{1}{N_{\mathrm{t}}} \sum_{k=1}^{N_{\mathrm{t}}} \left| \frac{\mu(t_k) - \widehat{\mu}(t_k)}{\sigma(t_k)} \right|, \quad \text{for} \quad \mu(t) = E[X(t)] = 0 \tag{3.32}$$

The average relative error upon the standard deviation function (Std.D-ARE) of non-stationary stochastic process $X(t)$ is defined as

$$\varepsilon_{\mathrm{m},2} = \frac{1}{N_{\mathrm{t}}} \sum_{k=1}^{N_{\mathrm{t}}} \left| \frac{\sigma(t_k) - \widehat{\sigma}(t_k)}{\sigma(t_k)} \right| \tag{3.33}$$

where $\mu(t_k)$ and $\sigma(t_k)$ denote the values of mean and standard deviation (Std.D) of the target non-stationary stochastic process at the k-th time instant, respectively. $\widehat{\mu}(t_k)$ and $\widehat{\sigma}(t_k)$ denote the values of mean and Std.D of the simulated process at the k-th time instant, respectively. N_{t} represents the number of time instants involved in representative time-histories.

Figures 3.3 and 3.4 present comparisons of the mean and Std.D of the representative time-history sets of the non-stationary seismic acceleration processes generated through the dimension-reduction simulation methods, against the corresponding

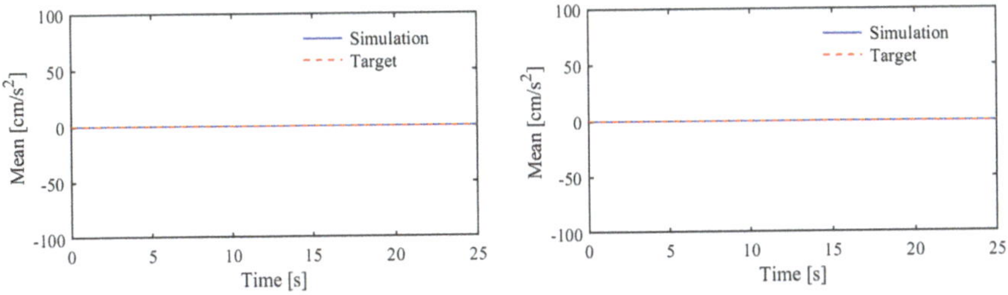

(a) DR-RA (b) DR-RP

Fig. 3.3 Comparison of the mean between the simulation values and the corresponding target values

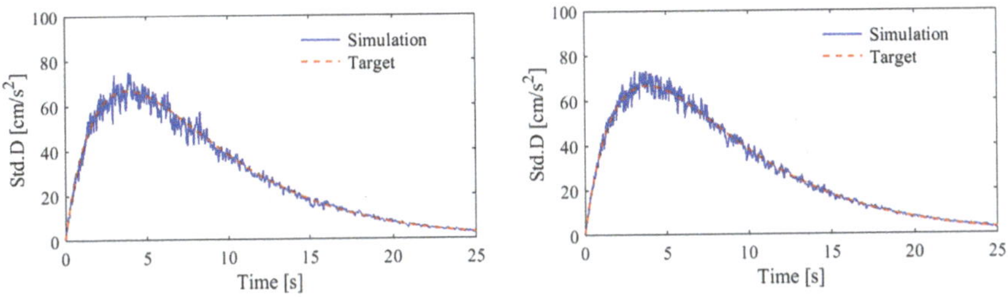

(a) DR-RA (b) DR-RP

Fig. 3.4 Comparison of the Std. D between the simulation values and the corresponding target values

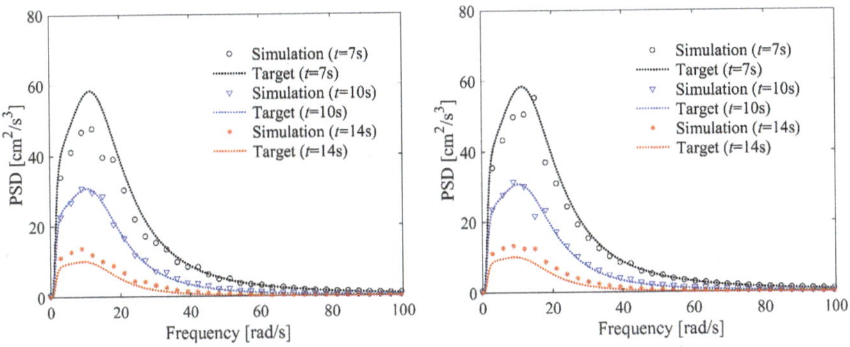

(a) DR-RA (b) DR-RP

Fig. 3.5 Comparison of the PSD at typical time instants between the simulation values and the corresponding target values

target values. Figure 3.5 compares the PSD of the simulated non-stationary seismic acceleration processes at typical time instants with the corresponding target values. It is evident from these figures that all the simulated values align well with the corresponding target ones in terms of the mean, Std.D and PSD, thereby confirming the validity of the proposed dimension-reduction simulation methods.

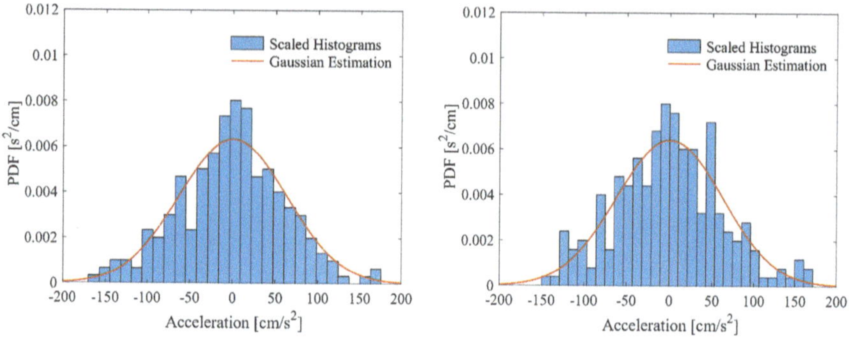

(c) DR-RA (d) DR-RP

Fig. 3.6 PDFs of the simulated non-stationary ground motion acceleration processes at time instant $t = 5$ s applying the dimension-reduction methods

Table 3.2 Comparisons of the accuracy and efficiency between the dimension-reduction simulation methods and the conventional Monte Carlo methods

Simulation method		Mean-ARE (%)	Std.D-ARE (%)	Time expense (s)
Dimension-reduction methods $n_{sel} = 233$	DR-RA	2.24×10^{-4}	4.66	0.21
	DR-RP	1.27×10^{-3}	3.68	0.11
Monte Carlo methods $n_{sel} = 233$	MC-RA	4.81	3.66	0.63
	MC-RP	5.58	3.86	0.62
Monte Carlo methods $n_{sel} = 1000$	MC-RA	2.48	1.86	0.73
	MC-RP	2.64	1.85	0.68

Figure 3.6 presents the PDFs of the representative time-history sets of non-stationary seismic acceleration processes generated by the dimension-reduction simulation methods at the time instant $t = 5$ s, employing both statistical method and Gaussian estimation technique. As depicted in the figures, irrespective of whether the defined random variables conform to Gaussian distribution (DR-RA) or non-Gaussian distribution (DR-RP), the simulated truncated random variables exhibit Gaussian distribution characteristics.

Table 3.2 presents the comparisons of the accuracy and efficiency between the dimension-reduction simulation methods and the conventional Monte Carlo methods including the conventional random amplitude method (MC-RA) and the conventional random phase method (MC-RP). It is found that the dimension-reduction simulation methods necessitate the generation of just over 200 representative time-histories, achieving an average relative error of less than 5% for both the mean and Std.D, thereby satisfying engineering accuracy demands. However, the conventional Monte Carlo methods usually require the generation of 1000 samples to attain an average relative error below 5% for both metrics, further validating the precision of the dimension-reduction approaches. From the perspective of computational efficiency, the conventional Monte Carlo methods require a larger number of samples to be generated, resulting in a longer total time. In addition, the time needed for structural time-history analysis remains constant under one single earthquake action. However, a smaller number of input ground motion samples translates to less time required for the overall dynamic time-history analysis of engineering structures. Consequently, the dimension-reduction simulation methods significantly enhance the efficiency of structural dynamic response analysis. In summary, the numerical examples fully verify the accuracy and efficiency of the proposed dimension-reduction simulation methods.

3.4.3 Non-Stationary Ground Motion Processes Consistent with Design Response Spectrum

In the following numerical examples, based on the Chinese "Code for Seismic Design of Buildings" (GB50011-2010 2016), the dimension-reduction simulation methods are employed to generate the representative time-history set of non-stationary seismic acceleration processes that are consistent with the design response spectrum. Commonly, there is a certain fitting error between the average response spectrum $\overline{S}_a(T_0, \zeta)$ of the representative time-history set of seismic acceleration processes generated by the dimension-reduction simulation methods and the design response spectrum $S_a(T_0, \zeta)$. To comply with the seismic design code, it is often necessary to iteratively refine all the generated representative time-histories. To this end, the two relative errors of the seismic acceleration response spectrum are defined as outlined below.

Average Relative Error

$$\varepsilon_m = \frac{1}{N_f}\sum_{k=1}^{N_f}\left|\frac{S_a(T_{0,k},\zeta) - \overline{S}_a(T_{0,k},\zeta)}{S_a(T_{0,k},\zeta)}\right| \le [\varepsilon_m] \tag{3.34}$$

where $T_{0,k}$ represents the k-th control point of the structural natural vibration period T_0. ζ denotes the structural damping ratio, which is taken as $\zeta = 0.05$. N_f denotes the number of discrete points for the structural natural vibration period within 6 s, say $N_f = 6/\Delta T_0$. ΔT_0 denotes the discrete interval for the structural natural vibration period, generally determined by $\Delta T_0 = 0.01$ s. $[\varepsilon_m]$ denotes the allowable average relative error, usually taken as $[\varepsilon_m] = 5\%$.

Maximum Relative Error

$$\varepsilon_{\mathrm{m}}=\frac{1}{N_{\mathrm{f}}}\sum_{k=1}^{N_{\mathrm{f}}}\left|\frac{S_{\mathrm{a}}(T_{0,k},\zeta)-\overline{S}_{\mathrm{a}}(T_{0,k},\zeta)}{S_{\mathrm{a}}(T_{0,k},\zeta)}\right|\leq[\varepsilon_{\mathrm{m}}] \tag{3.35}$$

where $[\varepsilon_{\max}]$ denotes the allowable maximum relative error, generally taken as $[\varepsilon_{\max}]=10\%$.

The average relative error ε_{m} reflects the overall fitting degree between the average response spectrum $\overline{S}_{\mathrm{a}}(T_0,\zeta)$ and the design response spectrum $S_{\mathrm{a}}(T_0,\zeta)$, while the maximum relative error $\varepsilon_{\max}$ reflects the local fitting degree between the two.

When the aforementioned two relative errors cannot be satisfied, iterative correction of the EPSD needs to be conducted, such that (Liu et al. 2017):

$$S_X(\omega,t)|_{k+1}=\begin{cases}S_X(\omega,t), & 0<\omega\leq\omega_{\mathrm{c}}\\ S_X(\omega,t)|_k\times\dfrac{S_{\mathrm{a}}(T_0,\zeta)^2}{\overline{S}_{\mathrm{a}}(T_0,\zeta)^2|_k}, & \omega>\omega_{\mathrm{c}}\end{cases},\quad k=0,1,\cdots,K \tag{3.36}$$

where $S_X(\omega,t)|_{k+1}$ and $S_X(\omega,t)|_k$ are the two-sided EPSD after k+1-th and k-th iteration, respectively. $\overline{S}_{\mathrm{a}}(T_0,\zeta)^2|_k$ denotes the square of the average response spectrum at the k-th iteration, in which $T_0=2\pi/\omega$. ω_{c} denotes the cut-off frequency, it can be taken as $\omega_{\mathrm{c}}=1.05$ rad/s.

Substitute Eq. (3.36) into either Eq. (3.24) or Eq. (3.27) to produce the representative time-history set for non-stationary seismic acceleration processes, compute the average response spectrum $\overline{S}_{\mathrm{a}}(T_0,\zeta)$, and verify Eqs. (3.34) and (3.35) until both two relative errors satisfy the criteria.

In the numerical examples, based on the representative time-history sets of non-stationary seismic acceleration processes generated in Sect. 3.4.2 using the dimension-reduction simulation methods based on random amplitude and random phase, the aforementioned direct iteration method for EPSD is further employed to generate the representative time-history sets of seismic acceleration processes that are consistent with the design response spectrum. Therefore, the parameter values for simulating seismic acceleration processes are all the same.

Figure 3.7 presents the representative time-histories of non-stationary seismic acceleration processes, generated through DR-RP and DR-RA, each encompassing iterations at both the 0th and 3rd iteration. It is evident from the figures that these representative time-histories exhibit typical non-stationary characteristics.

Figure 3.8 depicts the EPSD for Soil site II and Seismic group two. Specifically, Fig. 3.8a illustrates the outcome of the 0th iteration, whereas Figs. 3.8b, c display the result after the 3rd iteration using the dimension-reduction simulation methods based on random amplitude and random phase, respectively. It can be seen that the main frequency range of the EPSD in the 3rd iteration has become smaller, while the

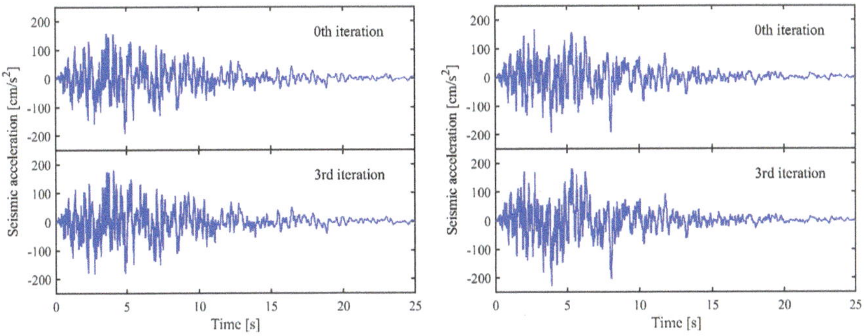

(a) DR-RA (b) DR-RP

Fig. 3.7 Representative time-histories of non-stationary seismic acceleration processes

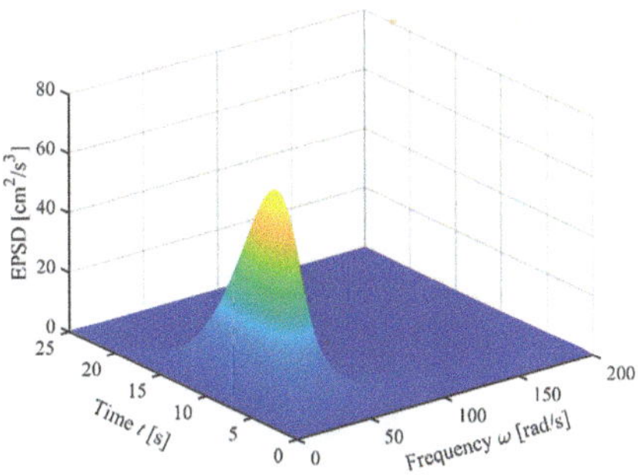

(a) 0th iteration

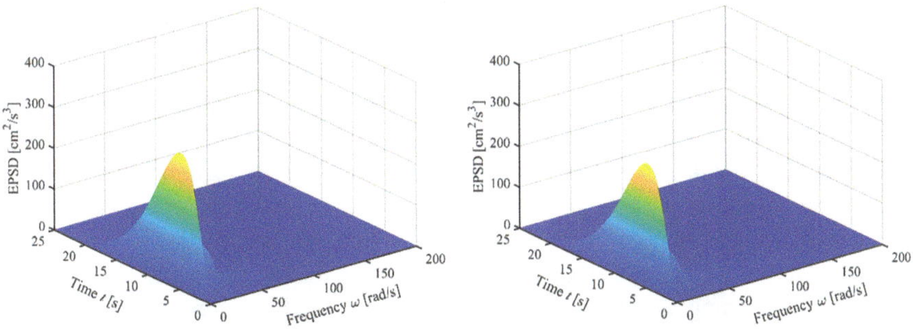

(b) 3rd iteration (DR-RA) (c) 3rd iteration (DR-RP)

Fig. 3.8 The EPSD for Soil site II and Seismic group two

spectral values have increased, indicating that the seismic energy is more concentrated within a certain range.

Figure 3.9 illustrates the comparison of the acceleration response spectrum of the representative time-history set generated by the dimension-reduction simulation methods with the design response spectrum. Observing the figures, it is evident

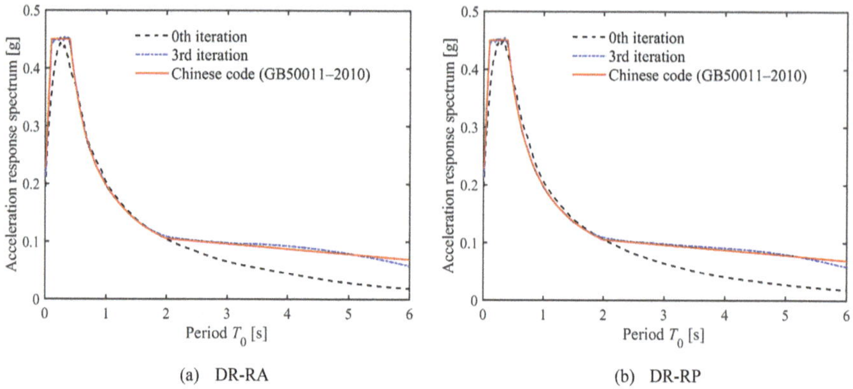

Fig. 3.9 Comparison of the simulated acceleration response spectrum with the design response spectrum

Table 3.3 Relative errors of response spectrum iteration (%)

Relative errors	Iteration times	Dimension-reduction simulation methods	
		DR -RA	DR-RP
ε_{m}	0	11.73	10.85
	3	0.88	0.73
ε_{max}	0	69.83	70.44
	3	6.85	6.03

that the results of the 0th iteration deviate significantly from the design response spectrum in the long-period range, attributed to the deliberate enhancement of spectral values in this range within the design response spectrum. Conversely, the results of the 3rd iteration demonstrate an exceptionally good alignment with the design response spectrum, which can fully comply with the requirements of structural seismic design.

Table 3.3 presents the relative errors between the average response spectrum of the seismic representative time-history sets generated by the dimension-reduction simulation methods and the design response spectrum. As can be seen from the table, after 3 iterations, both the average and maximum relative errors are within the allowable error range. This verifies the effectiveness of the dimension-reduction simulation methods combing with the iteration of EPSD for generating ground motions fitting with design response spectrum proposed in this book.

3.5 Conclusion Remarks

This chapter derives the O-SRM for 1D-1V non-stationary stochastic processes and traces back to the theoretical basis and essential origins of the conventional Monte Carlo methods and dimension-reduction methods based on frequency domain analysis for simulating non-stationary stochastic processes. Specifically, both the conventional Monte Carlo methods and dimension-reduction methods are derived from the O-SRM, and they are special cases of the O-SRM in essence. The dimension-reduction methods based on random orthogonal functions are proposed, facilitating precise simulation of non-stationary stochastic seismic ground motion processes with merely 1 to 2 elementary random variables. Numerical examples have verified the effectiveness of the dimension-reduction simulation methods, and some remarks are summarized as follows.

1. Sufficient and necessary conditions for dimension-reduction simulation methods. The simulation methods for 1D-1V non-stationary stochastic processes based on the O-SRM includes both the conventional Monte Carlo methods and the dimension-reduction simulation methods proposed in this book. They both belong to frequency domain analysis and require satisfaction of sufficient and necessary conditions. Among them, the basic conditions defined in Eq. (3.7) satisfied by the set of standard orthogonal random variables are the necessary conditions for simulation, which are the same for both conventional Monte Carlo methods and dimension-reduction methods. However, in the dimension-reduction methods, it is typically required to transform the predefined set of standard orthogonal random variables into the target set of standard orthogonal random variables via a deterministic one-to-one mapping. This deterministic mapping relationship constitutes a sufficient condition for the dimension-reduction approaches, whereas the sufficient condition for the conventional Monte Carlo methods hinges on the mutual independence of the random variables. Evidently, the sufficient conditions for these two methods are different. Actually, by defining alternative forms of random orthogonal functions that satisfy the basic conditions, various simulation methods can also be obtained.
2. The non-Gaussian nature of dimension-reduction simulation methods. In the conventional Monte Carlo simulation methods, regardless of whether it is the random amplitude method or the random phase method, the simulated random processes conform to either Gaussian processes or asymptotically Gaussian processes. However, in the dimension-reduction simulation method based on random amplitude, although the set of standard orthogonal random variables $\{R_k, I_k\}_{k=1}^{N}$ is a set of Gaussian random variables with the same distribution, the premise that their joint probability distribution is Gaussian cannot be presupposed. At this time, the independence and orthogonality of the random variables are not equivalent, thus it cannot be guaranteed that $\{R_k, I_k\}_{k=1}^{N}$ is a set of mutually independent random variables, and as a result it cannot be ensured that the simulated stochastic processes are Gaussian processes. Furthermore, in the dimension-reduction simulation method based on random phase, the set of standard orthogonal random variables $\{R_k, I_k\}_{k=1}^{N}$ is a set of non-Gaussian random variables with the same distribution. However, since they are not a set of mutually independent random variables, the simulated stochastic processes are generally not asymptotic Gaussian processes. In fact, engineering random

dynamic events, such as earthquake ground motions, are typically non-Gaussian processes. Consequently, employing the proposed dimension-reduction methods to simulate stochastic seismic ground motions is reasonable and effective.

3. Physical interpretation of dimension-reduction simulation methods. In the dimension-reduction methods for simulating 1D-1V non-stationary stochastic processes, the construction of random orthogonal function forms plays an important role, which is diverse and flexible. By defining different forms of random orthogonal functions, this book proposes two types of dimension-reduction methods (based on random amplitude and random phase), which exhibit different degrees of randomness. Therefore, by introducing different forms of random orthogonal functions (random constraints), efficient dimension-reduction of stochastic processes can be achieved, similar to the Rayleigh-Ritz method in deterministic structural dynamic systems, which involves reducing a high-dimensional structural system to a low-dimensional one by introducing shape vectors or basis vectors (deterministic constraints). Hence, the dimension-reduction ideas in stochastic dynamic systems and deterministic dynamic systems share the same physical significance.
4. Full probabilistic nature of dimension-reduction simulation methods. In the dimension-reduction methods for simulating 1D-1V non-stationary stochastic processes, due to the extremely small number of required elementary random variables (usually merely 1 to 2), it is sufficient to generate just a few hundred representative time-histories to achieve satisfactory accuracy. In contrast, the conventional Monte Carlo methods typically require hundreds or thousands of random variables, and the number of samples generated can reach thousands or even tens of thousands. Benefiting from this, the dimension-reduction methods can utilize advanced NTM for selecting a representative point set of the elementary random variables. Each representative point and each corresponding representative time-history generated have an assigned probability, and the probabilistic information of the representative time-history set is complete. Therefore, the dimension-reduction methods are fully probabilistic approaches, which is essentially consistent with the third-generation structural design theory (such as probability density evolution theory). It lays the foundation for achieving full probabilistic analysis of random seismic dynamic response and seismic reliability evaluation of complex engineering structures.

References

Housner GW (1947) Characteristics of strong-motion of earthquakes. Bull Seismol Soc Am 37(1): 19–31

Hua LK, Wang Y (1981) Applications of number theory to numerical analysis. Springer, Berlin

Li J, Chen JB (2007) The number theoretical method in response analysis of non-linear stochastic structures. Comput Mech 39(6):693–708

Liu ZJ, Liu W, Peng YB (2016) Random function based spectral representation of stationary and non-stationary stochastic processes. Prob Eng Mech 45:115–126

Liu ZJ, Liu ZH, Liu W (2017) Probability model of fully non-stationary ground motion with the target response spectrum compatible. J Vibr Shock 36(2):32–38. (in Chinese)
Liu ZJ, Chen JB, Peng YB (2022) Dynamics of structures. China Architecture & Building Press, Beijing. (in Chinese)
Liu ZX, Liu ZJ, Ruan XX et al (2025) Advances in dimension-reduction methods for simulating univariate and multivariate non-stationary stochastic processes via spectral representation. Prob Eng Mech 79:103720
Priestley MB (1965) Evolutionary spectra and non-stationary processes. J R Stat Soc Ser B Stat Methodol 27(3):204–237
Shinozuka M, Deodatis G (1991) Simulation of stochastic processes by spectral representation. Appl Mech Rev 44(4):191–204

Chapter 4
Dimension-Reduction Representation of Univariate Stochastic Ground Motions: Time-Domain Analysis Method

Abstract This chapter addresses three typical time-domain analysis methods for simulating univariate stochastic processes, say the Karhunen-Loève decomposition, filtered white noise process models in physics and time-domain representation of stochastic processes in mathematics. In particular, the filtered white-noise process models in physics focus on two types: the single-filtered white-noise process model and the double-filtered white-noise process model. For the time-domain representation of stochastic processes in mathematics, emphasis is placed on a unified time-domain representation form applicable to both stationary and non-stationary stochastic processes. Furthermore, by employing the constraint form of random orthogonal functions, the dimension-reduction simulation of non-stationary seismic acceleration processes can be accomplished using merely one elementary random variable.

Keywords Non-stationary stochastic processes · Karhunen-Loève decomposition · Double-filtered white noise process model · Time-domain representation

In the first three chapters, the Fourier transform relationship between the PSD function and the auto-correlation function of a stationary stochastic process has been introduced, as depicted in Eqs. (1.78) and (1.79), known as the Wiener-Khinchin formula. However, different Fourier transform pairs are frequently utilized across various literatures. To highlight the discrepancies in theoretical derivations among different Fourier transform pairs, the time-domain analysis method presented in this chapter specifically employs the Fourier transform pair that differs from those used in the preceding three chapters, as below:

$$\begin{cases} S(\omega) = \int_{-\infty}^{\infty} R(\tau)\,\mathrm{e}^{-\mathrm{i}\omega\tau}\mathrm{d}\tau \\ R(\tau) = \dfrac{1}{2\pi}\int_{-\infty}^{\infty} S(\omega)\,\mathrm{e}^{\mathrm{i}\omega\tau}\mathrm{d}\omega \end{cases} \tag{4.1}$$

Z. Liu et al., *Dimension-reduction Representation of Stochastic Ground Motion and Engineering Applications*, https://doi.org/10.1007/978-981-92-0168-6_4

It is worth noting that employing different Fourier transform pairs does not influence the final simulation outcomes. The primary aim of this arrangement is to facilitate readers' better understanding and flexible application of various Fourier transform pairs.

4.1 Karhunen-Loève Decomposition

In Sect. 1.1.4, through introducing eigen decomposition of matrix, the correlation structure of random vectors can be easily decomposed, thereby converting the correlated random vectors into a linear superposition form of mutually uncorrelated random variables. In fact, a stochastic process characterized by continuous time parameters can be viewed as an infinite-dimensional random vector. Consequently, this concept can also be extended to the expansion of stochastic processes, enabling the representation of a stochastic process as a linear superposition form of mutually uncorrelated random variables.

For a real-valued zero mean stochastic process $X(t)$ which is indexed on the bounded interval $t \in [0,\ T]$, suppose it is a second-order moment process. Then similar to Eq. (1.39), the auto-correlation function $R(t_1, t_2)$ of $X(t)$ can be decomposed as the following series form:

$$R(t_1,t_2)=\sum_{i=1}^{\infty}\lambda_i f_i(t_1)f_i(t_2)\approx\sum_{i=1}^{N}\lambda_i f_i(t_1)f_i(t_2) \tag{4.2}$$

where λ_i and $f_i(t)$ represent the i-th eigenvalue and eigenfunction of the auto-correlation function $R(t_1, t_2)$, respectively, which satisfy the following second Fredholm integral equation:

$$\int_0^T R(t_1,t_2)f_i(t_1)\mathrm{d}t_1=\lambda_i f_i(t_2) \tag{4.3}$$

In fact, the deterministic eigenfunction set comprises a set of standard orthogonal functions defined on the interval $[0,\ T]$, satisfying the following orthogonality conditions:

$$\int_0^T f_i(t)f_j(t)\mathrm{d}t=\delta_{ij} \tag{4.4}$$

Therefore, akin to Eq. (1.42), the zero-mean stochastic process can be decomposed as the following series:

$$X(t) = \sum_{i=1}^{\infty} \sqrt{\lambda_i} \xi_i f_i(t) \approx \sum_{i=1}^{N} \sqrt{\lambda_i} \xi_i f_i(t) \tag{4.5}$$

where $\{\xi_i\}_{i=1}^{N}$ is a set of standard orthogonal random variables, fulfilling the following basic conditions:

$$E[\xi_i] = 0, \quad E[\xi_i \xi_j] = \delta_{ij} \tag{4.6}$$

In Eq. (4.5), the mean square relative error of the finite series can be defined as:

$$\varepsilon(N) = 1 - \frac{\sum_{i=1}^{N} \lambda_i}{\int_0^T R(t,t) \mathrm{d}t} \tag{4.7}$$

Equation (4.5) represents the **Karhunen-Loève (K-L) decomposition** of the stochastic process $X(t)$, sometimes also referred to as the **Proper Orthogonal Decomposition (POD)**. In fact, according to the K-L decomposition of the stochastic process, say Eq. (4.5), it is easy to define its auto-correlation function. Then, applying the orthogonality of the eigenfunctions in Eq. (4.4), Eq. (4.2) can thus be obtained. This demonstrates that the stochastic process obtained via the **K-L decomposition** is statistically equivalent to the original stochastic process in terms of second-order moment. Obviously, the K-L decomposition applies to both stationary and non-stationary stochastic processes.

Generally, the eigenvalues in Eq. (4.3) are, at most, countable and strictly positive. Therefore, the eigenvalues λ_i can be arranged in a descending order. It has been demonstrated that the truncated finite K-L decomposition series in Eq. (4.5) for stochastic processes are optimal, i.e., when the same number of series terms N is considered, the mean square relative error is minimized (Ghanem and Spanos 1991). Just as in the decomposition of the correlation structure of a random vector, the foremost task is to conduct eigen-decomposition on the covariance matrix to acquire its eigenvalues and eigenvectors. When using the K-L decomposition, the primary task is to obtain the eigenvalues λ_i and standard orthogonal eigenfunctions $f_i(t)$ of the auto-correlation function $R(t_1, t_2)$ by solving the Fredholm integral function, say Eq. (4.3). However, except for very few cases, it is extremely difficult to directly obtain an analytical solution for the Fredholm integral equation. To this end, numerical solution methods are typically employed, and the implementation steps are as follows (Liu et al. 2017):

1. Choose a set of standard orthogonal functions $\{\psi_k(t)\}_{k=1}^{\infty}$ and utilize the first M terms to compute the eigenvalues and eigenfunctions of the auto-correlation function. Typically, M should be greater than or equal to the number of finite series terms in the K-L decomposition, say $M \geqslant N$.

2. Define the correlation matrix $\boldsymbol{R}$:

$$\boldsymbol{R} = \left[\rho_{ij}\right]_{M \times M} \tag{4.8}$$

where ρ_{ij} are the elements of the correlation matrix, and they can be calculated as follows:

$$\rho_{ij} = \int_0^T \int_0^T R(t_1, t_2)\psi_i(t_1)\psi_j(t_2)\mathrm{d}t_1\mathrm{d}t_2, \quad i,j = 1, 2, \cdots, M \tag{4.9}$$

Obviously, the correlation matrix $\boldsymbol{R}$ is a real symmetric matrix, with its eigenvalues $\widetilde{\lambda}_i$ typically being positive real numbers and its eigenvectors $\boldsymbol{\Phi}_i$ all being real vectors. The eigenvalues $\widetilde{\lambda}_i$ and eigenvectors $\boldsymbol{\Phi}_i$ can be obtained through the following eigenvalue issue:

$$\boldsymbol{R\Phi}_i = \widetilde{\lambda}_i \boldsymbol{\Phi}_i, \quad i = 1, 2, \cdots, M \tag{4.10}$$

3. Calculate the standard orthogonal eigenfunctions, such that:

$$\widetilde{f}_i(t) = \boldsymbol{\Phi}_i^{\mathrm{T}} \boldsymbol{\Psi}(t) = \sum_{k=1}^{M} \psi_k(t) \cdot \phi_{ki}, \quad i = 1, 2, \cdots, M \tag{4.11}$$

where $\boldsymbol{\Psi}(t) = \{\psi_1(t), \cdots, \psi_k(t), \cdots, \psi_M(t)\}^{\mathrm{T}}$. ϕ_{ki} is the k-th element of the eigenvector $\boldsymbol{\Phi}_i$, say $\boldsymbol{\Phi}_i = \{\phi_{1i}, \cdots, \phi_{ki}, \cdots, \phi_{Mi}\}^{\mathrm{T}}$. In fact, $\widetilde{f}_i(t)$ defined by Eq. (4.11) satisfies the standard orthogonal condition of Eq. (4.4):

$$\begin{aligned} \int_0^T \widetilde{f}_i(t)\widetilde{f}_j(t)\mathrm{d}t &= \boldsymbol{\Phi}_i^{\mathrm{T}} \left(\int_0^T \boldsymbol{\Psi}(t)\boldsymbol{\Psi}^{\mathrm{T}}(t)\,\mathrm{d}t \right) \boldsymbol{\Phi}_j \\ &= \boldsymbol{\Phi}_i^{\mathrm{T}} \boldsymbol{I} \boldsymbol{\Phi}_j = \boldsymbol{\Phi}_i^{\mathrm{T}} \boldsymbol{\Phi}_j = \delta_{ij} \end{aligned} \tag{4.12}$$

4. Finally, the first N eigenvalues λ_i and eigenfunctions $f_i(t)$ in Eq. (4.3) are approximated as:

$$\lambda_i \approx \widetilde{\lambda}_i, \quad f_i(t) \approx \widetilde{f}_i(t), \quad i = 1, 2, \cdots, N \tag{4.13}$$

Apparently, when $M \to \infty$, the eigenvalues $\widetilde{\lambda}_i$ in Eq. (4.10) and the eigenfunctions $\widetilde{f}_i(t)$ defined in Eq. (4.11) are equivalent to the eigenvalues λ_i and eigenfunctions $f_i(t)$ in Eq. (4.3), respectively.

In the numerical analysis application, for the deterministic standard orthogonal function sets, the trigonometric orthonormal basis might be a preferred alternative, given by:

$$\psi_1(t) = \frac{1}{\sqrt{T}}, \psi_{2k}(t) = \frac{\sqrt{2}}{\sqrt{T}} \sin\left(\frac{2k\pi t}{T}\right), \psi_{2k+1}(t) = \frac{\sqrt{2}}{\sqrt{T}} \cos\left(\frac{2k\pi t}{T}\right), \quad k = 1, 2, \cdots \tag{4.14}$$

In the field of earthquake engineering, attention is always paid solely to the PSD function of stochastic ground motions, with relatively extensive research conducted in this regard. Nevertheless, specific explicit expressions for the auto-correlation function of stochastic ground motions are virtually nonexistent. To simulate non-stationary stochastic ground motions, we can adopt the following two methods.

On the one hand, for the case of considering just the intensity non-stationarity, it is commonly to first derive the auto-correlation function of the corresponding stationary seismic process from its PSD function, and then take into account the intensity modulation function to construct the auto-correlation function for intensity non-stationary stochastic ground motions. At this stage, the correlation matrix $\boldsymbol{R}$ can be computed based on Eq. (4.9), and the eigenvalues and eigenvectors can be derived using Eq. (4.10). On the other hand, for the case of considering both the intensity and frequency non-stationarities of stochastic ground motions, deriving an analytical expression for the auto-correlation function might be highly challenging. At this point, an alternative approach is to obtain the numerical solution of the auto-correlation function by applying the Fourier transform to the EPSD function of time-frequency fully non-stationary stochastic ground motions. Finally, by incorporating the proposed dimension-reduction method, non-stationary stochastic ground motion processes can be simulated applying Eq. (4.5).

4.2 Filtered White Noise Process Models in Physics

4.2.1 Single-Filtered White Noise Process Model

In seismic engineering, the single-filtered white noise process model is commonly employed to simulate earthquake acceleration process, which represents the outputs of a low-order linear system driven by the white noise process. Specifically, the earthquake acceleration stationary process $\overline{X}_1(t)$ observed on the ground surface can be represented by the absolute acceleration of a single-degree-of-freedom (SDOF) linear filter of soil layer, i.e., as shown in Fig. 4.1.

The expression of $\overline{X}_1(t)$ and the motion differential equations are defined as follows:

$$\overline{X}_1(t) = \ddot{X}_{\mathrm{g}}(t) + w(t) \tag{4.15}$$

$$\ddot{X}_{\mathrm{g}}(t) + 2\zeta_{\mathrm{g}}\omega_{\mathrm{g}}\dot{X}_{\mathrm{g}}(t) + \omega_{\mathrm{g}}^2 X_{\mathrm{g}}(t) = -w(t) \tag{4.16}$$

Fig. 4.1 The physical model of single-filtered white noise process of $\overline{X}_1(t)$

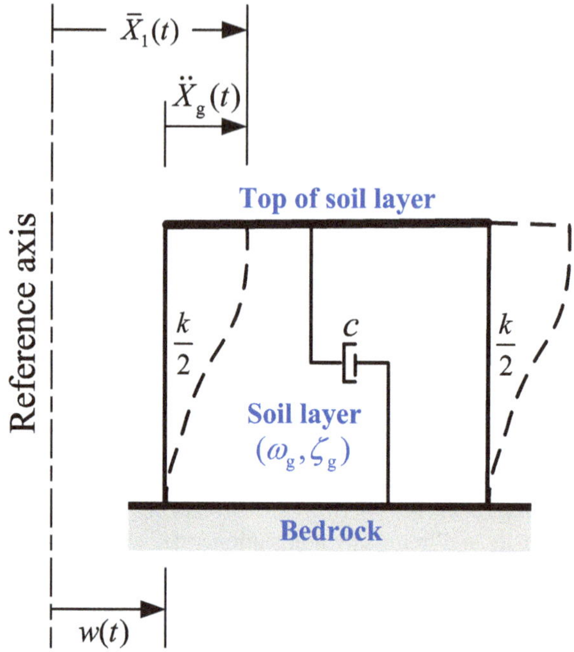

where $\ddot{X}_g(t)$, $\dot{X}_g(t)$, $X_g(t)$ denote the acceleration, velocity and displacement responses upon the top of soil layer (on the ground surface) relative to the bedrock, respectively. $w(t)$ represents a stationary white noise process with the two-sided PSD, say $S_w(\omega) = S_0$. ω_g and ζ_g refer to the dominant frequency and critical damping of the soil layer, respectively.

Considering the assumption of zero initial conditions and according to the Duhamel integral, the relative displacement response of the ground motion can be expressed as below:

$$X_g(t) = -\int_0^t h_{X_g}(t-\tau)w(\tau)\mathrm{d}\tau = -h_{X_g}(t) * w(t) \tag{4.17}$$

where $*$ denotes a convolutional symbol formally. $h_{X_g}(t)$ signifies the unit impulse response function of relative displacement, which is defined by:

$$h_{X_g}(t) = \frac{1}{\omega_D} e^{-\zeta_g \omega_g t} \sin(\omega_D t), \quad t \geq 0 \tag{4.18}$$

where $\omega_D = \omega_g \sqrt{1 - \zeta_g^2}$.

Based on the differential characteristics of convolution operation, the relative velocity response of the ground motion can be expressed as below:

$$\dot{X}_g(t) = -\int_0^t h_{\dot{X}_g}(t-\tau)w(\tau)\mathrm{d}\tau \tag{4.19}$$

where $h_{\dot{X}_g}(t)$ denotes the unit impulse response function of relative velocity, expressed as the following:

$$\begin{aligned} h_{\dot{X}_g}(t) &= h'_{X_g}(t) \\ &= \mathrm{e}^{-\zeta_g\omega_g t}\left[\cos(\omega_D t) - \frac{\zeta_g\omega_g}{\omega_D}\sin(\omega_D t)\right], \quad t \geq 0 \end{aligned} \tag{4.20}$$

According to Eqs. (4.15) and (4.16), the absolute acceleration of the ground motion can be expressed as follows:

$$\overline{X}_1(t) = -2\zeta_g\omega_g\dot{X}_g(t) - \omega_g^2 X_g(t) \tag{4.21}$$

Finally, through substituting Eqs. (4.17) and (4.19) into Eq. (4.21), the single-filtered white noise process model of earthquake stationary acceleration process $\overline{X}_1(t)$ as can be deduced, such that:

$$\overline{X}_1(t) = \int_0^t h_1(t-\tau)\cdot w(\tau)\mathrm{d}\tau \tag{4.22}$$

where the unit impulse response function $h_1(t)$ of absolute acceleration is given by:

$$\begin{aligned} h_1(t) &= 2\zeta_g\omega_g h_{\dot{X}_g}(t) + \omega_g^2 h_{X_g}(t) \\ &= \omega_g \mathrm{e}^{-\zeta_g\omega_g t}\left[\frac{\left(1-2\zeta_g^2\right)\omega_g}{\omega_D}\sin(\omega_D t) + 2\zeta_g\cos(\omega_D t)\right], \quad t \geq 0 \end{aligned} \tag{4.23}$$

In Eq. (4.22), the stationary white noise process $w(t)$ should satisfy the following conditions:

$$E[w(t)] = 0, \quad E[w(t)w(t')] = S_0\delta(t-t') \tag{4.24}$$

where $\delta(\cdot)$ denotes the Dirac function.

As a matter of fact, Eq. (4.22) denotes the time-domain model for single-filtered white noise process, and its corresponding frequency-domain model is the widely applied Kanai-Tajimi spectrum (K-T model) in the field of earthquake engineering, with its expression given as follows:

$$S_{\overline{X}_1}(\omega) = \frac{\omega_g^4 + 4\zeta_g^2\omega_g^2\omega^2}{\left(\omega^2 - \omega_g^2\right)^2 + 4\zeta_g^2\omega_g^2\omega^2} S_0 \tag{4.25}$$

Here it should be noted that the adopted Fourier transform relationship between the PSD function and the auto-correlation function in this section is represented by Eq. (4.1). Thus, the expression for the spectral intensity factor S_0 here should be the following:

$$S_0 = 2\pi \times \frac{a_{\max}^2}{r^2\pi\omega_g\left(2\zeta_g + \frac{1}{2\zeta_g}\right)} \tag{4.26}$$

which is different from Eq. (2.15).

It is noteworthy that Eq. (4.22) demonstrates that any real-valued stationary process, particularly those with second-order moments, can generally be expressed as a filtered white noise process. Furthermore, the function $h_1(t)$ is typically associated with the impulse response function of a linear time-invariant system. Therefore, a stationary process can commonly be viewed as the output response of a linear time-invariant system subjected to white noise process excitation. Moreover, the aforementioned results indicate that stationary processes hold equivalent representations in both the time and frequency domains.

4.2.2 *Double-Filtered White Noise Process Model*

Generally, the single-filtered white noise process model possesses clear physical significance, but it often tends to overestimate the energy of the low-frequency component of earthquake ground motions. In specific, the K-T model shows a non-zero value at zero frequency, which does not align with real-world conditions. To tackle this problem, numerous multi-filtered white noise process models have been proposed. Among these, the double-filtered white noise process model introduced by Clough and Penzien (C-P model) enjoys widespread engineering applications. In the C-P model, both the soil layer and the bedrock are treated as SDOF linear filters, thus constituting a doubled-filtered white noise process model, as shown in Fig. 4.2.

Consequently, the earthquake acceleration stationary process $\overline{X}_2(t)$ on the ground surface, relative to the bottom of bedrock, can be represented as the sum of the relative acceleration $\ddot{X}_g(t)$ of the soil layer and the relative acceleration $\ddot{X}_f(t)$ of the bedrock, such that:

$$\overline{X}_2(t) = \ddot{X}_g(t) + \ddot{X}_f(t) \tag{4.27}$$

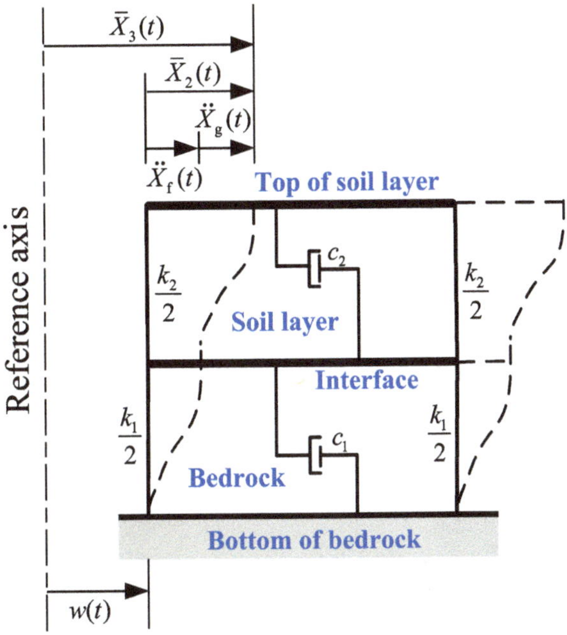

Fig. 4.2 The physical model of double-filtered white noise processes of $\overline{X}_2(t)$ and $\overline{X}_3(t)$

For the two SDOF linear filtering systems, say the soil layer and the bedrock, their motion differential equations are respectively defined as follows:

$$\ddot{X}_g(t) + 2\zeta_g\omega_g\dot{X}_g(t) + \omega_g^2 X_g(t) = -\ddot{X}_f(t) \tag{4.28}$$

$$\ddot{X}_f(t) + 2\zeta_f\omega_f\dot{X}_f(t) + \omega_f^2 X_f(t) = -w(t) \tag{4.29}$$

where $\ddot{X}_g(t)$, $\dot{X}_g(t)$, $X_g(t)$ denote the acceleration, velocity and displacement responses upon the top of soil layer relative to the bottom of soil layer (top of bedrock), respectively. $\ddot{X}_f(t)$, $\dot{X}_f(t)$, $X_f(t)$ represent the acceleration, velocity and displacement responses upon the top of bedrock relative to the bottom of bedrock, respectively. Additionally, ω_f and ζ_f refer to the dominant frequency and critical damping of the bedrock, respectively.

According to Eqs. (4.27) and (4.28), and considering the single-filtered case, the expression of $\overline{X}_2(t)$ can be derived as follows:

$$\begin{aligned}\overline{X}_2(t) &= -2\zeta_g\omega_g\dot{X}_g(t) - \omega_g^2 X_g(t)\\ &= \int_0^t h_1(t-\tau)\ddot{X}_f(\tau)\,\mathrm{d}\tau\end{aligned} \tag{4.30}$$

where $h_1(t)$ is consistent with that defined in Eq. (4.23).

Furthermore, based on Eq. (4.29), the absolute acceleration response at the top of bedrock can be defined as below:

$$\begin{aligned}\ddot{X}_{\mathrm{f}}(t)+w(t) &= -2\zeta_{\mathrm{f}}\omega_{\mathrm{f}}\dot{X}_{\mathrm{f}}(t)-\omega_{\mathrm{f}}^{2}X_{\mathrm{f}}(t)\\ &= \int_{0}^{t}h_{2}(t-\tau)w(\tau)\mathrm{d}\tau\end{aligned} \tag{4.31}$$

in which

$$h_{2}(t)=\omega_{\mathrm{f}}\mathrm{e}^{-\zeta_{\mathrm{f}}\omega_{\mathrm{f}}t}\left[\frac{\left(1-2\zeta_{\mathrm{f}}^{2}\right)\omega_{\mathrm{f}}}{\widetilde{\omega}_{\mathrm{D}}}\sin(\widetilde{\omega}_{\mathrm{D}}t)+2\zeta_{\mathrm{f}}\cos(\widetilde{\omega}_{\mathrm{D}}t)\right],\quad t\geq 0 \tag{4.32}$$

where $\widetilde{\omega}_{\mathrm{D}}=\omega_{\mathrm{f}}\sqrt{1-\zeta_{\mathrm{f}}^{2}}$.

Applying Eq. (4.31), the relative acceleration $\ddot{X}_{\mathrm{f}}(t)$ upon the top of bedrock with respect to the bottom of bedrock can be represented as:

$$\ddot{X}_{\mathrm{f}}(t)=-w(t)+\int_{0}^{t}h_{2}(t-\tau)w(\tau)\mathrm{d}\tau \tag{4.33}$$

As a result, by means of substituting Eq. (4.33) into Eq. (4.30), the integral expression for the earthquake acceleration stationary process $\overline{X}_{2}(t)$ relative to the bottom of bedrock can thus be achieved. It is noteworthy that due to the presence of the term '$-w(t)$' in Eq. (4.33), applying the Duhamel integral might lead to the emergence of a term including a single convolution. Finally, the expression of $\overline{X}_{2}(t)$ can be presented as follows (Liu et al. 2024):

$$\overline{X}_{2}(t)=-\int_{0}^{t}h_{1}(t-\tau)w(\tau)\mathrm{d}\tau+\int_{0}^{t}\int_{0}^{\tau}h_{1}(t-\tau)h_{2}(\tau-\tau')w(\tau')\mathrm{d}\tau'\mathrm{d}\tau,\quad 0\leq\tau'\leq\tau\leq t \tag{4.34}$$

In fact, Eq. (4.34) commonly denotes the time-domain model for stationary double-filtered white noise process, and the corresponding frequency-domain counterpart is represented by the widely applied C-P model, given by:

$$\begin{aligned}S_{\overline{X}_{2}}(\omega) &= \frac{\omega_{\mathrm{g}}^{4}+4\zeta_{\mathrm{g}}^{2}\omega_{\mathrm{g}}^{2}\omega^{2}}{\left(\omega^{2}-\omega_{\mathrm{g}}^{2}\right)^{2}+4\zeta_{\mathrm{g}}^{2}\omega_{\mathrm{g}}^{2}\omega^{2}}\cdot\frac{\omega^{4}}{\left(\omega^{2}-\omega_{\mathrm{f}}^{2}\right)^{2}+4\zeta_{\mathrm{f}}^{2}\omega_{\mathrm{f}}^{2}\omega^{2}}S_{0}\\ &= \frac{\omega^{4}}{\left(\omega^{2}-\omega_{\mathrm{f}}^{2}\right)^{2}+4\zeta_{\mathrm{f}}^{2}\omega_{\mathrm{f}}^{2}\omega^{2}}S_{\overline{X}_{1}}(\omega)\end{aligned} \tag{4.35}$$

From the above derivation, it is evident that there exist differences between the two time-domain filtered white noise process models presented in Eqs. (4.22) and (4.34).

The latter model, serving as a complement to the former, incorporates a double-filtered white noise process as a correction term to mitigate the issue of overestimating energy in the low-frequency range within the K-T model. Additionally, it should be noted that the double-filtered white noise process model comes in various forms, each accompanied by its unique motion differential equations. For the sake of simplicity, this book exclusively employs the C-P model for theoretical derivations and numerical example applications. Finally, Eq. (4.35) validates the equivalent frequency-domain representation of the double filter from the perspective of PSD function.

From Eqs. (4.27)–(4.29), it is evident that the aforementioned double-filtered white noise process model pertains to the earthquake acceleration stationary process observed on the ground surface relative to the bottom of bedrock. From this, we can also propose another double-filtered white noise process model, which is the absolute earthquake acceleration stationary process $\overline{X}_3(t)$ observed on the ground surface. This model is composed of three components, say the relative acceleration $\ddot{X}_g(t)$ of the soil layer, the relative acceleration $\ddot{X}_f(t)$ of the bedrock, and the white noise process $w(t)$ (acceleration), as illustrated in Fig. 4.2.

Then, the expression of $\overline{X}_3(t)$ and the motion differential equations for the two SDOF linear filtering systems, say the soil layer and the bedrock, are presented as follows:

$$\begin{cases} \overline{X}_3(t) = \ddot{X}_g(t) + \ddot{X}_f(t) + w(t) \\ \ddot{X}_g(t) + 2\zeta_g\omega_g\dot{X}_g(t) + \omega_g^2 X_g(t) = -\left[\ddot{X}_f(t) + w(t)\right] \\ \ddot{X}_f(t) + 2\zeta_f\omega_f\dot{X}_f(t) + \omega_f^2 X_f(t) = -w(t) \end{cases} \tag{4.36}$$

Through a derivation analogous to that of Eqs. (4.30) and (4.31), one can readily obtain the double-filtered white noise expression for $\overline{X}_3(t)$, followed by:

$$\overline{X}_3(t) = \int_0^t \int_0^\tau h_1(t-\tau)h_2(\tau-\tau')w(\tau')\mathrm{d}\tau'\mathrm{d}\tau, \quad 0 \le \tau' \le \tau \le t \tag{4.37}$$

In fact, the frequency-domain model corresponding to Eq. (4.37) can thus be represented by the following PSD function (Lai et al. 1995):

$$\begin{aligned} S_{\overline{X}_3}(\omega) &= \frac{\omega_g^4 + 4\zeta_g^2\omega_g^2\omega^2}{\left(\omega^2 - \omega_g^2\right)^2 + 4\zeta_g^2\omega_g^2\omega^2} \cdot \frac{\omega_f^4 + 4\zeta_f^2\omega_f^2\omega^2}{\left(\omega^2 - \omega_f^2\right)^2 + 4\zeta_f^2\omega_f^2\omega^2} S_0 \\ &= \frac{\omega_f^4 + 4\zeta_f^2\omega_f^2\omega^2}{\left(\omega^2 - \omega_f^2\right)^2 + 4\zeta_f^2\omega_f^2\omega^2} S_{\overline{X}_1}(\omega) \end{aligned} \tag{4.38}$$

From the two aforementioned double-filtered white noise process models, i.e., Eqs. (4.34) and (4.37), it is evident that they share commonalities and differences. Specifically, Eq. (4.37) constitutes merely the double integral part of Eq. (4.34).

From the expressions of their PSD functions, both of them share the same K-T model component. However, the energy of $S_{\overline{X}_3}(\omega)$ presented in Eq. (4.38) does not vanish at zero frequency, a characteristic similar to that of the K-L model, which is unreasonable in fact. Incidentally, the parameter of the PSD function $S_{\overline{X}_2}(\omega)$ in Eq. (4.35) should satisfy the condition $\omega_{\mathrm{f}} \ll \omega_{\mathrm{g}}$, whereas for the parameter of the PSD function $S_{\overline{X}_3}(\omega)$ in Eq. (4.38) should fulfill the condition $\omega_{\mathrm{f}} > \omega_{\mathrm{g}}$.

To sum up, in practical engineering, the C-P model can fulfill engineering application requirements in most cases, thus it might be typically the preferred choice.

4.2.3 *Discrete Representation of Filtered White Noise Processes*

The Single-Filtered Case for Stationary Process

To carry out the simulation of stochastic process, the single-filtered white noise process model of earthquake acceleration stationary process $\overline{X}_1(t)$ should be reformulated using piecewise integration. Consequently, Eq. (4.22) can be rewritten as follows:

$$\overline{X}_1(t) = \sum_{i=1}^{n} \int_{t_{i-1}}^{t_i} h_1(t-\tau)w(\tau)\mathrm{d}\tau + \int_{t_n}^{t} h_1(t-\tau)w(\tau)\mathrm{d}\tau \tag{4.39}$$

where $t \in [t_n, t_{n+1})$, $n = 1, 2, \cdots, N$; $t_k = k\Delta t$, $k = 0, 1, \cdots, N$. Δt denotes the time interval, thus there exists $\mathrm{int}(T/\Delta t) = N$ and $\mathrm{int}(t/\Delta t) = n$. T indicates the simulation duration of the ground motions.

Assume $h_1(t-\tau)$ to be a constant within a short time interval $t_{i-1} \le \tau \le t_i$, and disregard the last term in Eq. (4.39), then $\overline{X}_1(t)$ can be approximated in the following form (Rezaeian and Der Kiureghian 2008):

$$\overline{X}_1(t) \approx \sum_{i=1}^{n} h_1(t-t_i) \int_{t_{i-1}}^{t_i} w(\tau)\mathrm{d}\tau \tag{4.40}$$

Here, we can define:

$$W_i = \int_{t_{i-1}}^{t_i} w(\tau)\mathrm{d}\tau \tag{4.41}$$

According to Eq. (4.24), then it can be handily verified that W_i ($i = 1, 2, \cdots, N$) constitutes a set of zero-mean orthogonal random variables, which satisfies the following conditions:

$$E[W_i] = \int_{t_{i-1}}^{t_i} E[w(\tau)]\mathrm{d}\tau = 0 \tag{4.42a}$$

$$\begin{aligned} E\left[W_i W_j\right] &= E\left[\int_{t_{i-1}}^{t_i} w(\tau)\mathrm{d}\tau \int_{t_{j-1}}^{t_j} w(\tau')\mathrm{d}\tau'\right] \\ &= \int_{t_{i-1}}^{t_i}\int_{t_{j-1}}^{t_j} E[w(\tau)w(\tau')]\,\mathrm{d}\tau'\mathrm{d}\tau \\ &= S_0 \int_{t_{i-1}}^{t_i}\int_{t_{j-1}}^{t_j} \delta(\tau-\tau')\,\mathrm{d}\tau'\mathrm{d}\tau \\ &= S_0 \Delta t \delta_{ij} \end{aligned} \tag{4.42b}$$

Furthermore, the orthogonal random variables W_i ($i = 1, 2, \cdots, N$) can be standardized and denoted as $V_i = W_i/\sqrt{S_0\Delta t}$. As a result, Eq. (4.40) can be rewritten as follows:

$$\overline{X}_1(t) \approx \sqrt{S_0\Delta t}\sum_{i=1}^{n} h_1(t-t_i)\,V_i\ ,\ \ t\in[t_n, t_{n+1}),\quad n = 1, 2, \cdots, N \tag{4.43}$$

where $\{V_i\}_{i=1}^{N}$ represents a set of standardized orthogonal random variables that satisfy the following basic conditions:

$$E[V_i] = 0,\quad E\left[V_i V_j\right] = \delta_{ij} \tag{4.44}$$

It is noteworthy that in Eq. (4.43), the time interval Δt is relatively sensitive to the computational accuracy. To eliminate the influence of the time interval, the standard deviation $\sigma_{\overline{X}_1}$ of the zero-mean stationary stochastic process $\overline{X}_1(t)$ can be further introduced. As a matter of fact, based on Eq. (4.43), the expression for the standard deviation can be readily derived as follows:

$$\sigma_{\overline{X}_1} = \sqrt{S_0\Delta t\sum_{j=1}^{n} h_1^2(t-t_j)} \tag{4.45}$$

Thus, by substituting Eq. (4.45) into Eq. (4.43), the standardized representation of $\overline{X}_1(t)$ can be obtained as follows (Liu et al. 2024):

$$\overline{X}_1(t) \approx \sigma_{\overline{X}_1}\sum_{i=1}^{n} u_i(t)\,V_i\ ,\quad t\in[t_n, t_{n+1}),\quad n = 1, 2, \cdots, N \tag{4.46}$$

where $u_i(t)$ represents the standardized impulse response function, which can be expressed as follows:

$$u_i(t) = \frac{h_1(t - t_i)}{\sqrt{\sum_{j=1}^{n} h_1^2(t - t_j)}} \tag{4.47}$$

The Double-Filtered Case for Stationary Process

Similarly, in the double-filtered white noise process model for earthquake acceleration stationary process $\overline{X}_2(t)$, it is also necessary to discretize $\overline{X}_2(t)$ with respect to time t. In this issue, the discrete form of the first term $\int_0^t h_1(t-\tau)w(\tau)\mathrm{d}\tau$ on the right-hand side of Eq. (4.34) remains the same as the single-filtered case. For the double integration in the second term, the outer integration can first be discretized. Subsequently, according to Eq. (4.39) to Eq. (4.41), a similar discretization treatment can be extended to the inner integration $\int_0^{t_i} h_2(t_i-\tau')w(\tau')\mathrm{d}\tau'$, yielding the following expression for the second term:

$$\begin{aligned} \int_0^t \int_0^\tau h_1(t-\tau)h_2(\tau-\tau')w(\tau')\mathrm{d}\tau'\mathrm{d}\tau &\approx \Delta t \times \sum_{i=1}^{n} \left[h_1(t-t_i) \int_0^{t_i} h_2(t_i-\tau')w(\tau')\mathrm{d}\tau' \right] \\ &= \Delta t \times \sum_{i=1}^{n} \left[h_1(t-t_i) \sum_{j=1}^{i} h_2(t_i-t_j) W_j \right] \end{aligned} \tag{4.48}$$

where $t \in [t_n, t_{n+1})$, $n = 1, 2, \cdots, N$.

Similarly, by means of standardizing the set of orthogonal random variables $\{W_j\}_{j=1}^N$ in Eq. (4.48), the set of standardized orthogonal random variables $\{V_j\}_{j=1}^N$ can be obtained. Then Eq. (4.34) can be rewritten in the following discrete form (Liu et al. 2024):

$$\overline{X}_2(t) \approx \sqrt{S_0 \Delta t} \left\{ \Delta t \times \sum_{i=1}^{n} \left[h_1(t-t_i) \sum_{j=1}^{i} h_2(t_i-t_j) V_j \right] - \sum_{i=1}^{n} h_1(t-t_i)\, V_i \right\} \tag{4.49}$$

where $t \in [t_n, t_{n+1})$, $n = 1, 2, \cdots, N; j \le i \le n$.

The Non-Stationary Stochastic Ground Motion Case

The aforementioned content presents the derivation of discrete expressions for the earthquake acceleration stationary process. However, in actuality, seismic ground motion constitutes a non-stationary stochastic process, generally displaying dual non-stationary characteristics in both intensity and frequency. To depict the full non-stationarities of seismic ground motion in both time and frequency domains, the deterministic intensity modulation function $f(t)$ and time-varying site parameters are employed. Thus, the fully non-stationary processes $X_k(t)$ $(k = 1, 2)$ of earthquake acceleration can be presented as follows:

$$X_k(t) = f(t)\overline{X}_k(t)\ , \quad k = 1,2 \tag{4.50}$$

where $X_1(t)$ denotes the standardized single-filtered non-stationary earthquake acceleration process, and $X_2(t)$ denotes the double-filtered non-stationary earthquake acceleration process.

In this section, the three-segment intensity modulation function defined in Eq. (2.64) is adopted:

$$f(t) = \begin{cases} (t/t_1)^2, & 0 \le t \le t_1 \\ 1, & t_1 < t < t_2 \\ \mathrm{e}^{-c(t-t_2)}, & t_2 \le t \le T \end{cases} \tag{4.51}$$

Thus, according to Eq. (4.46), the time-domain model for the fully non-stationary process $X_1(t)$ of earthquake acceleration upon single-filtered white noise can be presented as follows:

$$X_1(t) = \sigma_{\overline{X}_1} f(t) \sum_{i=1}^{n} \widehat{u}_i(t)\, V_i\ , t \in [t_n, t_{n+1}), \quad n = 1, 2, \cdots, N \tag{4.52}$$

in which

$$\widehat{u}_i(t) = \frac{h_1[t - t_i; \boldsymbol{\lambda}_1(t_i)]}{\sqrt{\sum_{j=1}^{n} h_1^2[t - t_j; \boldsymbol{\lambda}_1(t_j)]}} \tag{4.53}$$

where the standardized impulse response function $\widehat{u}_i(t)$ denotes a linear filter with time-varying frequency components, which governs the spectral non-stationarity of the filtering process. Here, the parameters $\boldsymbol{\lambda}_1(t) = [\omega_g(t), \zeta_g(t)]$ can employ the time-varying forms defined in Eq. (2.75) and Eq. (2.76), and their expressions are presented as follows:

$$\omega_g(t) = \overline{\omega}_g + \frac{d}{2} - d \times \left(\frac{t}{T}\right)^m, \quad \zeta_g(t) = \overline{\zeta}_g - \frac{e}{2} + e \times \left(\frac{t}{T}\right)^m \tag{4.54}$$

where $\overline{\omega}_g$ and $\overline{\zeta}_g$ respectively represent the deterministic dominant circular frequency and the deterministic damping of the soil site, of which the values can be specified according to Table 2.5. The parameter m can be set to 1 or 2. The parameters d and e can be referred to Table 2.8.

Similarly, the time-domain model for the fully non-stationary process $X_2(t)$ of earthquake acceleration upon double-filtered white noise can be expressed as follows (Liu et al. 2024):

$$X_2(t)=\sqrt{S_0\Delta t}f(t)\left\{\Delta t\times\sum_{i=1}^{n}h_1[t-t_i;\boldsymbol{\lambda}_1(t_i)]\sum_{j=1}^{i}h_2\left[t_i-t_j;\boldsymbol{\lambda}_2(t_j)\right]V_j-\sum_{i=1}^{n}h_1\left[t-t_i;\boldsymbol{\lambda}_1(t_i)\right]V_i\right\} \tag{4.55}$$

where $\boldsymbol{\lambda}_2(t)=[\omega_f(t),\zeta_f(t)]$ denotes the parameters of the bedrock, valued by $\omega_f(t)=0.1\omega_g(t)$ and $\zeta_f(t)=\zeta_g(t)$, respectively.

4.3 Time-Domain Representation of Stochastic Processes in Mathematics

4.3.1 Stationary Stochastic Processes

According to the Priestley's spectral decomposition theory, a real-valued zero-mean stationary stochastic process $\overline{X}(t)$ can be represented as the following Fourier-Stieltjes integral form:

$$\overline{X}(t)=\int_{-\infty}^{\infty}e^{i\omega t}dZ(\omega)=\int_{-\infty}^{\infty}e^{i\omega t}H(\omega)dU(\omega) \tag{4.56}$$

where $Z(\omega)$ and $U(\omega)$ denote the complex orthogonal increment processes with zero mean, and their increments should satisfy the following basic conditions:

$$E[dZ(\omega)]=0,\quad E[dU(\omega)]=0 \tag{4.57a}$$

$$dZ(-\omega)=dZ^*(\omega),\ \ dU(-\omega)=dU^*(\omega) \tag{4.57b}$$

$$E[dZ(\omega)dZ^*(\omega')]=\frac{1}{2\pi}S_{\overline{X}}(\omega)\,\delta_{\omega\omega'}\,d\omega \tag{4.57c}$$

$$E[dU(\omega)dU^*(\omega')]=\frac{1}{2\pi}\delta_{\omega\omega'}\,d\omega \tag{4.57d}$$

where $S_{\overline{X}}(\omega)$ denotes the two-sided PSD function of the stationary stochastic process $\overline{X}(t)$. $\delta_{\omega\omega'}$ stands for the Kronecker-delta.

It should be noted that $H(\omega)$ in Eq. (4.56) is a deterministic complex-valued function with its modulus fulfilling the following:

$$S_{\overline{X}}(\omega)=|H(\omega)|^2 \tag{4.58}$$

In addition, the function $H(\omega)$ possesses the property of conjugate symmetry, say $H(-\omega) = H^*(\omega)$, and it is not unique. In fact, any function $\widehat{H}(\omega) = g(\omega)H(\omega)$ can satisfy Eq. (4.58) in the case that $|g(\omega)| = 1$.

Generally, Eq. (4.56) is the so-called frequency-domain representation of a stationary stochastic process $\overline{X}(t)$. Furthermore, if $h(t)$ is defined as the inverse Fourier transform of $H(\omega)$, according to Eq. (4.1), the Fourier transform pair between $h(t)$ and $H(\omega)$ can be presented as follows:

$$\begin{cases} H(\omega) = \int_{-\infty}^{\infty} h(t)\mathrm{e}^{-\mathrm{i}\omega t}\mathrm{d}t \\ h(t) = \dfrac{1}{2\pi}\int_{-\infty}^{\infty} H(\omega)\mathrm{e}^{\mathrm{i}\omega t}\mathrm{d}\omega \end{cases} \tag{4.59}$$

It should be noted that there exists $h(t) = h^*(t)$, which indicates that the function $h(t)$ is a real-valued one.

Consequently, applying the first equation in Eq. (4.59), Eq. (4.56) can be rewritten as below:

$$\overline{X}(t) = \int_{-\infty}^{\infty} h(\tau)\int_{-\infty}^{\infty} \mathrm{e}^{\mathrm{i}\omega(t-\tau)}\mathrm{d}U(\omega)\mathrm{d}\tau \tag{4.60}$$

Here, we can define a stationary stochastic process as follows:

$$w(t) = \int_{-\infty}^{\infty} \mathrm{e}^{\mathrm{i}\omega t}\mathrm{d}U(\omega) \tag{4.61}$$

It is handy to verify that $w(\tau)$ is the so-called white noise process, of which the PSD function equals to 1. As a matter of fact, applying Eq. (4.57d), it can be further obtained that:

$$\begin{aligned} E[w(t)w^*(t')] &= E\left[\int_{-\infty}^{\infty} \mathrm{e}^{\mathrm{i}\omega t}\mathrm{d}U(\omega)\int_{-\infty}^{\infty} \mathrm{e}^{-\mathrm{i}\omega' t'}\mathrm{d}U^*(\omega')\right] \\ &= \int_{-\infty}^{\infty}\int_{-\infty}^{\infty} \mathrm{e}^{\mathrm{i}(\omega t-\omega' t')}E[\mathrm{d}U(\omega)\mathrm{d}U^*(\omega')] \\ &= \int_{-\infty}^{\infty}\int_{-\infty}^{\infty} \mathrm{e}^{\mathrm{i}(\omega t-\omega' t')}\left(\frac{1}{2\pi}\delta_{\omega\omega'}\mathrm{d}\omega\right) \\ &= \frac{1}{2\pi}\int_{-\infty}^{\infty} \mathrm{e}^{\mathrm{i}\omega(t-t')}\mathrm{d}\omega = \delta(t-t') \end{aligned} \tag{4.62}$$

Then, through substituting Eq. (4.61) into Eq. (4.60), the time-domain representation of a stationary process associated with the white noise process can thus be achieved as follows:

$$\overline{X}(t)=\int_{-\infty}^{\infty}h(\tau)w(t-\tau)\mathrm{d}\tau=\int_{-\infty}^{\infty}h(t-\tau)w(\tau)\mathrm{d}\tau \tag{4.63}$$

where $w(\tau)$ denotes the real-valued white noise process with zero mean, of which the correlation function and PSD function are respectively presented as follows:

$$R_w(\tau)=\delta(\tau),\quad S_w(\omega)=1 \tag{4.64}$$

Furthermore, we can also define an incremental process associated with the white noise process as follows:

$$\mathrm{d}W(\tau)=w(\tau)\mathrm{d}\tau \tag{4.65}$$

Thus, Eq. (4.63) can also be written as below (Spanos and Zeldin 1998):

$$\overline{X}(t)=\int_{-\infty}^{\infty}h(t-\tau)w(\tau)\mathrm{d}\tau=\int_{-\infty}^{\infty}h(t-\tau)\mathrm{d}W(\tau) \tag{4.66}$$

where $W(\tau)$ is a process with orthogonal increment satisfying the following conditions:

$$\begin{aligned}E[\mathrm{d}W(\tau)\mathrm{d}W^*(\tau')]&=E[w(\tau)w^*(\tau')]\mathrm{d}\tau'\mathrm{d}\tau\\&=\delta(\tau-\tau')\mathrm{d}\tau'\mathrm{d}\tau\\&=\delta_{\tau\tau'}\mathrm{d}\tau\end{aligned} \tag{4.67}$$

In fact, Eq. (4.66) shows that any stationary second-order process can be expressed as a filtered white noise process. Moreover, the function $h(t)$ can be associated with the impulse response of a linear system. Therefore, any stationary process can be treated as the response of a linear dynamic system driven by the white noise process.

Through investigation of the time-domain representations for stationary stochastic processes in physics and mathematics, it is evident that Eq. (4.22) and Eq. (4.66) are identical in the expression form. However, the two have different physical implications. Specifically, for the time-domain representation in physics, i.e., Eq. (4.22), the PSD of the white noise process is S_0, whereas for that in mathematics, i.e., Eq. (4.66), the PSD is 1. Additionally, in Eq. (4.22), the unit impulse response function $h_1(t)$ is derived from physical processes, thereby giving it a unique characteristic. In Eq. (4.66), the function $h(t)$ is mathematically derived, specifically from the second equation in Eq. (4.59), and thus exhibits generality. This demonstrates that Eq. (4.22) represents a special case of Eq. (4.66) in essence.

4.3.2 Non-Stationary Stochastic Processes

As previously stated, earthquake ground motion is inherently a non-stationary stochastic process, displaying dual non-stationary features in both intensity and frequency. At this point, we can employ the deterministic time-frequency modulation function $A(t, \omega)$ to characterize the non-stationary nature of seismic motion. In the same way, a real-valued zero-mean non-stationary stochastic process $X(t)$ can be represented by the following Fourier-Stieltjes integral:

$$X(t) = \int_{-\infty}^{\infty} A(t,\omega)\mathrm{e}^{\mathrm{i}\omega t}\mathrm{d}Z(\omega) = \int_{-\infty}^{\infty} \mathrm{e}^{\mathrm{i}\omega t}A(t,\omega)H(\omega)\mathrm{d}U(\omega) \tag{4.68}$$

where $U(\omega)$ denotes the complex orthogonal increment process with zero mean, which satisfies Eq. (4.57d). The deterministic function $H(\omega)$ satisfies Eq. (4.58).

Here, the EPSD function $S_X(\omega, t)$ of the non-stationary stochastic process $X(t)$ employs the model defined in Eq. (2.72), such that:

$$S_X(\omega,t) = |A(t,\omega)|^2 S_{\overline{X}}(\omega) \tag{4.69}$$

Further, we can define $\widetilde{H}(t,\omega) = A(t,\omega)H(\omega)$, then $S_X(\omega, t)$ can be defined as follows:

$$S_X(\omega,t) = \left|\widetilde{H}(t,\omega)\right|^2 \tag{4.70}$$

For each given time instant t, if we define $\widetilde{h}(t,\tau)$as the inverse Fourier transform of $\widetilde{H}(t,\omega)$, the Fourier transform pair between $\widetilde{H}(t,\omega)$ and $\widetilde{h}(t,\tau)$ can thus be presented as below:

$$\begin{cases} \widetilde{H}(t,\omega) = \int_{-\infty}^{\infty} \widetilde{h}(t,\tau)\mathrm{e}^{-\mathrm{i}\omega\tau}\mathrm{d}\tau \\ \widetilde{h}(t,\tau) = \dfrac{1}{2\pi}\int_{-\infty}^{\infty} \widetilde{H}(t,\omega)\mathrm{e}^{\mathrm{i}\omega\tau}\mathrm{d}\omega \end{cases} \tag{4.71}$$

Note that the function $\widetilde{H}(t,-\omega) = \widetilde{H}^*(t,\omega)$, thus there exists $\widetilde{h}(t,\tau) = \widetilde{h}^*(t,\tau)$, which denotes that the function $\widetilde{h}(t,\tau)$ is a real-valued one.

Then, by means of substituting the first equation of Eq. (4.71) into Eq. (4.68), the non-stationary stochastic process $X(t)$ can be rewritten as follows:

$$X(t) = \int_{-\infty}^{\infty} \widetilde{h}(t,\tau)\int_{-\infty}^{\infty} \mathrm{e}^{\mathrm{i}\omega(t-\tau)}\mathrm{d}U(\omega)\mathrm{d}\tau = \int_{-\infty}^{\infty} \widetilde{h}(t,t-\tau)\int_{-\infty}^{\infty} \mathrm{e}^{\mathrm{i}\omega\tau}\mathrm{d}U(\omega)\mathrm{d}\tau \tag{4.72}$$

Finally, through substituting Eq. (4.61) and Eq. (4.65) into Eq. (4.72), we can obtain the time-domain representation for a non-stationary stochastic process, which is analogous to Eq. (4.66) for a stationary stochastic process, as detailed below (Liu et al. 2025):

$$X(t) = \int_{-\infty}^{\infty} \widetilde{h}(t, t-\tau) w(\tau) \mathrm{d}\tau = \int_{-\infty}^{\infty} \widetilde{h}(t, t-\tau) \mathrm{d}W(\tau) \tag{4.73}$$

It is worth mentioning that Eqs. (4.56), (4.68) and Eqs. (4.66), (4.73) obtain two equivalent representations for both stationary and non-stationary stochastic processes in frequency-domain (Eqs. (4.56), (4.68)) and time-domain (Eqs. (4.66), (4.73)), respectively. This effectively verifies that the time-domain analysis methods hold equal importance as the frequency-domain analysis strategies. Discretization of these equations yields two methods of synthesizing stochastic processes. In fact, the discretization of Eqs. (4.56) and (4.68) can achieve the SRM. As for the discretization of Eqs. (4.66) and (4.73), there are two strategies in common. One strategy is similar to the treatment described in Sect. 4.2.3, as illustrated in Eq. (4.43). And the other can yield the linear filter method.

4.3.3 *Discretization Strategy of Time-Domain Representation for Non-Stationary Processes*

In practical engineering, the time parameter of a non-stationary stochastic process $X(t)$ is defined within a certain range, say $t \in (0, T]$, where T denotes the duration of the stochastic process $X(t)$. Therefore, in the case of $t < 0$, there exists $X(t) = 0$. Meanwhile, note that in the time-domain representation for non-stationary stochastic processes, say Eq. (4.73), when $t < \tau$, there exists $\widetilde{h}(t, t-\tau) = 0$. Consequently, the infinite integral expression in Eq. (4.73) can be converted into the following finite integral form:

$$X(t) = \int_{0}^{t} \widetilde{h}(t, t-\tau) w(\tau) \mathrm{d}\tau = \int_{0}^{t} \widetilde{h}(t, t-\tau) \mathrm{d}W(\tau) \tag{4.74}$$

Then, for synthesizing non-stationary stochastic processes, three discretization strategies for the finite integral expression, say Eq. (4.74), are elaborated as follows.

Conventional Discretization Strategy

Similar to the discrete form of the stationary filtered white noise model presented in Sect. 4.2.3, namely Eq. (4.39), the time-domain representation for a non-stationary stochastic process, say Eq. (4.74), can also be discretized into the following conventional form:

$$X(t) = \sum_{i=1}^{n} \int_{t_{i-1}}^{t_i} \widetilde{h}(t, t-\tau) w(\tau) \mathrm{d}\tau + \int_{t_n}^{t} \widetilde{h}(t, t-\tau) w(\tau) \mathrm{d}\tau \tag{4.75}$$

Assume that $\widetilde{h}(t, t-\tau)$ remains a constant within a short time interval $t_{i-1} \le \tau \le t_i$, and disregard the last term in Eq. (4.75), then $X(t)$ can be approximated as:

$$X(t) \approx \sum_{i=1}^{n} \widetilde{h}(t, t-t_i) \int_{t_{i-1}}^{t_i} w(\tau) \mathrm{d}\tau \,, \quad t \in [t_n, t_{n+1}) \,, \quad n = 1, 2, \cdots, N \tag{4.76}$$

Similar to Eqs. (4.41) and (4.42), and note that the PSD function of the white noise process is $S_w(\omega) = 1$, as shown in Eq. (4.64), we can define a set of standard orthogonal random variables as follows:

$$V_i = \frac{1}{\sqrt{\Delta t}} \int_{t_{i-1}}^{t_i} w(\tau) \mathrm{d}\tau, \quad i = 1, 2, \cdots, N \tag{4.77}$$

As a result, substituting Eq. (4.77) into Eq. (4.76), the time-domain representation of a non-stationary stochastic process can be rewritten as follows (Liu et al. 2025):

$$X(t) \approx \sqrt{\Delta t} \sum_{i=1}^{n} \widetilde{h}(t, t-t_i) \, V_i, \ t \in [t_n, t_{n+1}) \,, \quad n = 1, 2, \cdots, N \tag{4.78}$$

In the same way, Eq. (4.78) can be expressed in the following standardized form:

$$X(t) \approx \sigma_X(t) \sum_{i=1}^{n} u_i(t) \, V_i, \ t \in [t_n, t_{n+1}] \,, \quad n = 1, 2, \cdots, N \tag{4.79}$$

where $\sigma_X(t)$ denotes the time-varying standard deviation of $X(t)$. The standardized linear filter $u_i(t)$ can be expressed as the following:

$$u_i(t) = \frac{\widetilde{h}(t, t-t_i)}{\sqrt{\sum_{j=1}^{n} \widetilde{h}^2(t, t-t_j)}} \tag{4.80}$$

Moving Average (MA) Model

As a matter of fact, there are various discretization strategies for the finite integral expression in Eq. (4.74). Thus, in addition to the conventional form presented in

Eq. (4.75), the following can also be employed to implement the discretization, such that:

$$X(t)=\int_{0}^{t-(q+1)\Delta t}\widetilde{h}(t,t-\tau)w(\tau)\mathrm{d}\tau+\int_{t-(q+1)\Delta t}^{t}\widetilde{h}(t,t-\tau)w(\tau)\mathrm{d}\tau \tag{4.81}$$

Obviously, for each given time instant t, the process $X(t)$ can be obtained by superimposing the response $\mathrm{d}W(\tau)\widetilde{h}(t,t-\tau)$ induced by the impulse excitation $\mathrm{d}W(\tau)\delta(t-\tau)$ $(0\le\tau\le t)$. In practical physical processes, the function $\widetilde{h}(t,t-\tau)$ commonly experiences a rapid decline as $t-\tau$ increases. It indicates that the contribution of the excitation diminishes as τ decreases, allowing this portion to be disregarded, yielding:

$$X(t)\approx\int_{t-(q+1)\Delta t}^{t}\widetilde{h}(t,t-\tau)w(\tau)\mathrm{d}\tau=\sum_{i=0}^{q}\int_{t-(i+1)\Delta t}^{t-i\Delta t}\widetilde{h}(t,t-\tau)w(\tau)\mathrm{d}\tau \tag{4.82}$$

Additionally, for each short time interval $\tau\in[t-(i+1)\Delta t,\ t-i\Delta t]$, the function $\widetilde{h}(t,t-\tau)$ can be considered as a constant, i.e., $\widetilde{h}(t,t-\tau)=\widetilde{h}(t,i\Delta t)$. Then, Eq. (4.82) can thus be further expressed as follows:

$$X(t)\approx\sum_{i=0}^{q}\widetilde{h}(t,i\Delta t)\int_{t-(i+1)\Delta t}^{t-i\Delta t}w(\tau)\mathrm{d}\tau \tag{4.83}$$

Subsequently, we can also define a set of standard orthogonal random variables as follows:

$$V_{t-i}=\frac{1}{\sqrt{\Delta t}}\int_{t-(i+1)\Delta t}^{t-i\Delta t}w(\tau)\mathrm{d}\tau \tag{4.84}$$

As a result, Eq. (4.83) can be rewritten as below:

$$X(t)\approx\sqrt{\Delta t}\sum_{i=0}^{q}\widetilde{h}(t,i\Delta t)V_{t-i} \tag{4.85}$$

In order to construct a standard Moving Average (MA) model, the continuous process $X(t)$ is firstly discretized into a time series X_t, i.e., $X_t=X(t)$. Meanwhile, let $\theta_i(t)=\widetilde{h}(t,i\Delta t)/\widetilde{h}(t,0)$, and define $\widetilde{V}_{t-i}=\sqrt{\Delta t}\widetilde{h}(t,0)V_{t-i}$. Then, Eq. (4.85) can be renewed as follows (Spanos and Zeldin 1998; Dahlhaus 1997):

$$X_t=\sum_{i=0}^{q}\theta_i(t)\widetilde{V}_{t-i}=\widetilde{V}_t+\theta_1(t)\widetilde{V}_{t-1}+\cdots+\theta_q(t)\widetilde{V}_{t-q} \tag{4.86}$$

Equation (4.86) is generally known as the MA model, in which a non-stationary time series can be represented applying a finite number of time-varying coefficients $\theta_i(t)$ and a set of mutually uncorrelated random variables $\{\widetilde{V}_{t-i}\}$ which satisfy the following orthogonality conditions:

$$E\left[\widetilde{V}_{t-i}\right]=0, E\left[\widetilde{V}_{t-i}\widetilde{V}_{t-j}\right]=\widetilde{h}^2(t,0)\Delta t\delta_{ij} \tag{4.87}$$

Auto Regression (AR) Model

In general, Eq. (4.86) can also be presented as the following alternative form based on the lag operator (Spanos and Zeldin 1998; Box et al. 2015):

$$X_t=\Theta(t,B)\widetilde{V}_t \tag{4.88}$$

where $\Theta(t,B) = 1 + \theta_1(t)B + \theta_2(t)B^2 + \cdots + \theta_q(t)B^q$ denotes a polynomial of degree q in the lag operator B. $B^i\,(i = 1,2,\cdots,q)$ denotes the lag operator defined by $B^i\widetilde{V}_t=\widetilde{V}_{t-i}$.

According to the properties of the MA model associated with non-stationary time series, for any given time instant t, the polynomial $\Theta(t,B)$ should satisfy the following condition (Box et al. 2015):

$$\Psi(t,B)\Theta(t,B)=1 \tag{4.89}$$

where $\Psi(t,B) = 1 - \varphi_1(t)B - \varphi_2(t)B^2 - \cdots - \varphi_p(t)B^p$ denotes a polynomial of degree p in the lag operator B.

Through combining Eqs. (4.88) and (4.89), and note that $B^iX_t = X_{t-i} = X(t - i\Delta t)$, thus we can obtain (Deodatis and Shinozuka 1988):

$$X_t=\sum_{i=1}^{p}\varphi_i(t)B^iX_t+\widetilde{V}_t=\sum_{i=1}^{p}\varphi_i(t)X_{t-i}+\widetilde{V}_t \tag{4.90}$$

Further, by means of introducing the standard orthogonal random variables V_t, i.e., $\widetilde{V}_t=\sqrt{\Delta t}\widetilde{h}(t,0)V_t=L(t)V_t$, Eq. (4.90) can be rewritten as follows:

$$X_t=\sum_{i=1}^{p}\varphi_i(t)X_{t-i}+L(t)V_t \tag{4.91}$$

where $L(t)=\sqrt{\Delta t}\widetilde{h}(t,0)$ is the time-varying gain coefficient of the noise term.

Apparently, the time series variable X_t of the non-stationary stochastic process $X(t)$ and the noise variable V_t satisfy the following orthogonality condition:

$$E[X_sV_t]=0,\quad s<t \tag{4.92}$$

It is worth noting that for non-stationary stochastic processes in practical engineering applications, when $t \leq 0$, there exists $X(t) = 0$. Therefore, it can generally be assumed that $X_t = 0$ ($t \leq 0$) in Eq. (4.91).

In fact, Eq. (4.91) represents the so-called Auto Regression (AR) model for non-stationary time series. It indicates that X_t at any given time instant is determined by all the previous information up to that moment and the noise information at that current time. What's more, the previous information can be regarded as deterministic, whereas the noise information at that instant is stochastic.

4.4 Novel Dimension-Reduction Simulation Methods

Similar to the frequency-domain analysis methods, for synthesizing non-stationary stochastic processes, regardless of whether the standard orthogonal random variables $\{\xi_i\}_{i=1}^N$ in the K-L decomposition (Eq. (4.5)), $\{V_i\}_{i=1}^N$ in the filtered white noise process models in physics (Eqs. (4.52) and (4.55)), or $\{V_i\}_{i=1}^N$ in the time-domain representation in mathematics (Eqs. (4.79), (4.85) and (4.91)), the probability distributions of them are not specified. Consequently, the aforementioned formulas cannot be directly utilized for simulation. However, the conventional time-domain analysis methods all belong to the framework of the Monte Carlo methods, which still confront significant challenges caused by ultra-high-dimensional randomness.

To address this issue, a novel dimension-reduction simulation approach is proposed, where $\{\xi_i\}_{i=1}^N$ and $\{V_i\}_{i=1}^N$ can be defined as random orthogonal functions of elementary random variables. As a result, the precise simulation of stochastic ground motion processes can be realized with just one elementary random variable applying the time-domain analysis methods. For the sake of simplicity, just the dimension-reduction simulation method of the K-L decomposition is adopted as an illustrative example since the random orthogonal function forms of other time-domain representation methods remain exactly the same.

As stated previously, in the K-L decomposition, say Eq. (4.5), the standard orthogonal random variables $\{\xi_i\}_{i=1}^N$ should satisfy the basic conditions defined in Eq. (4.6), which can be recognized as a necessary condition for simulation. Here, the dimension-reduction representation idea is introduced, thus we can define $\{\xi_i\}_{i=1}^N$ as the random orthogonal functions of elementary random variables, given by:

$$\xi_i = g_i(\Theta), \quad i = 1, 2, \cdots, N \tag{4.93}$$

where $g_i(\cdot)$ ($i = 1, 2, \cdots, N$) denotes the deterministic orthogonal functions, and Θ signifies the elementary random variable with known probability distribution, respectively.

Furthermore, Eq. (4.6) can be rewritten as follows:

$$E[\xi_i] = E[g_i(\Theta)] = \int_\Omega g_i(\theta) p_\Theta(\theta) \mathrm{d}\theta = 0 \tag{4.94a}$$

$$E[\xi_i \xi_j] = E[g_i(\Theta) g_j(\Theta)] = \int_\Omega g_i(\theta) g_j(\theta) p_\Theta(\theta) \mathrm{d}\theta = \delta_{ij} \tag{4.94b}$$

In fact, there are various kinds of random orthogonal functions that just require to satisfy the orthogonality conditions (necessary condition) defined in Eq. (4.94). Presented below are five kinds of random orthogonal functions, comprising four non-Gaussian types and one Gaussian one.

4.4.1 Random Orthogonal Functions for Non-Gaussian Random Variables

The random orthogonal functions for non-Gaussian random variables typically encompass polynomial and trigonometric function forms. Four specific formulations, three polynomial functions and one trigonometric function are addressed below for illustrative purpose.

1. For Legendre orthogonal polynomial function, the standard orthogonal random variables ξ_i can be defined as follows:

$$\xi_i = g_i(\Theta) = \sqrt{2i+1} P_i(\Theta), \quad i = 1, 2, \cdots, N \tag{4.95a}$$

where Θ denotes the uniformly distributed elementary random variable on the interval $(-1, 1)$. $P_i(\cdot)$ is the Legendre polynomial function, and its recurrence formulas are given by:

$$\begin{cases} P_{i+1}(x) = \dfrac{2i+1}{i+1} x P_i(x) - \dfrac{i}{i+1} P_{i-1}(x) \\ P_0(x) = 1, \quad P_1(x) = x \end{cases} \tag{4.95b}$$

2. For the first family of Chebyshev orthogonal polynomial functions, the standard orthogonal random variables ξ_i can be defined as:

$$\xi_i = g_i(\Theta) = \sqrt{2} T_i(\Theta), \quad i = 1, 2, \cdots, N \tag{4.96a}$$

where Θ denotes the elementary random variable on the interval $(-1, 1)$, which has the PDF $p_\Theta(\theta) = \frac{1}{\pi\sqrt{1-\theta^2}}$. $T_i(\cdot)$ is the first family of Chebyshev polynomial function, and its recurrence formulas are given by (Liu 1982):

$$\begin{cases} T_{i+1}(x) = 2xT_i(x) - T_{i-1}(x) \\ T_0(x) = 1, \quad T_1(x) = x \end{cases} \tag{4.96b}$$

Typically, the first family of Chebyshev orthogonal basis could also be rewritten as follows (Liu 1982):

$$\xi_i = g_i(\Theta) = \sqrt{2}T_i(\Theta) = \sqrt{2}\cos(i\arccos\Theta), \quad i = 1, 2, \cdots, N \tag{4.96c}$$

3. For the second family of Chebyshev orthogonal polynomial functions, the standard orthogonal random variables ξ_i can be defined as:

$$\xi_i = g_i(\Theta) = U_i(\Theta), \quad i = 1, 2, \cdots, N \tag{4.97a}$$

where Θ denotes the elementary random variable on the interval $(-1, 1)$, and its PDF is $p_\Theta(\theta) = \frac{2}{\pi}\sqrt{1-\theta^2}$. $U_i(\cdot)$ is the second family of Chebyshev polynomial function, and its recurrence formulas are given by (Liu 1982):

$$\begin{cases} U_{i+1}(x) = 2xU_i(x) - U_{i-1}(x) \\ U_0(x) = 1, \quad U_1(x) = 2x \end{cases} \tag{4.97b}$$

Similarly, the second family of Chebyshev orthogonal basis could also be rewritten as (Liu 1982):

$$\xi_i = g_i(\Theta) = U_i(\Theta) = \frac{\sin[(i+1)\arccos\Theta]}{\sqrt{1-\Theta^2}}, \quad i = 1, 2, \cdots, N \tag{4.97c}$$

4. For the trigonometric function form, the sine function series is presented here as an example, with the standard orthogonal random variables ξ_i defined as below:

$$\xi_i = g_i(\Theta) = \sqrt{2}\sin(i\Theta + \varphi), \quad i = 1, 2, \cdots, N \tag{4.98}$$

where Θ denotes the uniformly distributed elementary random variable on the interval $(0, 2\pi)$, and the parameter φ is usually valued by $\pi/4$, which can also be recognized as the Hartley orthogonal basis in this case.

It is worth further elaborating that the random orthogonal functions for non-Gaussian random variables possess various forms. Among the orthogonal polynomial functions, in addition to the aforementioned Legendre, the first and the second families of Chebyshev polynomial formulations, the commonly used ones also include Hermite and Laguerre polynomial functions (Liu et al. 2017). For trigonometric function forms, besides the sine function series mentioned above, the cosine function series and the sine-cosine function series can also be alternative, where the sine-cosine function series could consist a complete set of orthogonal functions (Liu et al. 2017).

4.4.2 Random Orthogonal Functions for Gaussian Random Variables

In order to gain a collection of standard orthogonal random variables $\{\xi_i\}_{i=1}^{N}$ with the same Gaussian distribution, the equal-probability-inverse-transformation method presented in Sect. 3.3.1 can be utilized here. For convenience, just the typical Gaussian-sine-function-series, which is transformed from the non-Gaussian-sine-function-series defined in Eq. (4.98), is addressed as follows:

$$\xi_i = \Phi^{-1}\left[\frac{1}{2} + \frac{1}{\pi}\arcsin(\sin(i\Theta + \varphi))\right], \quad i = 1, 2, \cdots, N \tag{4.99}$$

where Θ denotes the uniformly distributed elementary random variable on the interval $(0, 2\pi)$, and the parameter φ is usually valued by $\pi/4$, which is also regarded as the Gaussian-Hartley-orthogonal-basis in this case.

The five random orthogonal functions employed in this section and the corresponding PDFs of the elementary random variables are shown in Table 4.1.

It should be clarified that during the simulation process, in the K-L decomposition, the standard orthogonal random variables generated by the random orthogonal functions defined in Eqs. (4.95)–(4.99) can be directly substituted into Eq. (4.5) for generating stochastic processes. However, in other time-domain analysis methods, the initial random variables produced by random orthogonal functions should be transformed into the target random variables through a deterministic one-to-one mapping, which acts as a sufficient condition for simulation. The specific transformation details can be found in Sect. 3.3.1. It is just the technical benefit of the random orthogonal functions-based representation that the series of terms in the K-L

Table 4.1 Typical orthogonal functions and probability density function of the elementary random variables

Standard orthogonal random variables	Type of functions	Orthogonal functions	PDF of elementary random variables	Support
Non-Gaussian type	Polynomial functions	Legendre polynomials	Uniform	$(-1, 1)$
		The 1st Chebyshev polynomials	$\frac{1}{\pi\sqrt{1-\theta^2}}$	$(-1, 1)$
		The 2nd Chebyshev polynomials	$\frac{2}{\pi}\sqrt{1-\theta^2}$	$(-1, 1)$
	Trigonometric functions	Sine function series (Hartley orthogonal basis)	Uniform	$(0, 2\pi)$
Gaussian type	Trigonometric functions	Sine function series (Hartley orthogonal basis)	Uniform	$(0, 2\pi)$

decomposition are arranged in a descending order of eigenvalues, consistent with the order of the standard orthogonal random variables defined by random orthogonal functions. While the series of terms in other time-domain analysis methods are arranged according to the discrete terms of impulse response functions.

4.5 Numerical Examples

4.5.1 Numerical Examples of Karhunen-Loève Decomposition

In this section, the C-P model defined in Eq. (4.35) is used to simulate the stationary stochastic ground motion acceleration process $\overline{X}_{\mathrm{g}}(t)$. According to the Wiener-Khintchine theorem, the auto-correlation function of the stationary ground motion acceleration process, as an inverse Fourier transform of Eq. (4.35), is given by:

$$R_{\overline{X}_{\mathrm{g}}}(t_2-t_1)=R_{\overline{X}_{\mathrm{g}}}(\tau)=\frac{1}{2\pi}\int_{-\infty}^{\infty}S_{\overline{X}_{\mathrm{g}}}(\omega)\mathrm{e}^{\mathrm{i}\omega\tau}\mathrm{d}\omega=\mathrm{i}S_0\sum_{j=1}^{4}R_j(\tau) \tag{4.100}$$

where S_0 denotes the intensity factor of power spectrum, i.e., Eq. (4.26), and i denotes the imaginary unit, i.e., $\mathrm{i}=\sqrt{-1}$, respectively. The expressions of $R_j(\tau)$ are given by (Liu et al. 2017):

$$R_1(\tau)=\frac{\omega_{\mathrm{g}}^4+4\zeta_{\mathrm{g}}^2\omega_{\mathrm{g}}^2\omega_1^2}{(\omega_1-\omega_2)(\omega_1-\omega_3)(\omega_1-\omega_4)}\times\frac{\omega_1^4\exp(\mathrm{i}|\tau|\omega_1)}{\left(\omega_1^2-\omega_{\mathrm{f}}^2\right)^2+4\zeta_{\mathrm{f}}^2\omega_{\mathrm{f}}^2\omega_1^2} \tag{4.101a}$$

$$R_2(\tau)=\frac{\omega_{\mathrm{g}}^4+4\zeta_{\mathrm{g}}^2\omega_{\mathrm{g}}^2\omega_2^2}{(\omega_2-\omega_1)(\omega_2-\omega_3)(\omega_2-\omega_4)}\times\frac{\omega_2^4\exp(\mathrm{i}|\tau|\omega_2)}{\left(\omega_2^2-\omega_{\mathrm{f}}^2\right)^2+4\zeta_{\mathrm{f}}^2\omega_{\mathrm{f}}^2\omega_2^2} \tag{4.101b}$$

$$R_3(\tau)=\frac{\omega_{\mathrm{g}}^4+4\zeta_{\mathrm{g}}^2\omega_{\mathrm{g}}^2\omega_5^2}{\left(\omega_5^2-\omega_{\mathrm{g}}^2\right)^2+4\zeta_{\mathrm{g}}^2\omega_{\mathrm{g}}^2\omega_5^2}\times\frac{\omega_5^4\exp(\mathrm{i}|\tau|\omega_5)}{(\omega_5-\omega_6)(\omega_5-\omega_7)(\omega_5-\omega_8)} \tag{4.101c}$$

$$R_4(\tau)=\frac{\omega_{\mathrm{g}}^4+4\zeta_{\mathrm{g}}^2\omega_{\mathrm{g}}^2\omega_6^2}{\left(\omega_6^2-\omega_{\mathrm{g}}^2\right)^2+4\zeta_{\mathrm{g}}^2\omega_{\mathrm{g}}^2\omega_6^2}\times\frac{\omega_6^4\exp(\mathrm{i}|\tau|\omega_6)}{(\omega_6-\omega_5)(\omega_6-\omega_7)(\omega_6-\omega_8)} \tag{4.101d}$$

where

$$\omega_1=\omega_{\mathrm{g}}\sqrt{1-\zeta_{\mathrm{g}}^2}+\mathrm{i}\zeta_{\mathrm{g}}\omega_{\mathrm{g}},\quad \omega_2=-\omega_{\mathrm{g}}\sqrt{1-\zeta_{\mathrm{g}}^2}+\mathrm{i}\zeta_{\mathrm{g}}\omega_{\mathrm{g}},\omega_3=\omega_1^*,\omega_4=\omega_2^*,$$

$$\omega_5=\omega_\mathrm{f}\sqrt{1-\zeta_\mathrm{f}^2}+\mathrm{i}\zeta_\mathrm{f}\omega_\mathrm{f},\quad \omega_6=-\omega_\mathrm{f}\sqrt{1-\zeta_\mathrm{f}^2}+\mathrm{i}\zeta_\mathrm{f}\omega_\mathrm{f},\omega_7=\omega_5^*,\omega_8=\omega_6^*.$$

Furthermore, the non-stationary ground motion acceleration process could be gained through applying a deterministic modulation function to the stationary acceleration process, given by:

$$X_\mathrm{g}(t)=f(t)\overline{X}_\mathrm{g}(t) \tag{4.102}$$

where $X_\mathrm{g}(t)$ denotes non-stationary seismic acceleration process. $f(t)$ denotes the deterministic intensity modulation function, which is defined as Eq. (2.60), such that:

$$f(t)=\left[\frac{t}{c}\exp\left(1-\frac{t}{c}\right)\right]^d \tag{4.103}$$

where c denotes average arrival time of peak ground acceleration, and the parameter d is used to control the shape of $f(t)$, respectively. In this example, $c=4$ s and $d=1$.

The auto-correlation function of non-stationary seismic acceleration process can thus be expressed as:

$$R_{X_\mathrm{g}}(t_1,t_2)=f(t_1)f(t_2)R_{\overline{X}_\mathrm{g}}(t_2-t_1) \tag{4.104}$$

Meanwhile, the EPSD function of non-stationary seismic acceleration process $X_\mathrm{g}(t)$ is expressed as Eq. (2.71), such that:

$$S_{X_\mathrm{g}}(\omega,t)=f^2(t)S_{\overline{X}_\mathrm{g}}(\omega) \tag{4.105}$$

According to statistical relationship of stochastic processes, the mean and variance of a non-stationary seismic acceleration process are derived as follows, respectively:

$$E\left[X_\mathrm{g}(t)\right]=f(t)E\left[\overline{X}_\mathrm{g}(t)\right]=0 \tag{4.106}$$

$$\sigma_{X_\mathrm{g}}^2=R_{X_\mathrm{g}}(t,t)=\frac{1}{2\pi}\int_{-\infty}^{\infty}S_{X_\mathrm{g}}(\omega,t)\,\mathrm{d}\omega=\frac{1}{2\pi}f^2(t)\int_{-\infty}^{\infty}S_{\overline{X}_\mathrm{g}}(\omega)\,\mathrm{d}\omega \tag{4.107}$$

In this numerical example, the parameters of the stationary PSD model are valued by $\omega_\mathrm{g}=5\pi$ rad/s, $\zeta_\mathrm{g}=0.6$, $\omega_\mathrm{f}=0.5\pi$ rad/s, $\zeta_\mathrm{f}=0.6$. As for the parameters in the intensity factor of the power spectrum, say S_0 in Eq. (4.26), r denotes the peak factor and is valued by 3.0 for deep cohesionless soil, and the PGA is determined as $a_{\max}=0.2\mathrm{g}=200$ cm/s^2, respectively. The item number of K-L decomposition is defined as $N=600$, and the number of involved eigenfunctions is taken as $M=600$, respectively. The duration of ground motion acceleration process is defined as

$T = 20\,\text{s}$, and the time interval of representative samples is valued by $\Delta t = 0.01$ s, respectively.

In addition, it should be noted that when generating seismic ground motion acceleration process using the proposed dimension-reduction simulation methods, in the case that the Legendre polynomials, the Gaussian and non-Gaussian Hartley orthogonal basis are employed for the random orthogonal function, since the elementary random variable follows an equal probability distribution, the representative point set can be achieved using the operation of equal spatial distribution on the corresponding interval. However, in the case that the first and second families of Chebyshev polynomials are applied for the random orthogonal function, the equal-probability-inverse-transformation method could be utilized for transforming the initial representative point set to the target representative point set, of which the details can be found in Sect. 3.3.1. Then, trough substituting the standard orthogonal random variables into the corresponding simulation formula, the representative time-histories of non-stationary seismic acceleration process can thus be obtained. Finally, $n_{\text{sel}} = 626$ representative time-histories are generated using the dimension-reduction simulation method based on the K-L decomposition, of which the numerical results are presented in Figs. 4.3, 4.4, and 4.5.

Figure 4.3 depicts the representative time-histories of non-stationary ground motion processes simulated applying the Legendre polynomials, the non-Gaussian and Gaussian Hartley orthogonal basis, respectively. It is evident that significant

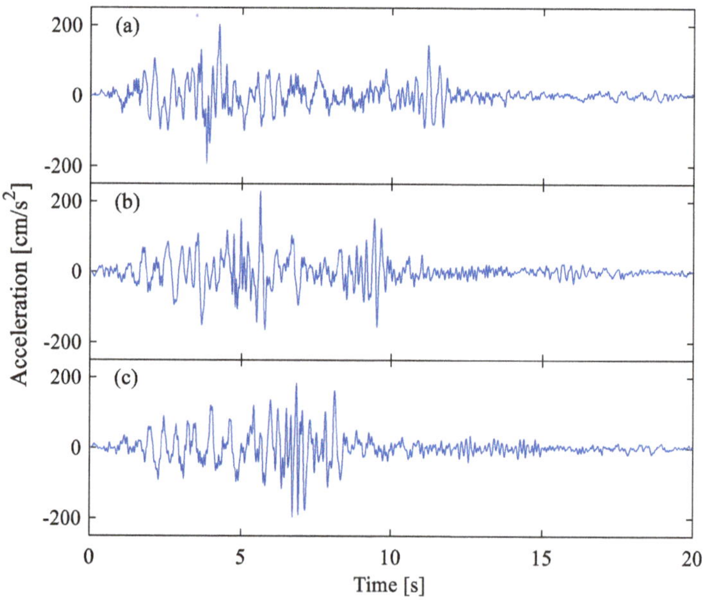

Fig. 4.3 Representative time-histories of non-stationary seismic acceleration processes generated by dimension-reduction simulation methods with random orthogonal functions: (**a**) Legendre polynomials; (**b**) non-Gaussian Hartley orthogonal basis; and (**c**) Gaussian Hartley orthogonal basis

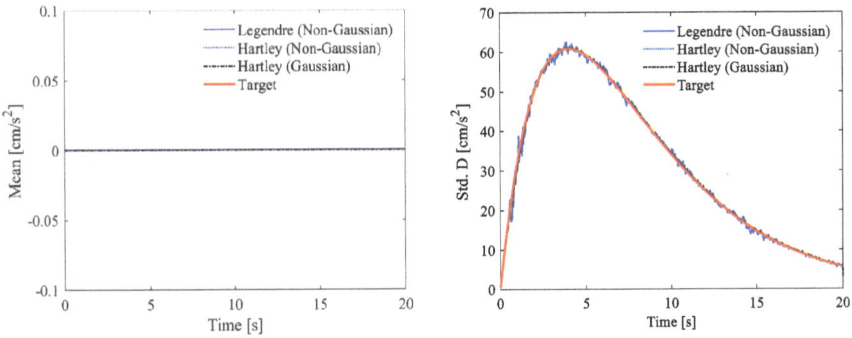

(a) mean (b) standard deviation

Fig. 4.4 Comparison of mean and Std. D between target statistics and those of simulated non-stationary seismic acceleration processes

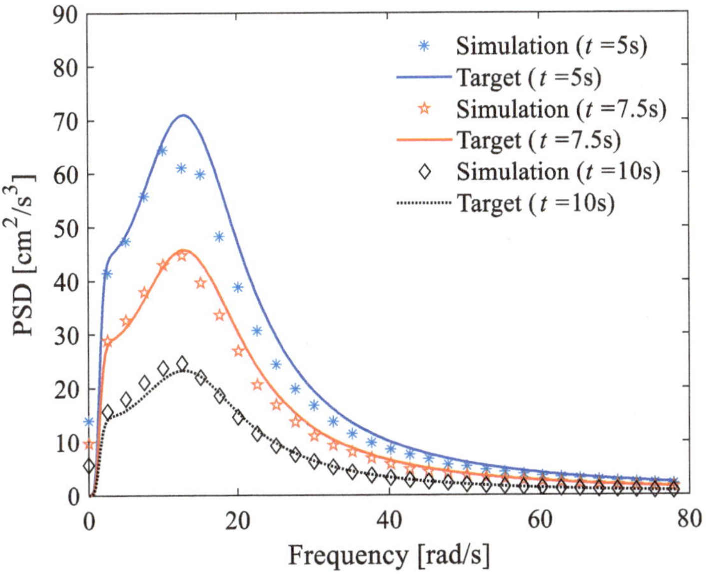

Fig. 4.5 Comparison of the PSD between the simulated non-stationary seismic acceleration processes by the updated scheme with the non-Gaussian Hartley orthogonal basis with the corresponding targets at three typical time instants

non-stationary characteristics emerged in the acceleration time-histories. One might notice that when employing the updated K-L decomposition with three different random orthogonal functions, the discrepancies among the representative time-histories appear obviously, particularly in case of the time lengths of large-amplitude segments vary significantly. Indeed, even with the same random orthogonal function, notable differences persist among the representative time-histories.

Nevertheless, the statistical processes derived from the three different random orthogonal functions remain consistent.

Comparison of the mean value and standard deviation (Std. D) between the target statistics and those of simulated non-stationary seismic acceleration processes with the three different random orthogonal functions are described in Fig. 4.4. It can be easily found that the K-L decomposition, updated with random orthogonal functions of the Legendre polynomials, the non-Gaussian and Gaussian Hartley orthogonal bases, all yielded satisfactory simulation results, with second-order statistics matching the corresponding target values precisely.

Figure 4.5 presents the comparison of the PSD between the simulated non-stationary seismic acceleration processes by the updated K-L decomposition using the non-Gaussian Hartley orthogonal basis at the time instants of 5 s, 7.5 s and 10 s with the corresponding target values. It is evident that the simulated PSD exhibits a high degree of agreement with the corresponding targets, particularly in the amplitude decay stage after 10 s. As a matter of fact, employing other random orthogonal functions, such as Legendre polynomials and Gaussian Hartley orthogonal basis, could also achieve satisfactory fitting results.

4.5.2 Numerical Examples of Filtered White Noise Process Models in Physics

In this section, the employed EPSD function of fully non-stationary seismic acceleration process is defined as Eq. (2.73), given by:

$$S_{X_g}(\omega, t) = f^2(t) S(\omega, t) \tag{4.108}$$

where $f(t)$ has been given in Eq. (4.51), and the generalized PSD (time-varying PSD) function $S(\omega, t)$ is defined as Eq. (2.74), of which the parameters $\boldsymbol{\lambda}_1(t) = [\omega_g(t), \zeta_g(t)]$ have been given in Eq. (4.54), and the expression of $S_0(t)$ can be found in Eq. (2.78).

In the previous numerical examples of the updated K-L decomposition, the Legendre polynomials, the non-Gaussian and Gaussian Hartley orthogonal basis were employed for generating non-stationary ground motion acceleration processes. It can be seen that all the three random orthogonal functions achieve satisfactory results. Therefore, for simplicity, in the dimension-reduction simulation using the filtered white noise process model, just the non-Gaussian Hartley orthogonal basis is utilized for investigation purpose. Here, different soil sites have been considered in detail, including rock, deep cohesionless soil, and soft soil, which are classified by the equivalent shear wave velocity of the soil layers and the thickness of the site coverage layers.

In this case, the time interval is set as $\Delta t = 0.01$ s and the upper cut-off frequency is valued by $\omega_u = 100\pi$ rad/s. The seismic fortification intensity is supposed as 8 degree and the corresponding PGA is $a_{max} = 200$ cm/s^2. According to the research

Table 4.2 Model parameter values

	Soil conditions		
Parameter	Rock	Deep cohesionless soil	Soft soil
$\overline{\omega}_g$ (rad/s)	8π	5π	2.4π
$\overline{\zeta}_g$	0.60	0.72	0.85
d (1/s)	7	5	3
e	0.2	0.2	0.2
m	2	2	2
t_1 (s)	2	2.5	3
t_2 (s)	12	15	18
c	0.38	0.30	0.25
T (s)	20	25	30
r	3.1	3.0	2.9

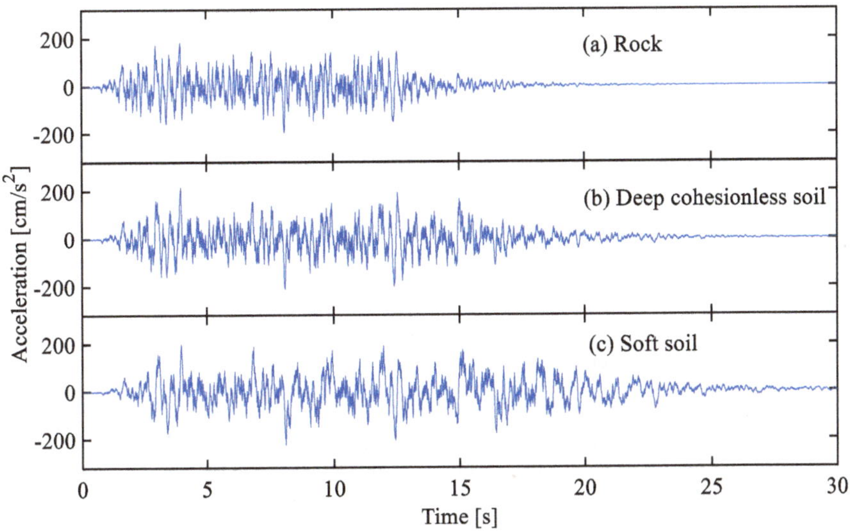

Fig. 4.6 Representative time-histories generated by the updated double-filtered white noise process model with the non-Gaussian Hartley orthogonal basis for different soil sites

(Seya et al. 1993; Deodatis 1996), the specific parameter values of different soil sites are provided in Table 4.2. Finally, $n_{\text{sel}} = 610$ representative time-histories of non-stationary seismic ground motions are generated applying the updated double-filtered white noise process model.

Figure 4.6 displays the seismic ground motion representative time-histories generated through the updated double-filtered white noise process model for different soil sites. It is evident from the figures that these representative time-histories emerge typical non-stationary characteristics. Obviously, the acceleration processes of different site categories show remarkable variations in the peak arrival time, duration of strong motion, and attenuation rate. It should be clarified that for rock and deep cohesionless soil sites, the duration for simulating ground motions is

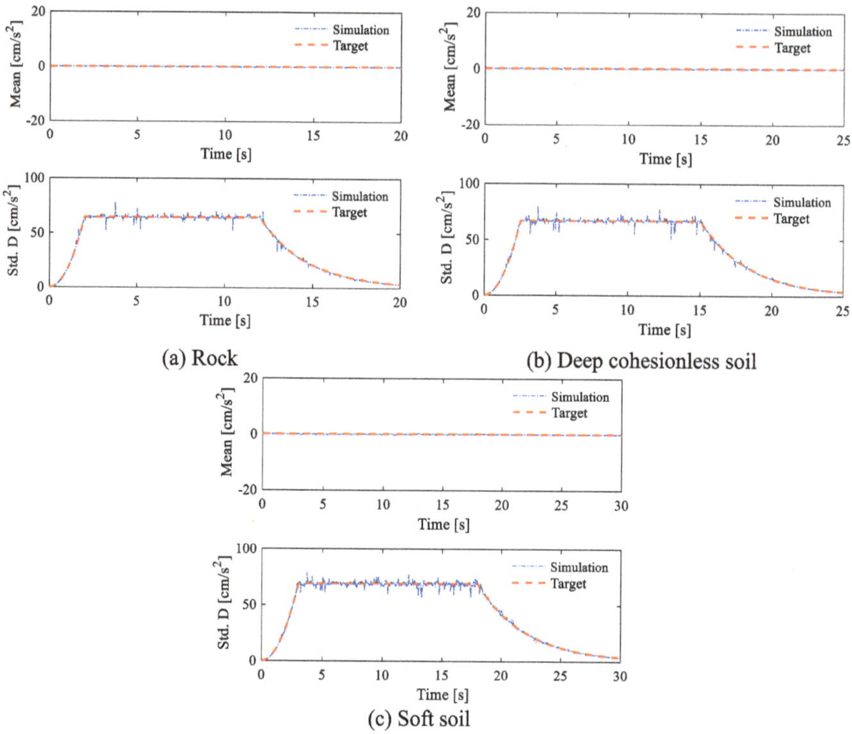

Fig. 4.7 Comparison of mean and Std. D between target statistics and those of simulated non-stationary seismic acceleration processes for different soil sites

calculated as 20 s and 25 s, respectively, as specified in Table 4.2. However, in order to more distinctly present the variations in acceleration time-histories across different site categories, all the representative time-histories are plotted for a duration of 30 s. Clearly, the intensity for rock and deep cohesionless soil sites drops to nearly zero after 20 s and 25 s.

Figure 4.7 illustrates the comparison of the mean and Std. D between the representative time-history sets of ground motion acceleration processes for different soil sites generated using the updated double-filtered white noise process model and their corresponding target values. It is evident from the figures that the simulated values align closely with the target ones for all the three site categories.

Figure 4.8 displays the 3D views of the EPSD function associated with the generated representative time-history sets of non-stationary ground motion acceleration processes for different soil sites using the improved double-filtered white noise process model. It is evident from the figures that the PSD function values vary simultaneously with both time and frequency, effectively proving that the applied EPSD model can capture the dual non-stationary characteristics of seismic ground motions in both intensity and frequency at the same time.

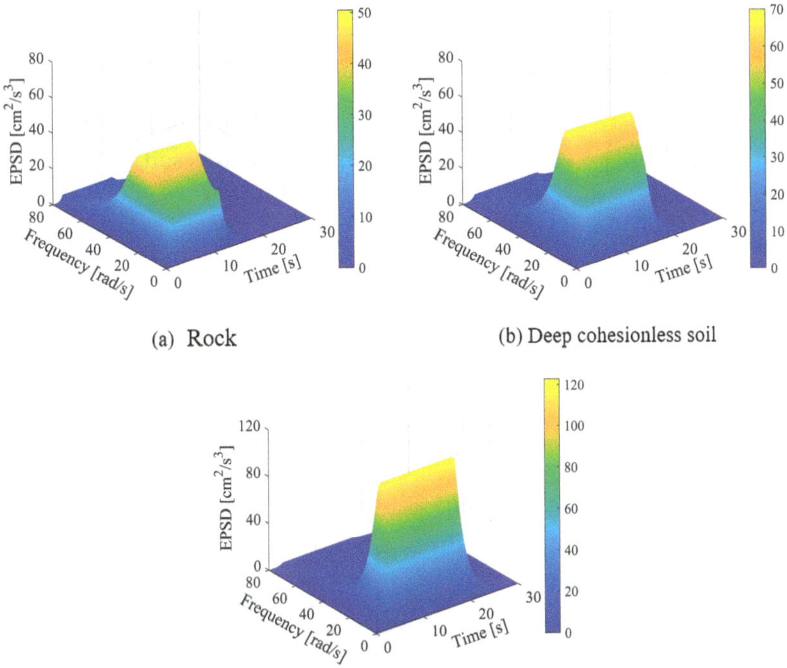

(a) Rock

(b) Deep cohesionless soil

(c) Soft soil

Fig. 4.8 EPSD of the simulated representative time-history sets for different soil sites

Figure 4.9 presents a comparison of the PSD function between the simulated non-stationary ground motion acceleration representative time-history sets for the three different soil site and their corresponding target values at the time instant $t = 10$ s. It is evident from the figure that the simulated values for all the three site categories exhibit a desirable fit with their corresponding target ones.

4.5.3 Numerical Examples of Time-Domain Representation of Non-Stationary Processes in Mathematics

In this example, the time-frequency modulation function defined in Eq. (2.69) is employed to comprehensively describe the non-stationary features of strong motions in both intensity and frequency, such that:

$$A(t, \omega) = \frac{\exp(-at) - \exp\left[-\left(c \times \left|\omega - \omega_g\right| + b\right) \times t\right]}{\exp(-at^*) - \exp\left[-\left(c \times \left|\omega - \omega_g\right| + b\right) \times t^*\right]}, \quad t > 0, \quad \omega > 0 \quad (4.109a)$$

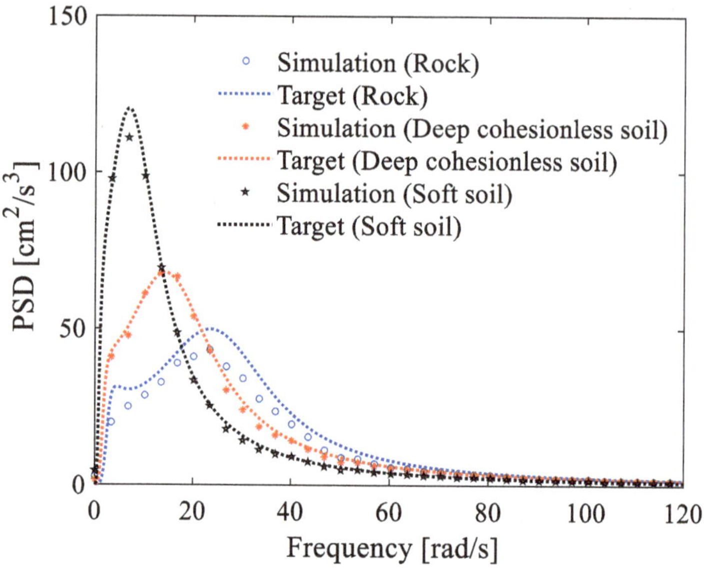

Fig. 4.9 Comparison of PSD between the simulated representative time-history sets and the corresponding targets for different soil sites at the time instant $t = 10$ s

$$t^* = \frac{\ln\left(c\times\left|\omega-\omega_{\mathrm{g}}\right|+b\right)-\ln(a)}{c\times\left|\omega-\omega_{\mathrm{g}}\right|+(b-a)},\quad \omega>0 \tag{4.109b}$$

where the parameters a, b and c satisfy the conditions $b = a+0.001$ and $c = 0.005$.

The C-P model defined in Eq. (4.35) is also adopted here. According to Eq. (4.58), the expression of $H(\omega)$ can be handily obtained as follows:

$$H(\omega)=\sqrt{S_0}\cdot\frac{\omega_{\mathrm{g}}^2+2\mathrm{i}\zeta_{\mathrm{g}}\omega_{\mathrm{g}}\omega}{\left(\omega_{\mathrm{g}}^2-\omega^2\right)+2\mathrm{i}\zeta_{\mathrm{g}}\omega_{\mathrm{g}}\omega}\times\frac{\omega^2}{\left(\omega_{\mathrm{f}}^2-\omega^2\right)+2\mathrm{i}\zeta_{\mathrm{f}}\omega_{\mathrm{f}}\omega} \tag{4.110}$$

Then, we can get the frequency response function as below:

$$\widetilde{H}(t,\omega)=A(t,\omega)H(\omega) \tag{4.111}$$

Furthermore, applying the second formula of Eq. (4.71), the impulse response function can be achieved as the following:

$$\widetilde{h}(t,\tau)=\frac{1}{2\pi}\int_{-\infty}^{\infty}\widetilde{H}(t,\omega)\mathrm{e}^{\mathrm{i}\omega\tau}\mathrm{d}\omega \tag{4.112}$$

Table 4.3 Calculation parameters for simulation of non-stationary ground motions

Parameter	Value	Parameter	Value
Dominant frequency of the soil layer	$\omega_g = 5\pi$ rad/s	Damping ratio of the soil layer	$\zeta_g = 0.6$
Dominant frequency of the rock bed	$\omega_f = 0.5\pi$ rad/s	Damping ratio of the rock bed	$\zeta_f = 0.6$
Parameter of the time-frequency modulation function	$a = 0.15\ s^{-1}$	Number of representative samples	$n_{sel} = 233$
PGA	$a_{max} = 200\ cm/s^2$	Peak factor	$r = 3.0$
The entire duration	$T = 30$ s	Time step	$\Delta t = 0.01$ s
The upper cut-off frequency	$\omega_u = 100\pi$ rad/s	Frequency interval	$\Delta\omega = 0.209$ rad/s

Finally, the non-stationary ground motion acceleration process can be simulated utilizing Eq. (4.79), of which the time-varying standard deviation $\sigma_{X_g}(t)$ holding the expression as follows:

$$\sigma_{X_g}(t) = \sqrt{\frac{1}{2\pi}\int_{-\infty}^{\infty} S_{X_g}(\omega, t)\mathrm{d}\omega} = \sqrt{\frac{1}{2\pi}\int_{-\infty}^{\infty} \left|\widetilde{H}(t,\omega)\right|^2 \mathrm{d}\omega} \tag{4.113}$$

The specific parameters for simulation of non-stationary ground motion processes are listed in Table 4.3.

Here, it is worth noting that within the conventional discretization strategy for time-domain representation in mathematics, the EPSD function of non-stationary processes is inherently more complex than the PSD function of stationary processes. Consequently, deriving an analytical solution for the corresponding impulse response function $\widetilde{h}(t,\tau)$ is highly challenging. Instead, through employing the inverse fast Fourier transform method with the help of MATLAB software, numerical solutions can be efficiently obtained. Based on this, the representative time-histories of non-stationary ground motion acceleration processes can be simulated by the renewed time-domain analysis method in mathematics. The non-Gaussian Hartley orthogonal basis is also chosen as the random orthogonal function in this case.

Figure 4.10 displays the 50th, 100th, and 150th representative time-histories of the non-stationary ground motion acceleration processes generated by the renewed time-domain analysis approach in mathematics with the non-Gaussain Hartley orthogonal basis. It is evident from the figures that the acceleration time-histories show typical non-stationary features. Furthermore, notable differences exist among different representative time-histories, highlighting the significant randomness inherent in earthquake ground motions.

Figure 4.11 presents a comparison of the mean and Std. D between the representative time-histories of non-stationary seismic acceleration processes generated through the updated time-domain analysis approach in mathematics and the

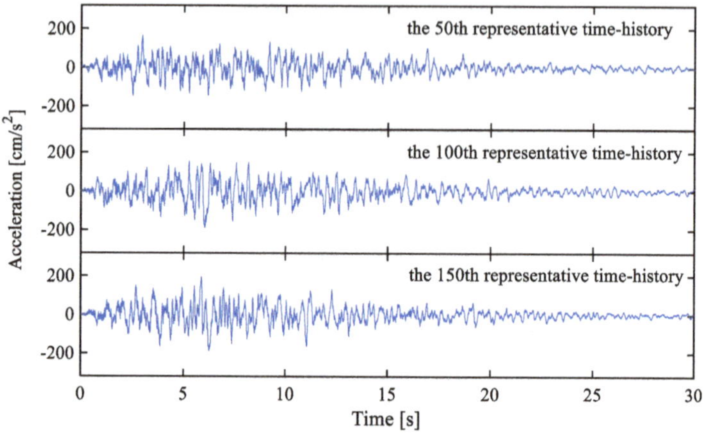

Fig. 4.10 The 50th, 100th and 150th representative time-histories of non-stationary seismic acceleration processes simulated by the renewed time-domain analysis method in mathematics

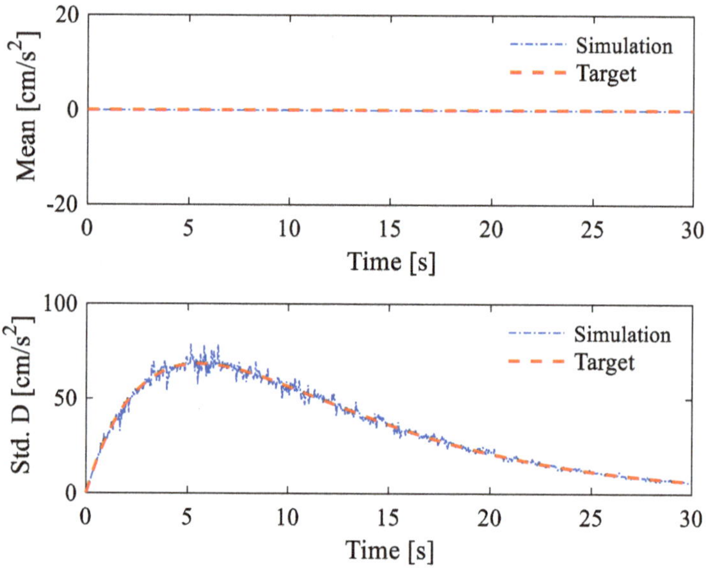

Fig. 4.11 Comparison of mean and Std. D between target statistics and those of simulated non-stationary seismic acceleration processes

corresponding target values. It is evident from the figure that the simulated values are both well aligned with the corresponding targets.

Figure 4.12 displays the 3D vies of the EPSD associated with the representative time-history set simulated by the renewed time-domain analysis method in mathematics. As shown in the figure, it is evident that the PSD value exhibits a clear trend of initially increasing and subsequently decreasing over time, with energy

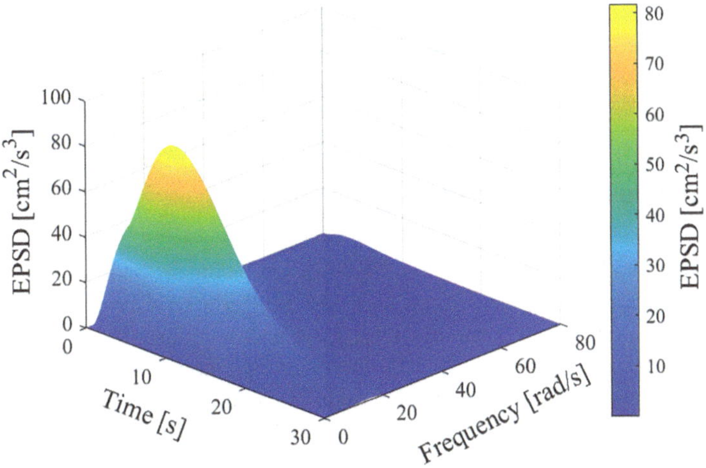

Fig. 4.12 EPSD of the representative time-history set simulated by the renewed time-domain analysis method in mathematics

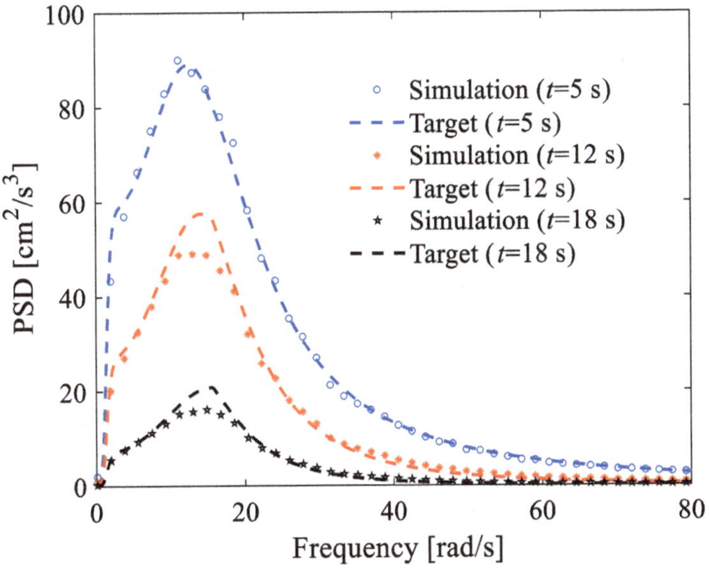

Fig. 4.13 Comparison of PSD between the simulated representative time-history set and the corresponding targets at three typical time instants

predominantly concentrated within a certain frequency range. Overall, the selected EPSD model can effectively reflect the full non-stationary characteristics of earthquake ground motions in both time and frequency domains.

Figure 4.13 presents a comparison of the PSD between the simulated non-stationary ground motion representative time-histories at three typical time instants, say $t = 5$ s, 12 s and 18 s and their corresponding target values. As can be clearly observed from the figure, all the simulated values at different time instants are well-fitted with their corresponding target ones.

4.6 Conclusion Remarks

This chapter addresses three typical kinds of time-domain analysis methods for simulating univariate stochastic processes, say the Karhunen-Loève decomposition, filtered white noise process models in physics and time-domain representation of stochastic processes in mathematics. Furthermore, by employing the constraint form of random orthogonal functions, the dimension-reduction simulation of non-stationary seismic acceleration processes can thus be accomplished using merely one elementary random variable. Through the detailed derivations presented in this chapter, the distinct characteristics of the aforementioned three typical time-domain analysis methods have been clarified, with the conclusion remarks drawn as follows:

1. The K-L decomposition is applicable not only to stationary stochastic processes, but also to non-stationary ones. In fact, the truncated finite K-L decomposition series in Eq. (4.5) for stochastic processes are optimal, i.e., when the same number of series terms N is considered, the mean square relative error is minimized. Moreover, the standard orthogonal random variables can be directly obtained through the definition of random orthogonal functions, without the requirement for deterministic one-to-one mapping transformation. Its disadvantage lies in the fact that calculating the correlation matrix is time-consuming. Especially for non-stationary stochastic processes in engineering, obtaining an explicit expression for their autocorrelation function is challenging, and calculating their eigenvalues and eigen functions would be quite time-consuming.
2. The filtered white noise process models in physics are commonly employed to simulate stationary stochastic seismic ground motions. Based on this, the non-stationary stochastic seismic ground motions can be constructed by further incorporating the intensity modulation functions and time-varying PSD parameters. The advantage of this method lies in its clear physical significance, where the impulse response function can be directly obtained through derivation of physical processes (motion equations). However, its drawback is that it can just simulate some specific physical processes, thereby lacking generality.
3. As for the time-domain representation of stochastic processes in mathematics, any stationary and non-stationary stochastic processes can be treated as the response of a linear dynamic system driven by the white noise process. For any stochastic process with second-order moment, once its PSD or EPSD is provided, an explicit or implicit expression for the impulse response function can be obtained. However, for time-frequency fully non-stationary stochastic processes, just the implicit numerical results of the impulse response function could be obtained in most cases.

References

Box GE, Jenkins GM, Reinsel GC et al (2015) Time series analysis: forecasting and control. Wiley, Hoboken, NJ

Dahlhaus R (1997) Fitting time series models to nonstationary processes. Ann Stat 25(1):1–37

Deodatis G (1996) Non-stationary stochastic vector processes: seismic ground motion applications. Prob Eng Mech 11(3):149–168

Deodatis G, Shinozuka M (1988) Auto-regressive model for nonstationary stochastic processes. J Eng Mech 114(11):1995–2012

Ghanem RG, Spanos PD (1991) Stochastic finite elements: a spectral approach. Springer, New York

Lai M, Ye TY, Li YM (1995) Bi-filtered white noise model of earthquake ground motion. Chin Civil Eng J 28(6):60–66. (in Chinese)

Liu ZK (1982) Orthogonal functions and its applications. National Defence Industry Press, Beijing. (in Chinese)

Liu ZJ, Liu ZX, Peng YB (2017) Dimension reduction of Karhunen-Loeve expansion for simulation of stochastic processes. J Sound Vib 408:168–189

Liu ZJ, Liu M, Xu BH, et al (2024) Time-domain dimension-reduction representation for stochastic ground motion utilizing filtered white noise. Prob Eng Mech 77:103678

Liu ZJ, Xu BH, Ruan XX et al (2025) Dimension-reduction representations for stationary and non-stationary stochastic processes in time domain. Mech Syst Signal Process 241:113541

Rezaeian S, Der Kiureghian A (2008) A stochastic ground motion model with separable temporal and spectral nonstationarities. Earthq Eng Struct Dyn 37(13):1565–1584

Seya H, Talbott ME, Hwang HHM (1993) Probabilistic seismic analysis of a steel frame structure. Prob Eng Mech 8(2):127–136

Spanos PD, Zeldin BA (1998) Monte Carlo treatment of random fields: a broad perspective. Appl Mech Rev 5(3):219–237

Chapter 5
Dimension-Reduction Representation of Multivariate Stochastic Ground Motions

Abstract This chapter deduces simulation methods for multivariate non-stationary stochastic processes (1D-nV non-stationary stochastic vector processes), including the spectral decomposition methods based on frequency-domain analysis and the methods based on time-domain analysis. For frequency-domain analysis methods, this chapter systematically derives the unified form of original spectral decomposition representation incorporating both the spectral representation method (SRM) and proper orthogonal decomposition (POD). Moreover, the conventional Monte Carlo methods and the dimension-reduction methods are both derived from the original spectral decomposition, implying that both can be regarded as special cases of the original spectral decomposition. For time-domain analysis methods, this chapter introduces the conventional time-domain representation method and the TVP-VAR model. Further, through introducing the constraint form of random orthogonal functions associated with the standard orthogonal random vector, the multivariate stochastic processes can be precisely represented with merely 1–3 elementary random variables. Meanwhile, the chapter also briefly describes the dimension-reduction simulation method for multi-point and multi-component stochastic earthquake ground motions.

Keywords Non-stationary stochastic vector processes · Original spectral decomposition representation · Conventional Monte Carlo methods · Dimension-reduction simulation methods · Frequency-domain analysis methods · Time-domain analysis methods · Multi-point and multi-component ground motions

In seismic analysis of engineering structures, one-dimensional multi-variate (1D-nV) non-stationary stochastic vector process is typically employed to simulate the potential non-uniform seismic excitations at various supports of large-span structures, such as long-span bridges and dams. Essentially, the 1D-nV stochastic vector process constitutes a discrete spatio-temporal stochastic field. The primary simulation techniques encompass frequency-domain and time-domain analysis methods. This chapter mainly introduces the basic theories of these methods, and also covers the advancements in dimension-reduction simulation methods for non-stationary stochastic vector process based on the above analytical theories. It

Z. Liu et al., *Dimension-reduction Representation of Stochastic Ground Motion and Engineering Applications*, https://doi.org/10.1007/978-981-92-0168-6_5

should be clarified that, to avoid confusion, the Fourier transform pair between the PSD and the correlation function of the stochastic process in this section adopts Eqs. (1.78) and (1.79), rather than Eq. (4.1).

5.1 Original Spectral Decomposition Representation of Non-Stationary Stochastic Vector Processes

5.1.1 Unified Form of Original Spectral Decomposition Representation

Suppose that $\boldsymbol{X}(t) = [X_1(t), X_2(t), \cdots, X_n(t)]^{\mathrm{T}}$ is a 1D-nV non-stationary stochastic vector process, of which the components $X_i(t)$ $(i = 1, 2, \cdots, n)$ are real-valued non-stationary stochastic processes with zero-mean. According to the spectral decomposition representation theory, $\boldsymbol{X}(t)$ can be presented by Fourier-Stieltjes integral as follows (Priestley 1965):

$$\boldsymbol{X}(t) = \int_{-\infty}^{\infty} \boldsymbol{A}(t, \omega) \mathrm{e}^{\mathrm{i}\omega t} \mathrm{d}\boldsymbol{Z}(\omega) \tag{5.1}$$

where the time-frequency modulation function matrix $\boldsymbol{A}(t, \omega) = \mathrm{diag}\ [A_1(t, \omega), A_2(t, \omega), \cdots, A_n(t, \omega)]$ is a nonnegative real diagonal matrix and satisfies $\boldsymbol{A}(t, -\omega) = \boldsymbol{A}(t, \omega)$. $\boldsymbol{Z}(\omega) = [Z_1(\omega), Z_2(\omega), \cdots, Z_n(\omega)]^{\mathrm{T}}$ is a zero-mean 1D-nV complex orthogonal incremental process, of which the increment is $\mathrm{d}\boldsymbol{Z}(\omega) = \boldsymbol{Z}(\omega + \mathrm{d}\omega) - \boldsymbol{Z}(\omega)$ and obeys basic conditions as below:

$$E[\mathrm{d}\boldsymbol{Z}(\omega)] = \boldsymbol{0}_{n \times 1}, \ \ \mathrm{d}\boldsymbol{Z}(-\omega) = \mathrm{d}\boldsymbol{Z}^*(\omega) \tag{5.2a}$$

$$E\left[\mathrm{d}\boldsymbol{Z}(\omega)\mathrm{d}\boldsymbol{Z}^{*\mathrm{T}}(\omega')\right] = \boldsymbol{S}(\omega)\delta_{\omega\omega'}\,\mathrm{d}\omega \tag{5.2b}$$

where the superscript "*" and "T" denote complex conjugate and matrix transpose, respectively. $\boldsymbol{S}(\omega)$ is the two-sided PSD matrix of the corresponding stationary stochastic vector process. In general, $\boldsymbol{S}(\omega)$ is a nonnegative Hermitian matrix.

Here, define the complex increment $\mathrm{d}\boldsymbol{Z}(\omega)$ as follows:

$$\mathrm{d}\boldsymbol{Z}(\omega) = \mathrm{d}\boldsymbol{U}(\omega) - \mathrm{i}\mathrm{d}\boldsymbol{V}(\omega) \tag{5.3}$$

where $\mathrm{d}\boldsymbol{U}(\omega)$ and $\mathrm{d}\boldsymbol{V}(\omega)$ are both real-valued increments.

According to Eq. (5.2a), the real-valued increments $\mathrm{d}\boldsymbol{U}(\omega)$ and $\mathrm{d}\boldsymbol{V}(\omega)$ should satisfy the following conditions:

$$E[\mathrm{d}\boldsymbol{U}(\omega)] = E[\mathrm{d}\boldsymbol{V}(\omega)] = \mathbf{0}_{n\times 1} \tag{5.4a}$$

$$\mathrm{d}\boldsymbol{U}(-\omega) = \mathrm{d}\boldsymbol{U}(\omega), \quad \mathrm{d}\boldsymbol{V}(-\omega) = -\mathrm{d}\boldsymbol{V}(\omega) \tag{5.4b}$$

Similarly, substituting Eq. (5.3) and Eq. (5.4b) into Eq. (5.1), and considering Euler's formula, the real-valued finite series form of Eq. (5.1) can be obtained, given by:

$$\begin{aligned}
\boldsymbol{X}(t) &= \int_{-\infty}^{\infty} \boldsymbol{A}(t,\omega)\mathrm{e}^{\mathrm{i}\omega t}[\mathrm{d}\boldsymbol{U}(\omega) - \mathrm{i}\mathrm{d}\boldsymbol{V}(\omega)] \\
&= \int_{-\infty}^{0} \boldsymbol{A}(t,\omega)\mathrm{e}^{\mathrm{i}\omega t}[\mathrm{d}\boldsymbol{U}(\omega) - \mathrm{i}\mathrm{d}\boldsymbol{V}(\omega)] + \int_{0}^{\infty} \boldsymbol{A}(t,\omega)\mathrm{e}^{\mathrm{i}\omega t}[\mathrm{d}\boldsymbol{U}(\omega) - \mathrm{i}\mathrm{d}\boldsymbol{V}(\omega)] \\
&= \int_{0}^{\infty} \boldsymbol{A}(t,\omega)\mathrm{e}^{-\mathrm{i}\omega t}[\mathrm{d}\boldsymbol{U}(\omega) + \mathrm{i}\mathrm{d}\boldsymbol{V}(\omega)] + \int_{0}^{\infty} \boldsymbol{A}(t,\omega)\mathrm{e}^{\mathrm{i}\omega t}[\mathrm{d}\boldsymbol{U}(\omega) - \mathrm{i}\mathrm{d}\boldsymbol{V}(\omega)] \\
&= 2\int_{0}^{\infty} \boldsymbol{A}(t,\omega)[\cos(\omega t)\,\mathrm{d}\boldsymbol{U}(\omega) + \sin(\omega t)\,\mathrm{d}\boldsymbol{V}(\omega)] \\
&\approx 2\sum_{k=1}^{N} \boldsymbol{A}(t,\omega_k)[\cos(\omega_k t)\Delta\boldsymbol{U}(\omega_k) + \sin(\omega_k t)\Delta\boldsymbol{V}(\omega_k)]
\end{aligned} \tag{5.5}$$

For the non-stationary stochastic vector process $\boldsymbol{X}(t)$, its full probabilistic information is determined by the EPSD matrix $\boldsymbol{S_X}(\omega, t)$. Commonly, $\boldsymbol{S_X}(\omega, t) = \boldsymbol{A}(t, \omega)\boldsymbol{S}(\omega)$ $\boldsymbol{A}(t, \omega)$ is a nonnegative definite Hermitian matrix. Here assume that the PSD matrix $\boldsymbol{S}(\omega)$ can be decomposed orthogonally as follows (Li and Kareem 1991):

$$\boldsymbol{S}(\omega) = \boldsymbol{Q}(\omega)\boldsymbol{Q}^{*\mathrm{T}}(\omega) \tag{5.6}$$

where $\boldsymbol{Q}(\omega)$ is an $n\times n$-order complex matrix.

Considering Eq. (5.2b), Eq. (5.3) and Eq. (5.6), the complex increment $\Delta\boldsymbol{Z}(\omega_k)$ can be further defined as:

$$\Delta\boldsymbol{Z}(\omega_k) = \Delta\boldsymbol{U}(\omega_k) - \mathrm{i}\Delta\boldsymbol{V}(\omega_k) = \boldsymbol{Q}(\omega_k)\boldsymbol{P}_k\sqrt{\Delta\omega} \tag{5.7}$$

in which

$$E[\boldsymbol{P}_k] = \mathbf{0}_{n\times 1}, \quad E\left[\boldsymbol{P}_k\boldsymbol{P}_m^{*\mathrm{T}}\right] = \boldsymbol{I}_{n\times n}\delta_{km} \tag{5.8}$$

where $\boldsymbol{I}_{n\times n}$ is an $n\times n$-order unit matrix.

Further, the complex matrix $\boldsymbol{Q}(\omega_k)$ and the complex random vector $\boldsymbol{P}_k$ are defined as follows:

$$\boldsymbol{Q}(\omega_k)=\boldsymbol{E}(\omega_k)-\mathrm{i}\boldsymbol{F}(\omega_k),\quad \boldsymbol{P}_k=\boldsymbol{\mathcal{R}}_k-\mathrm{i}\,\boldsymbol{\mathcal{I}}_k \tag{5.9}$$

Substituting Eq. (5.9) into Eq. (5.7), the following can be obtained:

$$\begin{cases}\Delta\boldsymbol{U}(\omega_k)=[\boldsymbol{E}(\omega_k)\boldsymbol{\mathcal{R}}_k-\boldsymbol{F}(\omega_k)\boldsymbol{\mathcal{I}}_k]\sqrt{\Delta\omega}\\ \Delta\boldsymbol{V}(\omega_k)=[\boldsymbol{E}(\omega_k)\boldsymbol{\mathcal{I}}_k+\boldsymbol{F}(\omega_k)\boldsymbol{\mathcal{R}}_k]\sqrt{\Delta\omega}\end{cases} \tag{5.10}$$

Finally, through substituting Eq. (5.10) into Eq. (5.5), the unified form of spectral decomposition representation for 1D-nV non-stationary stochastic vector process $\boldsymbol{X}(t)$ can be written as follows (Di Paola and Zingales 2000; Liu et al. 2025):

$$\begin{aligned}\boldsymbol{X}(t)\approx 2\sum_{k=1}^{N}\sqrt{\Delta\omega}\boldsymbol{A}(t,\omega_k)\{&\boldsymbol{E}(\omega_k)[\boldsymbol{\mathcal{R}}_k\cos(\omega_k t)+\boldsymbol{\mathcal{I}}_k\sin(\omega_k t)]\\ &+\boldsymbol{F}(\omega_k)[\boldsymbol{\mathcal{R}}_k\sin(\omega_k t)-\boldsymbol{\mathcal{I}}_k\cos(\omega_k t)]\}\end{aligned} \tag{5.11}$$

where $\boldsymbol{\mathcal{R}}_k=(\mathcal{R}_{1k},\mathcal{R}_{2k},\cdots,\mathcal{R}_{nk})^{\mathrm{T}}$ and $\boldsymbol{\mathcal{I}}_k=(\mathcal{I}_{1k},\mathcal{I}_{2k},\cdots,\mathcal{I}_{nk})^{\mathrm{T}}$ are zero-mean real-valued orthogonal random vectors satisfying the following orthogonality conditions:

$$E[\boldsymbol{\mathcal{R}}_k]=E[\boldsymbol{\mathcal{I}}_k]=\boldsymbol{0}_{n\times 1},\quad E\left[\boldsymbol{\mathcal{R}}_k\boldsymbol{\mathcal{I}}_m^{\mathrm{T}}\right]=\boldsymbol{0}_{n\times n},\ E\left[\boldsymbol{\mathcal{R}}_k\boldsymbol{\mathcal{R}}_m^{\mathrm{T}}\right]=E\left[\boldsymbol{\mathcal{I}}_k\boldsymbol{\mathcal{I}}_m^{\mathrm{T}}\right]=\frac{1}{2}\boldsymbol{I}_{n\times n}\delta_{km} \tag{5.12}$$

In the spectral decomposition representation of Eq. (5.11), there are usually two approaches. One is to apply Eq. (5.6) to perform direct orthogonal decomposition of the PSD matrix $\boldsymbol{S}(\omega)$. Besides, the PSD matrix $\boldsymbol{S}(\omega)$ can also be firstly decomposed as $\boldsymbol{S}(\omega)=\boldsymbol{B}(\omega)\boldsymbol{\gamma}(\omega)\boldsymbol{B}(\omega)$, where $\boldsymbol{B}(\omega)=\mathrm{diag}\left[\sqrt{S_1(\omega)},\sqrt{S_2(\omega)},\cdots,\sqrt{S_n(\omega)}\right]$, in which $S_i(\omega)$ $(i=1,2,\cdots,n)$ signifies the two-sided stationary PSD function corresponding to the stochastic process $X_i(t)$, and $\boldsymbol{\gamma}(\omega)$ denotes the coherency function matrix, which is a nonnegative definite Hermitian matrix, respectively. In this case, orthogonal decomposition of $\boldsymbol{\gamma}(\omega)$ can be implemented as $\boldsymbol{\gamma}(\omega)=\boldsymbol{\Gamma}(\omega)\boldsymbol{\Gamma}^{*\mathrm{T}}(\omega)$, thereby obtaining $\boldsymbol{Q}(\omega)=\boldsymbol{B}(\omega)\boldsymbol{\Gamma}(\omega)$.

Here, if we define $\boldsymbol{\Gamma}(\omega)=\boldsymbol{C}(\omega)-\mathrm{i}\boldsymbol{D}(\omega)$, then the expressions $\boldsymbol{E}(\omega)=\boldsymbol{B}(\omega)\boldsymbol{C}(\omega)$ and $\boldsymbol{F}(\omega)=\boldsymbol{B}(\omega)\boldsymbol{D}(\omega)$ can be obtained. At the same time, considering the standardized processing of the random variables, i.e. $R_{rk}=\sqrt{2}\mathcal{R}_{rk}$ and $I_{rk}=\sqrt{2}\mathcal{I}_{rk}$, and according to Eq. (5.11), the unified form of spectral decomposition representation for the i-th component process $X_i(t)$ can be derived as follows (Liu et al. 2025):

$$\begin{aligned}X_i(t)\approx\sum_{r=1}^{n}\sum_{k=1}^{N}\sqrt{2S_{X_i}(\omega_k,t)\Delta\omega}\{&C_{ir}(\omega_k)[R_{rk}\cos(\omega_k t)+I_{rk}\sin(\omega_k t)]\\ &+D_{ir}(\omega_k)[R_{rk}\sin(\omega_k t)-I_{rk}\cos(\omega_k t)]\}\end{aligned} \tag{5.13}$$

where $S_{X_i}(\omega,t)=A_i^2(t,\omega)S_i(\omega)$ denotes the two-sided EPSD function of the non-stationary stochastic process $X_i(t)$. $C_{ir}(\omega)$ and $D_{ir}(\omega)$ are the elements at the i-th row and r-th column of the real matrices $\boldsymbol{C}(\omega)$ and $\boldsymbol{D}(\omega)$, respectively.

In Eq. (5.13), $\{R_{rk}, I_{rk}\}$ is a set of zero-mean standard orthogonal random variables that satisfy the following basic conditions:

$$E[R_{rk}]=E[I_{rk}]=0,\quad E[R_{rk}I_{sm}]=0,\quad E[R_{rk}R_{sm}]=E[I_{rk}I_{sm}]=\delta_{rs}\delta_{km} \tag{5.14}$$

where r, s = 1, 2, ⋯, n and k, m = 1, 2, ⋯, N, respectively.

Obviously, as can be seen from Eqs. (5.12) and (5.14), the specific probability distributions for the set of orthogonal random vectors $\{\mathcal{R}_k, \mathcal{I}_k\}$ and the set of standard orthogonal random variables $\{R_{rk}, I_{rk}\}$ have not been provided. Consequently, Eqs. (5.11) and (5.13) cannot be utilized directly for simulating the 1D-nV non-stationary stochastic vector process $\boldsymbol{X}(t)$. Therefore, Eq. (5.11) can be designated as the unified form of original spectral decomposition representation for the 1D-nV non-stationary stochastic vector process $\boldsymbol{X}(t)$, whereas Eq. (5.13) denotes the unified form of original spectral decomposition representation for the i-th component process $X_i(t)$ of $\boldsymbol{X}(t)$.

5.1.2 *Two Classical Cases of Original Spectral Decomposition Representation*

For the spectral decomposition representation of non-stationary stochastic vector processes, there are typically two classical approaches: spectral representation method (SRM) and proper orthogonal decomposition (POD). The sole distinction between them lies in the techniques utilized for spectral decomposition. Here, one can directly perform the orthogonal decomposition on either the PSD matrix $\boldsymbol{S}(\omega)$ or the coherence function matrix $\gamma(\omega)$. For the sake of simplicity, the following will focus solely on the orthogonal decomposition of the PSD matrix $\boldsymbol{S}(\omega)$ as an illustrative example.

Cholesky Decomposition

If $\boldsymbol{S}(\omega)$ is a positive definite Hermitian matrix, the Cholesky decomposition can be implemented as follows:

$$\boldsymbol{S}(\omega)=\boldsymbol{H}(\omega)\boldsymbol{H}^{*\mathrm{T}}(\omega) \tag{5.15}$$

where $\boldsymbol{H}(\omega)$ denotes a lower triangular matrix.

If $\boldsymbol{S}(\omega)$ is a nonnegative definite Hermitian matrix, the improved Cholesky decomposition, say the $\boldsymbol{L\mathcal{D}L}^{*\mathrm{T}}$ decomposition, can be performed as below:

$$
\begin{aligned}
\boldsymbol{S}(\omega) &= \boldsymbol{L}(\omega)\boldsymbol{\mathcal{D}}(\omega)\boldsymbol{L}^{*\mathrm{T}}(\omega) \\
&= \left[\boldsymbol{L}(\omega)\sqrt{\boldsymbol{\mathcal{D}}(\omega)}\right]\left[\boldsymbol{L}(\omega)\sqrt{\boldsymbol{\mathcal{D}}(\omega)}\right]^{*\mathrm{T}} \\
&= \boldsymbol{H}(\omega)\boldsymbol{H}^{*\mathrm{T}}(\omega)
\end{aligned}
\tag{5.16}
$$

Therefore, it is easy to see that the decomposed spectral matrix $\boldsymbol{Q}(\omega)$ in Eq. (5.6) is exactly $\boldsymbol{H}(\omega)$ of the Cholesky decomposition in Eqs. (5.15) and (5.16). In this case, Eqs. (5.11) and (5.13) applying the Cholesky decomposition can be termed the original spectral representation method (O-SRM) for non-stationary stochastic vector process.

Eigen Decomposition

If $\boldsymbol{S}(\omega)$ is a positive or nonnegative definite Hermitian matrix, the following eigen decomposition can be implemented:

$$
\boldsymbol{S}(\omega) = \boldsymbol{\Phi}(\omega)\boldsymbol{\Lambda}(\omega)\boldsymbol{\Phi}^{*\mathrm{T}}(\omega)
\tag{5.17}
$$

where $\boldsymbol{\Lambda}(\omega)$ denotes a diagonal matrix, of which the elements are the eigenvalues $\lambda_i(\omega)$ $(i = 1, 2, \cdots, n)$ of the matrix $\boldsymbol{S}(\omega)$. $\boldsymbol{\Phi}(\omega)$ denotes the eigenmatrix of $\boldsymbol{S}(\omega)$, and its columns $\boldsymbol{\phi}_i(\omega)$ $(i = 1, 2, \cdots, n)$ are the normalized eigenvectors, i.e., $\boldsymbol{\phi}_i^{*\mathrm{T}}(\omega)\boldsymbol{\phi}_j(\omega) = \delta_{ij}$, respectively.

Obviously, the expression of $\boldsymbol{Q}(\omega)$ in Eq. (5.6) is $\boldsymbol{Q}(\omega) = \left[\sqrt{\lambda_1}\boldsymbol{\phi}_1(\omega), \sqrt{\lambda_2}\boldsymbol{\phi}_2(\omega), \cdots, \sqrt{\lambda_n}\boldsymbol{\phi}_n(\omega)\right]$. Similarly, on this occasion, Eqs. (5.11) and (5.13) employing the eigen decomposition can be regarded as the original proper orthogonal decomposition (O-POD) for non-stationary stochastic vector process.

By inspecting Eqs. (5.15)–(5.17), it is handy to find that the difference between the O-SRM and the O-POD for non-stationary stochastic vector process is mainly reflected in the different decomposition techniques of the PSD matrix $\boldsymbol{S}(\omega)$. This suggests that by utilizing various spectral decomposition strategies on the PSD matrix $\boldsymbol{S}(\omega)$ or the coherence function matrix $\boldsymbol{\gamma}(\omega)$, we can derive different spectral decomposition representations.

5.2 Conventional Monte Carlo Methods

To simulate non-stationary stochastic vector processes, it is necessary to provide the specific probability distribution for the set of standard orthogonal random variables $\{R_{rk}, I_{rk}\}$ in Eq. (5.13). Analogous to the frequency domain analysis approach for simulating the univariate non-stationary stochastic processes expounded in Chap. 3, traditionally, there exist two prevalent methodologies. One involves defining $\{R_{rk},$

$I_{rk}\}$ as a set of mutually independent standard Gaussian random variables, thereby yielding the conventional random amplitude simulation formula. Alternatively, $\{R_{rk}, I_{rk}\}$ can also be defined as a set of mutually independent random phase expressed in trigonometric functions, resulting in the conventional random phase simulation formula. Regardless of the specific form in which the set of standard orthogonal random variables $\{R_{rk}, I_{rk}\}$ is defined, it should satisfy the basic condition (also recognized as necessary conditions) presented in Eq. (5.14). Essentially, both the conventional random amplitude method and the conventional random phase method belong to the family of conventional Monte Carlo methods. This section introduces the conventional Monte Carlo methods for simulating non-stationary stochastic vector processes, clarifying the relationship and origin of the conventional random amplitude method and the conventional random phase method.

5.2.1 *Conventional Random Amplitude Method (Non-deterministic Spectral Amplitude Method)*

In the original spectral decomposition representation of the i-th non-stationary component process $X_i(t)$, i.e., Eq. (5.13), if $\{R_{rk}, I_{rk}\}$ $(r = 1, 2, \cdots, n; k = 1, 2, \cdots, N)$ is defined as a set of mutually independent standard Gaussian random variables, the conventional random amplitude simulation formula, also recognized as the non-deterministic spectral amplitude method, can be expressed as follows:

$$X_i^{(1)}(t) = \sum_{r=1}^{n}\sum_{k=1}^{N}\sqrt{2S_{X_i}(\omega_k, t)\Delta\omega}\,\mathcal{A}_{rk}G_{ir}(\omega_k)\cos[\omega_k t - \varphi_{ir}(\omega_k) - \alpha_{rk}] \tag{5.18}$$

in which

$$R_{rk} = \mathcal{A}_{rk}\cos\alpha_{rk}, \quad I_{rk} = \mathcal{A}_{rk}\sin\alpha_{rk} \tag{5.19}$$

where $X_i^{(1)}(t)$ is the simulated i-th non-stationary component process based on the conventional random amplitude method. Note that $\mathcal{A}_{rk}$ denotes the random amplitude and α_{rk} denotes random phase, respectively. Therefore, the conventional random amplitude method considers the randomness of both amplitude and phase in essence.

It should also be clarified that since $\{R_{rk}, I_{rk}\}$ $(r = 1, 2, \cdots, n; k = 1, 2, \cdots, N)$ is a set of mutually independent standard Gaussian random variables, according to Eq. (5.19), it can be proved that $\{\mathcal{A}_{rk}, \alpha_{rk}\}$ is also a set of mutually independent random variables, with $\mathcal{A}_{rk}$ obeying the Rayleigh distribution, i.e., $p_{\mathcal{A}_{rk}}(x) = x\exp(-x^2/2)$, $x \geqslant 0$ and α_{rk} following the uniform distribution over the interval $(0, 2\pi]$, respectively. Conversely, if $\{\mathcal{A}_{rk}, \alpha_{rk}\}$ $(r = 1, 2, \cdots, n; k = 1, 2, \cdots, N)$ is a set of mutually independent random variables, where $\mathcal{A}_{rk}$ holds the Rayleigh distribution, i.e., $p_{\mathcal{A}_{rk}}(x) = x\exp(-x^2/2)$, $x \geqslant 0$ and α_{rk} follows the uniform distribution over the interval (0, 2π], based on Eq. (5.19), it can also be verified that $\{R_{rk}, I_{rk}\}$ is a set of mutually independent standard Gaussian random variables. Obviously, the form of the random variables defined in Eq. (5.19) can completely satisfy the basic conditions involved in Eq. (5.14).

Furthermore, we can also define $\mathcal{A}_{rk} = \sqrt{-2\ln \beta_{rk}}$, and $\{\beta_{rk}, \alpha_{rk}\}$ is also a set of independent random variables, where β_{rk} obeys the uniform distribution over the interval (0, 1] and α_{rk} follows the uniform distribution over the interval (0, 2π], respectively. Accordingly, an alternative form of the random amplitude method can then be expressed as follows:

$$X_i^{(1)}(t) = 2\sum_{r=1}^{n}\sum_{k=1}^{N}\sqrt{-S_{X_i}(\omega_k, t)\Delta\omega \ln \beta_{rk}}\, G_{ir}(\omega_k)\cos[\omega_k t - \varphi_{ir}(\omega_k) - \alpha_{rk}] \tag{5.20}$$

In Eqs. (5.18) and (5.20), we have:

$$G_{ir}(\omega_k) = \sqrt{C_{ir}^2(\omega_k) + D_{ir}^2(\omega_k)} \tag{5.21a}$$

$$\varphi_{ir}(\omega_k) = \begin{cases} \arctan\dfrac{D_{ir}(\omega_k)}{C_{ir}(\omega_k)}, & C_{ir}(\omega_k) > 0 \\ -\dfrac{\pi}{2} \text{ or } \dfrac{\pi}{2}, & C_{ir}(\omega_k) = 0 \\ \pi + \arctan\dfrac{D_{ir}(\omega_k)}{C_{ir}(\omega_k)}, & C_{ir}(\omega_k) < 0 \end{cases} \tag{5.21b}$$

where the defined interval of the deterministic phase angle $\varphi_{ir}(\omega_k)$ is $\left[-\frac{\pi}{2}, \frac{3\pi}{2}\right)$.

5.2.2 Conventional Random Phase Method (Deterministic Spectral Amplitude Method)

In the original spectral decomposition representation of the i-th non-stationary component process $X_i(t)$, i.e., Eq. (5.13), if $\{R_{rk}, I_{rk}\}$ ($r = 1, 2, \cdots, n$; $k = 1, 2, \cdots, N$) is defined as a set of trigonometric functions with mutually independent random phase, the conventional random phase simulation formula, also recognized as the deterministic spectral amplitude method, can be obtained as follows (Deodatis 1996):

$$X_i^{(2)}(t) = 2\sum_{r=1}^{n}\sum_{k=1}^{N}\sqrt{S_{X_i}(\omega_k, t)\Delta\omega}\, G_{ir}(\omega_k)\cos[\omega_k t - \varphi_{ir}(\omega_k) - \alpha_{rk}] \tag{5.22}$$

in which

$$R_{rk} = \sqrt{2}\cos\alpha_{rk}, \quad I_{rk} = \sqrt{2}\sin\alpha_{rk} \tag{5.23}$$

where $X_i^{(2)}(t)$ is the simulated i-th non-stationary component process based on the conventional random phase method. $\{\alpha_{rk}\}$ is a set of mutually independent random phase uniformly distributed over the interval $(0, 2\pi]$. Apparently, the form of the random variables $\{R_{rk}, I_{rk}\}$ defined in Eq. (5.23) can fully satisfy the basic conditions included in Eq. (5.14).

Note that the sets of the independent random variables defined in Eq. (5.19) and Eq. (5.23) both completely satisfy the basic conditions involved in Eq. (5.14). As a result, both the random amplitude method (Eqs. (5.18) and (5.20)) and the random phase method (Eq. (5.22)) are deduced from the original spectral decomposition representation of the i-th non-stationary component process, i.e., Eq. (5.13). That is to say Eqs. (5.18), (5.20) and (5.22) are all the special cases of Eq. (5.13) in essence.

It is obvious that the when employing the random amplitude method (Eqs. (5.18) and (5.20)), the randomness degree of the simulated i-th non-stationary component process $X_i^{(1)}(t)$ is $2{\times}n{\times}N$. However, the randomness degree of $X_i^{(2)}(t)$ is just $n{\times}N$ when the random phase method (Eq. (5.22)) is applied, which is just half of that in $X_i^{(1)}(t)$. This clearly shows that different levels of randomness degree can be performed through defining various random orthogonal function forms for the set of standard orthogonal random variables within the original spectral decomposition representation.

5.3 Dimension-Reduction Simulation Methods

Through the derivation of the aforementioned conventional Monte Carlo method, it is not difficult to observe that if the set of standard orthogonal random variables in the original spectral decomposition representation is defined as different forms of random orthogonal functions, the simulation formula will exhibit varying levels of randomness degree. Hence, by incorporating suitable random orthogonal function forms into the set of standard orthogonal random variables within the original spectral decomposition representation, dimension-reduction simulation of non-stationary stochastic vector processes can be achieved.

5.3.1 Random Orthogonal Functions

For the 1D-nV non-stationary stochastic vector process, the set of standard orthogonal random variables $\{R_{rk}, I_{rk}\}$ in the original spectral decomposition representation of the i-th non-stationary component process, i.e., Eq. (5.13), can be defined as the following common form of random orthogonal function (Liu et al. 2018, 2019):

$$R_{rk} = g_{rk}(\boldsymbol{\Theta}), I_{rk} = h_{rk}(\boldsymbol{\Theta}), r = 1, 2, \cdots, n; \; k = 1, 2, \cdots, N \tag{5.24}$$

where $g_{rk}(\cdot)$ and $h_{rk}(\cdot)$ denote the deterministic orthogonal functions. $\boldsymbol{\Theta} = (\Theta_1, \Theta_2, \cdots, \Theta_q)$ denotes the q-dimensional elementary random vector, of which the joint PDF, say $p_{\boldsymbol{\Theta}}(\boldsymbol{\theta})$, is known.

Similarly, the defined form of Eq. (5.24) should satisfy the basic conditions involved in Eq. (5.14), which can be regarded as a necessary condition for simulation. Consequently, Eq. (5.14) can be rewritten as followings (Liu et al. 2018, Liu et al. 2019):

$$E[R_{rk}] = E[g_{rk}(\boldsymbol{\Theta})] = \int_{\Omega} g_{rk}(\boldsymbol{\theta}) p_{\boldsymbol{\Theta}}(\boldsymbol{\theta}) \mathrm{d}\boldsymbol{\theta} = 0 \tag{5.25a}$$

$$E[I_{rk}] = E[h_{rk}(\boldsymbol{\Theta})] = \int_{\Omega} h_{rk}(\boldsymbol{\theta}) p_{\boldsymbol{\Theta}}(\boldsymbol{\theta}) \mathrm{d}\boldsymbol{\theta} = 0 \tag{5.25b}$$

$$E[R_{rk} R_{sm}] = E[g_{rk}(\boldsymbol{\Theta}) g_{sm}(\boldsymbol{\Theta})] = \int_{\Omega} g_{rk}(\boldsymbol{\theta}) g_{sm}(\boldsymbol{\theta}) p_{\boldsymbol{\Theta}}(\boldsymbol{\theta}) \mathrm{d}\boldsymbol{\theta} = \delta_{rs} \delta_{km} \tag{5.25c}$$

$$E[I_{rk} I_{sm}] = E[h_{rk}(\boldsymbol{\Theta}) h_{sm}(\boldsymbol{\Theta})] = \int_{\Omega} h_{rk}(\boldsymbol{\theta}) h_{sm}(\boldsymbol{\theta}) p_{\boldsymbol{\Theta}}(\boldsymbol{\theta}) \mathrm{d}\boldsymbol{\theta} = \delta_{rs} \delta_{km} \tag{5.25d}$$

$$E[R_{rk} I_{sm}] = E[g_{rk}(\boldsymbol{\Theta}) h_{sm}(\boldsymbol{\Theta})] = \int_{\Omega} g_{rk}(\boldsymbol{\theta}) h_{sm}(\boldsymbol{\theta}) p_{\boldsymbol{\Theta}}(\boldsymbol{\theta}) \mathrm{d}\boldsymbol{\theta} = 0 \tag{5.25e}$$

where Ω denotes the distribution interval of the elementary random vector $\boldsymbol{\Theta}$.

Obviously, the forms of random orthogonal function (Eq. (5.24)) that satisfy the basic conditions involved in Eq. (5.25) are various, involving trigonometric functions, orthogonal polynomials, and the products of the above two. Here, several forms of random orthogonal functions realized as the products of trigonometric functions and orthogonal polynomials with two elementary random variables are presented in the following.

Form 1

$$\begin{cases} \overline{R}_{sm} = 2 \times T_s(\Theta_1) \times \cos(m\Theta_2 + \varphi) \\ \overline{I}_{sm} = 2 \times T_s(\Theta_1) \times \sin(m\Theta_2 + \varphi) \end{cases}, \quad s = 1, 2, \cdots, n; \quad m = 1, 2, \cdots, N \tag{5.26}$$

where φ is a deterministic parameter over the interval (0, 2π] generally valued by $\varphi = \pi/4$. Θ_1 and Θ_2 are mutually independent elementary random variables, and Θ_2 holds the uniform distribution on the interval (0, 2π]. Meanwhile, the PDF of the random variable Θ_1 is as below:

$$p_{\Theta_1}(\theta) = \frac{1}{\pi\sqrt{1 - \theta^2}}, \quad -1 < \theta < 1 \tag{5.27}$$

And $T_s(\cdot)$ denotes the first family of Chebyshev polynomial function, such that:

$$\begin{cases} T_{s+1}(x) = 2xT_s(x) - T_{s-1}(x) \\ T_0(x) = 1, \ T_1(x) = x \end{cases}, \quad s = 1, 2, \cdots, n \tag{5.28}$$

Form 2

$$\begin{cases} \overline{R}_{sm} = \sqrt{2} \times U_s(\Theta_1) \times \cos(m\Theta_2 + \varphi) \\ \overline{I}_{sm} = \sqrt{2} \times U_s(\Theta_1) \times \sin(m\Theta_2 + \varphi) \end{cases}, \quad s = 1, 2, \cdots, n; \quad m = 1, 2, \cdots, N \tag{5.29}$$

where φ is a deterministic parameter over the interval $(0, 2\pi]$ generally valued by $\varphi = \pi/4$. Θ_1 and Θ_2 are mutually independent elementary random variables, and Θ_2 holds the uniform distribution on the interval $(0, 2\pi]$. Meanwhile, the PDF of the random variable Θ_1 is as below:

$$p_{\Theta_1}(\theta) = \frac{2}{\pi}\sqrt{1 - \theta^2}, \quad -1 < \theta < 1 \tag{5.30}$$

And $U_s(\cdot)$ denotes the second family of Chebyshev polynomial function, such that:

$$\begin{cases} U_{s+1}(x) = 2xU_s(x) - U_{s-1}(x) \\ U_0(x) = 1, \ U_1(x) = 2x \end{cases}, \quad s = 1, 2, \cdots, n \tag{5.31}$$

Form 3

$$\begin{cases} \overline{R}_{sm} = \sqrt{2(1 + 2s)} \times P_s(\Theta_1) \times \cos(m\Theta_2 + \varphi) \\ \overline{I}_{sm} = \sqrt{2(1 + 2s)} \times P_s(\Theta_1) \times \sin(m\Theta_2 + \varphi) \end{cases}, \quad s = 1, 2, \cdots, n; \quad m = 1, 2, \cdots, N \tag{5.32}$$

where φ is a deterministic parameter over the interval $(0, 2\pi]$ generally valued by $\varphi = \pi/4$. Θ_1 and Θ_2 are mutually independent elementary random variables. Θ_1 holds the uniform distribution on the interval $(-1, 1)$, and Θ_2 is uniformly distributed over the interval $(0, 2\pi]$, respectively. $P_s(\cdot)$ denotes the Legendre polynomial function, such that:

$$\begin{cases} P_{s+1}(x) = \dfrac{2s + 1}{s + 1}xP_s(x) - \dfrac{s}{s + 1}P_{s-1}(x) \\ P_0(x) = 1, \ P_1(x) = x \end{cases}, \quad s = 1, 2, \cdots, n \tag{5.33}$$

It is handy to verify that the forms of the random orthogonal functions of Eqs. (5.26), (5.29) and (5.32) all can fully satisfy the basic conditions included in Eq. (5.25). Thus, all of them are sets of standard orthogonal random variables. Furthermore, $\{\overline{R}_{sm}, \overline{I}_{sm}\}$ defined in Eqs. (5.26), (5.29) and (5.32) are all required to be transformed into the set of standard orthogonal random variables $\{R_{rk}, I_{rk}\}$ in Eq. (5.13) through a deterministic one-to-one mapping relationship, say $\overline{R}_{sm} \rightarrow R_{rk}$ and $\overline{I}_{sm} \rightarrow I_{rk}$. Taking $\overline{R}_{sm} \rightarrow R_{rk}$ as an example, random variables with double subscript, i.e., R_{rk} and $\overline{R}_{sm}$, are defined as random variables with single subscript, i.e., R_p and $\overline{R}_q$, and there exists $R_p = R_{rk}$ and $\overline{R}_q = \overline{R}_{sm}$, where $p=(k-1)\times n+r$, $q=(m-1)\times n+s$. Then the deterministic one-to-one mapping can be readily implemented using the built-in functions of MATLAB toolbox, say rand ('state', 0) and randperm(·). Specifically, let temp = randperm($n\times N$) and $R_p = \overline{R}_q$, thus the one-to-one mapping relationship between p and q is determined by $q = \text{temp}(p)$. As a matter of fact, this deterministic mapping relationship performs as a sufficient condition for the dimension-reduction simulation methods of 1D-nV non-stationary stochastic vector process. In consequence, the target set of standard orthogonal random variables $\{R_{rk}, I_{rk}\}$ can be uniquely determined.

Finally, in response to the afore-mentioned two conventional Monte Carlo simulation methods, say the random amplitude method and the random phase method, we can also propose two corresponding dimension-reduction simulation methods, as explained below.

5.3.2 *Dimension-Reduction Simulation Method Via Random Amplitude*

For the novel dimension-reduction simulation method considering the randomness of both amplitude and phase, the set of standard orthogonal random variables $\{R_{rk}, I_{rk}\}$ can be expressed as below:

$$\begin{cases} \overline{R}_{sm} = \sqrt{-2\ln\left\{\frac{1}{2} + \frac{1}{\pi}\arcsin[\sin(m\Theta_1 + \varphi)]\right\}}\cos(s\Theta_2 + m\Theta_3) \\ \overline{I}_{sm} = \sqrt{-2\ln\left\{\frac{1}{2} + \frac{1}{\pi}\arcsin[\sin(m\Theta_1 + \varphi)]\right\}}\sin(s\Theta_2 + m\Theta_3) \end{cases}, \quad s = 1, 2, \cdots, n; \; m = 1, 2, \cdots, N \tag{5.34}$$

where Θ_1, Θ_2 and Θ_3 are three mutually independent elementary random variables, and all of them follow a uniform distribution over the interval $(0, 2\pi]$. φ is a constant in the interval $(0, 2\pi]$, and valued by $\varphi = \pi/4$ herein.

It is easy to verify that the set of standard orthogonal random variables $\{\overline{R}_{sm}, \overline{I}_{sm}\}$ defined in Eq. (5.34) fully satisfies the basic conditions included in Eq. (5.25), say the necessary conditions for simulation have been fulfilled. At the meantime, in Eq. (5.34), the deterministic one-to-one mapping relationship between $\{\overline{R}_{sm}, \overline{I}_{sm}\}$ and $\{R_{rk}, I_{rk}\}$ in Eq. (5.13) is $q = \text{temp}(p)$, serving as the sufficient condition required for simulation. The implementation method of this correspondence is the same as described in the previous section, and will not be elaborated here.

Therefore, by means of substituting Eq. (5.34) into the original spectral decomposition representation of the i-th non-stationary component process, i.e., Eq. (5.13), the dimension-reduction representation on the basis of random amplitude for simulating 1D-nV non-stationary stochastic vector process can be achieved, such that (Liu et al. 2025):

$$\widehat{X}_i^{(1)}(t)=\sum_{r=1}^{n}\sum_{k=1}^{N}\sqrt{2S_{X_i}(\omega_k,t)\Delta\omega}\mathcal{A}_{rk}(\Theta_1)G_{ir}(\omega_k)\cos[\omega_k t-\varphi_{ir}(\omega_k)-\alpha_{rk}(\Theta_2,\Theta_3)],\ \ i=1,2,\cdots,n \tag{5.35}$$

in which

$$\mathcal{A}_{rk}(\Theta_1)=\overline{\mathcal{A}}_{sm}(\Theta_1)=\sqrt{-2\ln\left\{\frac{1}{2}+\frac{1}{\pi}\arcsin[\sin(m\Theta_1+\varphi)]\right\}} \tag{5.36a}$$

$$\alpha_{rk}(\Theta_2,\Theta_3)=\overline{\alpha}_{sm}(\Theta_2,\Theta_3)=s\times\Theta_2+m\times\Theta_3 \tag{5.36b}$$

where $\widehat{X}_i^{(1)}(t)$ is the simulated i-th non-stationary component process presented by the dimension-reduction representation based on random amplitude method. The random variable Θ_1 describes the randomness of amplitude, as well as Θ_2 and Θ_3 describe the randomness of phase. The deterministic one-to-one mapping relationship between α_{rk} and $\overline{\alpha}_{sm}$ is the same as the mapping relationship between R_{rk} and $\overline{R}_{sm}$.

5.3.3 Dimension-Reduction Simulation Method Via Random Phase

For the novel dimension-reduction simulation method taking into account just the randomness of phase, the set of standard orthogonal random variables $\{R_{rk}, I_{rk}\}$ can be expressed as the random orthogonal functions with two elementary random variables, specifically:

$$\begin{cases}\overline{R}_{sm}=\sqrt{2}\cos(s\Theta_1+m\Theta_2+\varphi)\\ \overline{I}_{sm}=\sqrt{2}\sin(s\Theta_1+m\Theta_2+\varphi)\end{cases},\quad s=1,2,\cdots,n;\ \ m=1,2,\cdots,N \tag{5.37}$$

where Θ_1 and Θ_2 are mutually independent elementary random variables uniformly distributed over the interval $(0,2\pi]$. φ is a constant in the interval $(0,\ 2\pi]$ taken as $\varphi=\pi/4$ herein.

It is easy to verify that the set of standard orthogonal random variables $\{\overline{R}_{sm},\overline{I}_{sm}\}$ defined in Eq. (5.37) fully satisfy the basic conditions included in Eq. (5.25), say the necessary conditions for simulation have been fulfilled. Then, through substituting Eq. (5.37) into the original spectral decomposition representation of the i-th non-stationary component process, i.e., Eq. (5.13), the dimension-reduction representation on the basis of random phase for simulating 1D-nV non-stationary stochastic vector process can be achieved, such that (Liu et al. 2025):

$$\widehat{X}_i^{(2)}(t)=2\sum_{r=1}^{n}\sum_{k=1}^{N}\sqrt{S_{X_i}(\omega_k,t)\Delta\omega}\,G_{ir}(\omega_k)\cos[\omega_k t-\varphi_{ir}(\omega_k)-\alpha_{rk}(\boldsymbol{\Theta})],\quad i=1,2,\cdots,n \tag{5.38}$$

where $\widehat{X}_i^{(2)}(t)$ denotes the simulated i-th non-stationary component process presented by the dimension-reduction representation based on random phase. Apparently, there exists $\alpha_{rk}(\boldsymbol{\Theta})=\overline{\alpha}_{sm}(\boldsymbol{\Theta})=s\times\Theta_1+m\times\Theta_2+\varphi$. Similarly, the deterministic one-to-one mapping relationship between α_{rk} and $\overline{\alpha}_{sm}$ is the same as the mapping relationship between R_{rk} and $\overline{R}_{sm}$.

5.4 Time-Domain Representation of Non-Stationary Stochastic Vector Processes

Generally, for a real-valued 1D-nV non-stationary stochastic vector process $\boldsymbol{X}(t)=[X_1(t),X_2(t),\cdots,X_n(t)]^{\mathrm{T}}$ with zero-mean, all of its probabilistic information is determined by the EPSD function matrix or the correlation function matrix. In the previous sections of this chapter, we primarily consider the frequency-domain analysis method, which performs spectral decomposition on the PSD or the coherence function matrix and incorporates constraint forms of random orthogonal functions, thereby achieving dimension-reduction representation for 1D-nV non-stationary stochastic vector process. Besides, we can also employ the time-domain analysis method to achieve dimension-reduction representation of 1D-nV non-stationary stochastic vector process, which will be expounded in the following.

The correlation function matrix of a 1D-nV non-stationary stochastic vector process $\boldsymbol{X}(t)$ is defined as follows:

$$\boldsymbol{R}_X(t,t+\tau)=\begin{bmatrix} R_{11}(t,t+\tau) & R_{12}(t,t+\tau) & \cdots & R_{1n}(t,t+\tau) \\ R_{21}(t,t+\tau) & R_{22}(t,t+\tau) & \cdots & R_{2n}(t,t+\tau) \\ \vdots & \vdots & \ddots & \vdots \\ R_{n1}(t,t+\tau) & R_{n2}(t,t+\tau) & \cdots & R_{nn}(t,t+\tau) \end{bmatrix} \tag{5.39}$$

As Eqs. (2.100) and (2.101) defined in Chap. 2, the auto-correlation function $R_{ii}(t, t+\tau)$ of the component process $X_i(t)$ and the cross-correlation function $R_{ij}(t,t+\tau)$ between $X_i(t)$ and $X_j(t)$ ($j\neq i$) can be respectively expressed as follows (Deodatis 1996):

$$R_{ii}(t,t+\tau) = \int_{-\infty}^{\infty} A_i(t,\omega)A_i(t+\tau,\omega)\,S_i(\omega)\,\mathrm{e}^{\mathrm{i}\omega\tau}\mathrm{d}\omega, \quad i=1,2,\cdots,n \tag{5.40}$$

$$R_{ij}(t,t+\tau) = \int_{-\infty}^{\infty} A_i(t,\omega)A_j(t+\tau,\omega)\sqrt{S_i(\omega)S_j(\omega)}\gamma_{ij}(\omega)\mathrm{e}^{\mathrm{i}\omega\tau}\mathrm{d}\omega, \; i,j=1,2,\cdots,n; \; j\neq i \tag{5.41}$$

where $A_i(t, \omega)$ denotes the time-frequency modulation function of the component process $X_i(t)$. $S_i(\omega)$ is the PSD function of the corresponding stationary process of $X_i(t)$. $\gamma_{ij}(\omega)$ denotes the coherence function.

5.4.1 Conventional Time-Domain (CTD) Representation

Regarding Eq. (4.78) in Chap. 4, if the Fourier transform pair for the PSD and the correlation function defined in Eqs. (1.78) and (1.79) is employed, and the Fourier transform pair for the function $H(t, \omega)$ and $h(t, \tau)$ defined in Eq. (4.59) is applied simultaneously, then the discretized time-domain representation for any component process $X_i(t)$ within $\boldsymbol{X}(t)$ can be expressed as follows:

$$X_i(t) = \sqrt{2\pi\Delta t}\sum_{k=1}^{m} h_i(t, t-t_k)\,V_{ik}, \; t \in [t_m, t_{m+1}), \quad m=1,2,\cdots,N \tag{5.42}$$

The aforementioned equation can be rewritten as follows:

$$X_i(t) = \begin{pmatrix} X_i(t_1) \\ X_i(t_2) \\ \vdots \\ X_i(t_N) \end{pmatrix} = \sqrt{2\pi\Delta t}\begin{bmatrix} h_i(t_1,0) & & & \\ h_i(t_2,\Delta t) & h_i(t_2,0) & & \\ \vdots & \vdots & \ddots & \\ h_i(t_N,(N-1)\Delta t) & h_i(t_N,(N-2)\Delta t) & \cdots & h_i(t_N,0) \end{bmatrix}\begin{pmatrix} V_{i1} \\ V_{i2} \\ \vdots \\ V_{iN} \end{pmatrix} \tag{5.43}$$

Furthermore, Eq. (5.43) can also be written in the following matrix form:

$$X_i(t) = \sqrt{2\pi\Delta t}\boldsymbol{h}_i\,\boldsymbol{V}_i \tag{5.44}$$

where $\boldsymbol{h}_i$ represents an N-order lower triangular matrix. $\boldsymbol{V}_i = (V_{i1}, V_{i2}, \cdots, V_{iN})^{\mathrm{T}}$ is a standard orthogonal random vector, say $E[\boldsymbol{V}_i\boldsymbol{V}_i^{\mathrm{T}}] = \boldsymbol{I}$, where $\boldsymbol{I}$ denotes an N-order unit square matrix.

Then, the 1D-nV non-stationary stochastic vector process $\boldsymbol{X}(t)$ can be presented as below:

$$\boldsymbol{X}(t) = \begin{pmatrix} X_1(t) \\ X_2(t) \\ \vdots \\ X_n(t) \end{pmatrix} = \sqrt{2\pi\Delta t} \begin{bmatrix} \boldsymbol{h}_1 & & & \\ & \boldsymbol{h}_2 & & \\ & & \ddots & \\ & & & \boldsymbol{h}_n \end{bmatrix} \begin{pmatrix} \boldsymbol{V}_1 \\ \boldsymbol{V}_2 \\ \vdots \\ \boldsymbol{V}_n \end{pmatrix} = \sqrt{2\pi\Delta t}\, \boldsymbol{hV} \quad (5.45)$$

where $\boldsymbol{h}$ is an $(n \times N)$-order square matrix and $\boldsymbol{V}$ is an $(n{\times}N)$-dimensional random vector, respectively.

In fact, for each component process $X_i(t)$, $\boldsymbol{V}_i = (V_{i1}, V_{i2}, \cdots, V_{iN})^{\mathrm{T}}$ constitutes a standard orthogonal random vector. However, for two different component processes $X_i(t)$ and $X_j(t)$, the random vectors $\boldsymbol{V}_i$ and $\boldsymbol{V}_j$ are commonly not orthogonal, thus they need to be orthogonalized. To achieve this, we can introduce an $(n{\times}N)$-dimensional standard orthogonal random vector $\boldsymbol{W} = (W_1, W_2, \cdots, W_{n\times N})^{\mathrm{T}}$ and define $\boldsymbol{V} = \boldsymbol{ZW}$, then Eq. (5.45) can be rewritten as:

$$\boldsymbol{X}(t) = \sqrt{2\pi\Delta t}\,\boldsymbol{hV} = \sqrt{2\pi\Delta t}\,\boldsymbol{hZW} = \boldsymbol{KW} \quad (5.46)$$

where $\boldsymbol{W}$ denotes a standard orthogonal random vector satisfying $E[\boldsymbol{WW}^{\mathrm{T}}] = \boldsymbol{I}$, of which $\boldsymbol{I}$ is an $(n{\times}N)$-order unit square matrix. As a matter of fact, the matrix $\boldsymbol{K} = \sqrt{2\pi\Delta t}\ \boldsymbol{hZ}$ can be determined through the following relationship:

$$\begin{aligned} \boldsymbol{R_X}(t,t) &= E\left[\boldsymbol{X}(t)\boldsymbol{X}^{\mathrm{T}}(t)\right] \\ &= \boldsymbol{K}E\left[\boldsymbol{WW}^{\mathrm{T}}\right]\boldsymbol{K}^{\mathrm{T}} \\ &= \boldsymbol{KK}^{\mathrm{T}} \end{aligned} \quad (5.47)$$

It is worth noting that according to Eq. (5.39), $\boldsymbol{R_X}(t,t)$ represents an n-order square matrix related to the time parameter t. However, in Eq. (5.47), $\boldsymbol{R_X}(t,t)$ should be regarded as an $(n \times N)$-order square matrix. In fact, according to Eq. (5.43), for the ith component process $X_i(t) = [X_i(t_1), X_i(t_2), \cdots, X_i(t_N)]^{\mathrm{T}}$, any element of the matrix $\boldsymbol{R_X}(t,t)$ in Eq. (5.47) can be written as:

$$R_{ij}(t,t) = E\left[X_i(t)X_j^{\mathrm{T}}(t)\right] = \begin{bmatrix} R_{ij}(t_1,t_1) & R_{ij}(t_1,t_2) & \cdots & R_{ij}(t_1,t_N) \\ R_{ij}(t_2,t_1) & R_{ij}(t_2,t_2) & \cdots & R_{ij}(t_2,t_N) \\ \vdots & \vdots & \ddots & \vdots \\ R_{ij}(t_N,t_1) & R_{ij}(t_N,t_2) & \cdots & R_{ij}(t_N,t_N) \end{bmatrix} \quad (5.48)$$

It is obvious that the matrix $\boldsymbol{R_X}(t,t)$ in Eq. (5.47) is real and symmetric, which facilitates performing its orthogonal decomposition and thus makes it easier to solve the matrix $\boldsymbol{K}$. As a result, we can derive the expression of 1D-nV non-stationary stochastic vector process in terms of the standard orthogonal random vector $\boldsymbol{W}$ from Eq. (5.46) with the solution of $\boldsymbol{K}$. This provides the necessary condition for introducing the constraint form of random orthogonal functions to realize dimension-reduction simulation.

5.4.2 Time-Varying Parameter Vector Auto-Regressive (TVP-VAR) Model

The discrete time series of 1D-nV non-stationary stochastic vector process $\boldsymbol{X}(t)$ can be defined as $\boldsymbol{X}_t = [X_{t,\,1}, X_{t,\,2}, \cdots, X_{t,\,n}]^{\mathrm{T}}$, of which the components are also real-valued ones with zero-mean. In Chap. 4, the time-varying parameter auto-regressive (TVP-AR) model for univariate time series was provided in Eq. (4.91), which will be expanded to the time-varying parameter vector auto-regressive (TVP-VAR) model for multivariate time series, given by:

$$\boldsymbol{X}_t = \sum_{i=1}^{p} \boldsymbol{\varphi}_i(t)\boldsymbol{X}_{t-i} + \boldsymbol{L}(t)\boldsymbol{W}_t \tag{5.49}$$

where $\boldsymbol{\varphi}_i(t)$ $(i = 1, 2, \cdots, p)$ and $\boldsymbol{L}(t)$ are both the n-order square matrices associated with time t. $\boldsymbol{W}_t = (W_{t,\,1}, W_{t,\,2}, \cdots, W_{t,\,n})^{\mathrm{T}}$ denotes a random multivariate time series, and the components $W_{t,\,k}$ $(k = 1, 2, \cdots, n)$ are standard orthogonal random variables satisfying the following conditions:

$$E[W_{t,k}] = 0, \quad k = 1, 2, \cdots, n \tag{5.50a}$$

$$E[W_{s,l}W_{t,k}] = \delta_{kl}\delta_{ts}, \quad k, l = 1, 2, \cdots, n \tag{5.50b}$$

Similar to the univariate time series, the random multivariate time series $\boldsymbol{X}_t$ and $\boldsymbol{W}_t$ satisfy the following condition (Samaras et al. 1985):

$$E\left[\boldsymbol{X}_s\boldsymbol{W}_t^{\mathrm{T}}\right] = E\left[\boldsymbol{W}_t\boldsymbol{X}_s^{\mathrm{T}}\right] = \boldsymbol{0}_{n\times n}, \quad s < t \tag{5.51}$$

In Eq. (5.49), $\boldsymbol{\varphi}_i(t)$ and $\boldsymbol{L}(t)$ are two matrices to be determined, of which the specific solving methods are presented as follows.

At first, the matrix $\boldsymbol{\varphi}_i(t)\,(i = 1, 2, \cdots, p)$ can be calculated by solving a Yule-Walker equation (Box et al. 2015). The equation is constructed by post-multiplying both sides of Eq. (5.49) with $\boldsymbol{X}_{t-j}^{\mathrm{T}}$, and then taking expectations for $j = 1, 2, \cdots, p$. We thus obtain:

$$\boldsymbol{R}_X(t, t - j\Delta t) = \sum_{i=1}^{p} \boldsymbol{\varphi}_i(t)\boldsymbol{R}_X(t - i\Delta t, t - j\Delta t), \quad j = 1, 2, \cdots, p \tag{5.52}$$

Further, Eq. (5.52) can be written in the matrix form as follows:

$$[\boldsymbol{R}_X(t,t-\Delta t),\cdots,\boldsymbol{R}_X(t,t-p\Delta t)] = [\boldsymbol{\varphi}_1(t),\cdots,\boldsymbol{\varphi}_p(t)] \begin{bmatrix} \boldsymbol{R}_X(t-\Delta t,t-\Delta t) & \cdots & \boldsymbol{R}_X(t-\Delta t,t-p\Delta t) \\ \vdots & \ddots & \vdots \\ \boldsymbol{R}_X(t-p\Delta t,t-\Delta t) & \cdots & \boldsymbol{R}_X(t-p\Delta t,t-p\Delta t) \end{bmatrix} \tag{5.53}$$

where $\boldsymbol{R}_X(t-i\Delta t, t-j\Delta t)$ $(i,j = 1,2,\cdots, p)$ denotes the cross-correlation function matrix of multivariate time series $\boldsymbol{X}_{t-i}$ and $\boldsymbol{X}_{t-j}$, given by:

$$\boldsymbol{R}_X(t-i\Delta t,t-j\Delta t) = \begin{bmatrix} R_{11}(t-i\Delta t,t-j\Delta t) & \cdots & R_{1n}(t-i\Delta t,t-j\Delta t) \\ \vdots & \ddots & \vdots \\ R_{n1}(t-i\Delta t,t-j\Delta t) & \cdots & R_{nn}(t-i\Delta t,t-j\Delta t) \end{bmatrix} \tag{5.54}$$

where $R_{kl}(t-i\Delta t, t-j\Delta t)$ $(k,l = 1,2,\cdots,n)$ denotes the cross-correlation function of two stochastic time series $X_{t-i,\,k}$ and $X_{t-j,\,l}$.

Secondly, the square matrix $\boldsymbol{L}(t)$ in Eq. (5.49) can be derived from a system of simultaneous equations. The first equation is constructed by post-multiplying both sides of Eq. (5.49) with $\boldsymbol{X}_t^{\mathrm{T}}$, and then taking expectations. It is given by:

$$\boldsymbol{R}_X(t,t) = \sum_{i=1}^{p} \boldsymbol{\varphi}_i(t)\boldsymbol{R}_X(t-i\Delta t,t) + \boldsymbol{L}(t)\boldsymbol{R}_{WX}(t,t) \tag{5.55}$$

where $\boldsymbol{R}_{WX}(t, t)$ denotes the correlation function of $\boldsymbol{W}_t$ and $\boldsymbol{X}_t$. The second equation is constructed by post-multiplying both sides of Eq. (5.49) with $\boldsymbol{W}_t^{\mathrm{T}}$, and then taking expectations. It is given by:

$$\boldsymbol{R}_{XW}(t,t) = \boldsymbol{L}(t) \tag{5.56}$$

Combine Eqs. (5.55) and (5.56), the solution for the square matrix $\boldsymbol{L}(t)$ can be obtained as follows:

$$\boldsymbol{L}(t)\boldsymbol{L}^{\mathrm{T}}(t) = \boldsymbol{R}_X(t,t) - \sum_{i=1}^{p} \boldsymbol{\varphi}_i(t)\boldsymbol{R}_X(t-i\Delta t,t) \tag{5.57}$$

Finally, by means of solving Eqs. (5.53) and (5.57), the matrices $\boldsymbol{\varphi}_i(t)$ $(i = 1,2,\cdots,p)$ and $\boldsymbol{L}(t)$ can be obtained respectively, and then the 1D-nV non-stationary stochastic vector process can thus be represented by a set of standard orthogonal random variables applying Eq. (5.49). Clearly, this provides the necessary condition for introducing the constraint form of random orthogonal functions to achieve dimension-reduction simulation.

5.4.3 Dimension-Reduction Simulation Method

To accomplish dimension-reduction simulation of the 1D-nV non-stationary stochastic vector process $\boldsymbol{X}(t)$ based on the above-mentioned time-domain analysis methods, it is necessary to perform the dimension-reduction representation of the standard orthogonal random vectors $\boldsymbol{W}$ in Eq. (5.46) and $\boldsymbol{W}_t$ in Eq. (5.49) at first. This entails expressing all the random variables within $\boldsymbol{W}$ and $\boldsymbol{W}_t$ as orthogonal function forms of the elementary random variables. To this end, the forms of random orthogonal functions for the standard orthogonal random vectors $\boldsymbol{W}$ and $\boldsymbol{W}_t$ are constructed, respectively. As previously stated, the forms of random orthogonal functions are obvious. For simplicity, just the random orthogonal functions involving one and two elementary random variables are provided here for illustrative purpose.

As for the standard orthogonal random vector $\boldsymbol{W} = (W_1, W_2, \cdots, W_{n\times N})^{\mathrm{T}}$ in Eq. (5.46), its random orthogonal functions can be addressed as follows.

Form 1: one-random-variable form of the CTD Representation.

$$\overline{W}_i = \sqrt{2}\sin(i\times\Theta + \varphi);\quad i = 1, 2, \cdots, (n\times N) \tag{5.58}$$

where the random variable Θ is uniformly distributed over the interval $(0, 2\pi]$. φ is a constant in the interval $(0, 2\pi]$ commonly taken as $\varphi = \pi/4$.

Form 2: two-random-variables form of the CTD Representation.

In addition, we can also define $\boldsymbol{W} = (W_1, W_2, \cdots, W_{n\times N})^{\mathrm{T}} = (U_{11}, \ldots, U_{1N}; U_{21}, \ldots, U_{2N}; \cdots; U_{n1}, \ldots, U_{n\times N})^{\mathrm{T}}$, thus the random orthogonal function with two elementary random variables can be given as follows:

$$\overline{W}_i = U_{kl} = \sqrt{2}\sin(k\times\Theta_1 + l\times\Theta_2 + \varphi);\quad k = 1, 2, \cdots, n;\ l = 1, 2, \cdots, N \tag{5.59}$$

where $i = (k-1)\times N+l$; Θ_1 and Θ_2 are mutually independent elementary random variables uniformly distributed over the interval $(0, 2\pi]$. φ is a constant in the interval $(0, 2\pi]$ taken as $\varphi = \pi/4$ herein.

As for the standard orthogonal random vector $\boldsymbol{W}_t = (W_{t,\,1}, W_{t,\,2}, \cdots, W_{t,\,n})^{\mathrm{T}}$ in Eq. (5.49), first, the time variable t should be discretized, i.e., $t = t_1, t_2, \cdots, t_N$. Then $\boldsymbol{W}_t$ can be transformed into an $(N\times n)$-dimensional standard orthogonal random vector $\boldsymbol{W} = (W_{1,\,1}, \ldots, W_{1,\,n}; W_{2,\,1}, \ldots, W_{2,\,n}; \cdots; W_{N,\,1}, \ldots, W_{N,\,n})^{\mathrm{T}}$, of which the random orthogonal functions can be constructed as follows:

Form 1: one-random-variable form of the TVP-VAR Model.

$$\overline{W}_{k,l} = U_i = \sqrt{2}\sin(i\times\Theta + \varphi),\quad k = 1, 2, \cdots, N;\ l = 1, 2, \cdots, n \tag{5.60}$$

where $i = (k-1)\times n+l$; the random variable Θ is uniformly distributed over the interval $(0, 2\pi]$. φ is a constant in the interval $(0, 2\pi]$ commonly taken as $\varphi = \pi/4$.

Form 2: two-random-variables form of the TVP-VAR Model.

$$\overline{W}_{k,l} = \sqrt{2}\sin(k \times \Theta_1 + l \times \Theta_2 + \varphi), \quad k = 1, 2, \cdots, N; \; l = 1, 2, \cdots, n \tag{5.61}$$

where Θ_1 and Θ_2 are mutually independent elementary random variables uniformly distributed over the interval $(0, 2\pi]$. φ is a constant in the interval $(0, 2\pi]$ taken as $\varphi = \pi/4$ herein.

It is a straightforward work to verify that no matter the one-random-variable forms, say Eqs. (5.58) and (5.60), or the two-random-variables forms, say Eqs. (5.59) and (5.61), are all standard orthogonal random variable sets, which satisfy the necessary condition for simulation. Furthermore, these initial standard orthogonal random variables $\overline{W}_i$ or $\overline{W}_{k,l}$ require to be transformed into the target standard orthogonal random variables W_i or $W_{k,l}$ through a deterministic one-to-one mapping relationship, say $\overline{W}_i \rightarrow W_i$ or $\overline{W}_{k,l} \rightarrow W_{k,l}$, which serves as a sufficient condition for dimension-reduction simulation. The transformation can be implemented using the built-in functions of MATLAB toolbox as the same as described in previous sections.

5.5 Dimension-Reduction Representation of Multi-Point and Multi-Component Stochastic Ground Motions

The multi-point and multi-component non-stationary stochastic ground motions can be regarded as a non-stationary stochastic vector process $\boldsymbol{X}(t) = \left[\boldsymbol{X}_1^{\mathrm{T}}(t), \boldsymbol{X}_2^{\mathrm{T}}(t), \cdots, \boldsymbol{X}_n^{\mathrm{T}}(t)\right]^{\mathrm{T}}$, of which the component process is $\boldsymbol{X}_i(t) = [X_{ix}(t), X_{iy}(t), X_{iz}(t)]^{\mathrm{T}}$ $(i = 1, 2, \cdots, n)$ with zero-mean. Thus, $\boldsymbol{X}(t)$ can be extended as a 1D-(3n)V non-stationary stochastic vector process, of which the probability information is completely determined by the EPSD function matrix (two-sided spectrum) as follows:

$$\boldsymbol{S}_{\boldsymbol{X}}(\omega, t) = \begin{bmatrix} \boldsymbol{S}_{11}(\omega, t) & \boldsymbol{S}_{12}(\omega, t) & \cdots & \boldsymbol{S}_{1n}(\omega, t) \\ \boldsymbol{S}_{21}(\omega, t) & \boldsymbol{S}_{22}(\omega, t) & \cdots & \boldsymbol{S}_{2n}(\omega, t) \\ \vdots & \vdots & \ddots & \vdots \\ \boldsymbol{S}_{n1}(\omega, t) & \boldsymbol{S}_{n2}(\omega, t) & \cdots & \boldsymbol{S}_{nn}(\omega, t) \end{bmatrix} \tag{5.62}$$

In Eq. (5.62), the diagonal elements $\boldsymbol{S}_{ii}(\omega, t)$ $(i = 1, 2, \cdots, n)$ present the EPSD function matrix (two-sided spectrum) of the same or different components at the same action point i, such that:

$$\mathbf{S}_{ii}(\omega,t)=\begin{bmatrix} S_{ix}(\omega,t) & S_{ix,iy}(\omega,t) & S_{ix,iz}(\omega,t) \\ S_{iy,ix}(\omega,t) & S_{iy}(\omega,t) & S_{iy,iz}(\omega,t) \\ S_{iz,ix}(\omega,t) & S_{iz,iy}(\omega,t) & S_{iz}(\omega,t) \end{bmatrix},\quad i=1,2,\cdots,n \tag{5.63}$$

where the subscripts x and y represent two horizontal components of earthquake ground motions, and z denotes the vertical component, respectively. The diagonal elements are the auto-EPSD function of the same component at the i-th action point, and the off-diagonal elements are the cross-EPSD function of two different components at the i-th action point, respectively.

Besides, in Eq. (5.62), the off-diagonal elements $\mathbf{S}_{ij}(\omega,t)$ $(i,j=1,2,\cdots,n;\ i\neq j)$ are the EPSD function matrix (two-sided spectrum) of the same or different components at different action points, followed by:

$$\mathbf{S}_{ij}(\omega,t)=\begin{bmatrix} S_{ix,jx}(\omega,t) & S_{ix,jy}(\omega,t) & S_{ix,jz}(\omega,t) \\ S_{iy,jx}(\omega,t) & S_{iy,jy}(\omega,t) & S_{iy,jz}(\omega,t) \\ S_{iz,jx}(\omega,t) & S_{iz,jy}(\omega,t) & S_{iz,jz}(\omega,t) \end{bmatrix},\quad i,j=1,2,\cdots,n;\ i\neq j \tag{5.64}$$

where the diagonal elements are the cross-EPSD function of the same component at two different action points, and the off-diagonal elements signify the cross-EPSD function of two different components at two different action points, respectively.

Thus, by substituting Eqs. (5.63) and (5.64) into Eq. (5.62), the EPSD function matrix $\mathbf{S}_{\mathbf{X}}(\omega,t)$ can thus be achieved. Herein, the classical Matsushima model (Matsushima 1976) is applied to describe the correlation of different seismic components, given by (Liu et al. 2023):

$$S_{i\varepsilon,j\upsilon}(\omega,t)=\sigma\sqrt{S_{i\varepsilon}(\omega,t)\times S_{j\upsilon}(\omega,t)}\,\gamma_{ij}(\omega) \tag{5.65}$$

where the subscripts ε and υ denote translational seismic components x, y and z. σ is the correlation coefficient of Matsushima model. For two horizontal components, we have $\sigma=1$, and for one horizontal component and one vertical component, we have $\sigma=0.6$. $\gamma_{ij}(\omega)$ is the coherence function between the two different action points i and j, of which the specific models for engineering applications can be found in Sect. 2.5.2.

In the field of earthquake engineering, spectral decomposition can be directly performed on the EPSD function matrix to achieve the simulation of multi-point and multi-component stochastic ground motions. Additionally, it is more convenient to employ spectral decomposition techniques on the coherence function matrix. To this end, the coherence function matrix $\boldsymbol{\gamma}(\omega)$ of the 1D-(3n)V non-stationary stochastic vector process $\mathbf{X}(t)$ can be written as follows:

$$\boldsymbol{\gamma}(\omega)=\begin{bmatrix}\gamma_{11}(\omega) & \gamma_{12}(\omega) & \cdots & \gamma_{1n}(\omega)\\ \gamma_{21}(\omega) & \gamma_{22}(\omega) & \cdots & \gamma_{2n}(\omega)\\ \vdots & \vdots & \ddots & \vdots\\ \gamma_{n1}(\omega) & \gamma_{n2}(\omega) & \cdots & \gamma_{nn}(\omega)\end{bmatrix} \tag{5.66}$$

In Eq. (5.66), the diagonal elements $\boldsymbol{\gamma}_{ii}(\omega)$ $(i = 1, 2, \cdots, n)$ denote the coherence function matrix of three components at the same action point, according to Matsushima model and Eq. (5.65), $\boldsymbol{\gamma}_{ii}(\omega)$ can be defined as follows:

$$\gamma_{ii}(\omega)=\begin{bmatrix}1 & \gamma_{ix,iy}(\omega) & \gamma_{ix,iz}(\omega)\\ \gamma_{iy,ix}(\omega) & 1 & \gamma_{iy,iz}(\omega)\\ \gamma_{iz,ix}(\omega) & \gamma_{iz,iy}(\omega) & 1\end{bmatrix}=\begin{bmatrix}1 & 1 & 0.6\\ 1 & 1 & 0.6\\ 0.6 & 0.6 & 1\end{bmatrix},\quad i=1,2,\cdots,n \tag{5.67}$$

Besides, in Eq. (5.66), the off-diagonal elements $\boldsymbol{\gamma}_{ij}(\omega)$ $(i, j = 1, 2, \cdots, n;\ i \neq j)$ denote the coherence function matrix of three components at two different action points, according to Matsushima model and Eq. (5.65), $\boldsymbol{\gamma}_{ij}(\omega)$ can be defined as follows (Liu et al. 2023):

$$\gamma_{ij}(\omega)=\begin{bmatrix}\gamma_{ix,jx}(\omega) & \gamma_{ix,jy}(\omega) & \gamma_{ix,jz}(\omega)\\ \gamma_{iy,jx}(\omega) & \gamma_{iy,jy}(\omega) & \gamma_{iy,jz}(\omega)\\ \gamma_{iz,jx}(\omega) & \gamma_{iz,jy}(\omega) & \gamma_{iz,jz}(\omega)\end{bmatrix}=\gamma_{ij}(\omega)\begin{bmatrix}1 & 1 & 0.6\\ 1 & 1 & 0.6\\ 0.6 & 0.6 & 1\end{bmatrix},\quad i,j=1,2,\cdots,n;\ i\neq j \tag{5.68}$$

Finally, through substituting Eqs. (5.67) and (5.68) into Eq. (5.66), the coherence function matrix $\boldsymbol{\gamma}(\omega)$ can thus be obtained. Furthermore, based on the unified form of the original spectral decomposition representation of the non-stationary stochastic vector process, say Eq. (5.13), the Cholesky decomposition or eigen decomposition can be employed, followed by the introduction of the random orthogonal functions proposed in Sect. 5.3, to achieve dimension-reduction representation for multi-point and multi-component non-stationary stochastic seismic ground motions. The details can be found in the literature (Liu et al. 2023). Meanwhile, utilizing the EPSD function matrix $\boldsymbol{S}_{\boldsymbol{X}}(\omega, t)$ of multi-point and multi-component non-stationary stochastic seismic ground motions, we can further derive the corresponding correlation function matrix and subsequently conduct simulations using the time-domain representation of non-stationary stochastic vector processes outlined in Sect. 5.4.

5.6 Numerical Examples

In this numerical case, the frequency-domain analysis method is utilized to simulate the 1D-nV non-stationary ground motion stochastic vector process. The power spectral density (PSD) function employs the Clough-Penzien (C-P) model given in Eq. (2.10), such that:

$$S(\omega) = \frac{\omega_g^4 + 4\zeta_g^2\omega_g^2\omega^2}{\left(\omega^2 - \omega_g^2\right)^2 + 4\zeta_g^2\omega_g^2\omega^2} \times \frac{\omega^4}{\left(\omega^2 - \omega_f^2\right)^2 + 4\zeta_f^2\omega_f^2\omega^2} \times S_0 \tag{5.69}$$

where ω_g and ζ_g denote the predominant circular frequency and damping ratio of the site soil, respectively. ω_f and ζ_f denote the predominant circular frequency and damping ratio of the rock bed, respectively. S_0 is the spectral intensity defined in Eq. (2.15), followed by:

$$S_0 = \frac{a_{\max}^2}{r^2\left[\pi\omega_g\left(2\zeta_g + \frac{1}{2\zeta_g}\right)\right]} \tag{5.70}$$

where $a_{\max}$ denotes the mean of peak ground acceleration (PGA), which is taken by $a_{\max} = 200\text{cm/s}^2$ herein. r refers to the peak factor.

For the sake of reflecting the fully non-stationary characteristics of earthquake ground motions in both intensity and frequency, the time-frequency modulation function defined in Eq. (2.69) is applied, given by:

$$A(t,\omega) = \frac{\exp(-a\times t) - \exp\left[-\left(c\times\left|\omega-\omega_g\right| + b\right)\times t\right]}{\exp(-a\times t^*) - \exp\left[-\left(c\times\left|\omega-\omega_g\right| + b\right)\times t^*\right]}, \quad \omega > 0, t > 0 \tag{5.71}$$

in which

$$t^* = \frac{\ln\left(c\times\left|\omega-\omega_g\right| + b\right) - \ln(a)}{c\times\left|\omega-\omega_g\right| + b - a}, \quad \omega > 0 \tag{5.72}$$

where a, b and c are parameters that control the arrival time of the peak value and the rate of energy decay. Generally, we have $b = a+0.001$ and $c = 0.005$.

The dimension-reduction simulation of a 1D-4V non-stationary seismic ground motion stochastic vector process is taken as the numerical example. As depicted in Fig. 5.1, suppose there are four action points, namely A, B, C, and D, aligned along the propagation direction of the apparent velocity of seismic waves on the ground surface, with a distance of 200 m between every two adjacent action points. Suppose Points A and B are located in Soil site I_1, and Points C and D are located in Soil site

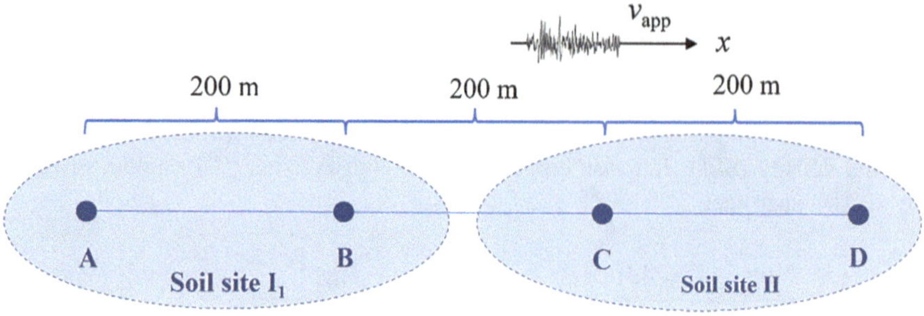

Fig. 5.1 Location of ground surface action points and site category of non-stationary seismic stochastic field

Table 5.1 Parameter values of the EPSD model

Soil sites	a (s^{-1})	ω_g (rad/s)	ζ_g	ω_f (rad/s)	ζ_f	r
I_1	0.30	8π	0.64	0.8π	0.64	3.0
II	0.20	5π	0.72	0.5π	0.72	2.88

II. Obviously, the seismic ground motions at the action points of A, B, C, and D can be considered as a 1D-4V non-stationary stochastic vector process.

The parameter values of the evolutionary PSD (EPSD) function are shown in Table 5.1.

Additionally, the spatial variability of seismic ground motion field generally includes lagged coherence effect, wave passage effect and local site effect. To fully capture the aforementioned spatial variation effects, the composite coherence function model defined in Eq. (2.104) is employed, given by:

$$\gamma_{ij}(\omega) = \gamma_{ij}^{\text{incoh}}(\omega) \cdot \gamma_{ij}^{\text{wave}}(\omega) \cdot \gamma_{ij}^{\text{site}}(\omega) = \gamma_{ij}^{\text{incoh}}(\omega) \cdot \exp\left[\mathrm{i}\left(\theta_{ij}^{\text{wave}}(\omega) + \theta_{ij}^{\text{site}}(\omega)\right)\right] \tag{5.73}$$

where $\gamma_{ij}^{\text{incoh}}(\omega)$ denotes lagged coherence effect, in fact, it is the modulus of the coherence function, say $|\gamma_{ij}(\omega)| = \gamma_{ij}^{\text{incoh}}(\omega)$. The phases $\theta_{ij}^{\text{wave}}$ and $\theta_{ij}^{\text{site}}$ denote the wave passage effect and local site effect, respectively.

The lagged coherence effect employs the Luco-Wang model defined in Eq. (2.108), such that:

$$\gamma_{ij}^{\text{incoh}}(\omega) = \exp\left[-\left(\eta \frac{\omega \xi_{ij}}{v_s}\right)^2\right] \tag{5.74}$$

where ξ_{ij} represents the distance between two action points (scalar). v_s denotes the shear wave velocity of seismic waves propagating through a medium (scalar). η denotes the partial coherence coefficient. According to the reference (Luco and Wong 1986), it takes $\frac{\eta}{v_s} = 2 \times 10^{-4}$ s/m herein.

The phase presenting the wave passage effect defined in Eq. (2.111) is applied, given by:

$$\theta_{ij}^{\text{w ave}}(\omega) = -\frac{j-i}{|j-i|} \times \frac{\omega \xi_{ij}}{v_{\text{app}}} \tag{5.75}$$

where the values of apparent wave velocity v_{app} are taken as $v_{\text{app}}^{\text{A-B}} = 850$, $v_{\text{app}}^{\text{A-C}} = 750$, $v_{\text{app}}^{\text{A-D}} = 750$, $v_{\text{app}}^{\text{B-C}} = 750$, $v_{\text{app}}^{\text{B-D}} = 750$, $v_{\text{app}}^{\text{C-D}} = 650$, of which the unit is m/s.

The phase reflecting the local site effect defined in Eq. (2.113) is utilized, such that:

$$\theta_{ij}^{\text{site}}(\omega) = \arctan \frac{\text{Im}\left[H_i(\omega) H_j^*(\omega)\right]}{\text{Re}\left[H_i(\omega) H_j^*(\omega)\right]} = \arctan \frac{\text{Im}\left[H_i(\omega) H_j(-\omega)\right]}{\text{Re}\left[H_i(\omega) H_j(-\omega)\right]} \tag{5.76}$$

where $H_i(\omega)$ and $H_j(\omega)$ denote the transfer functions of the absolute acceleration at action points i and j, respectively, of which the expressions are as follows:

$$H_i(\omega) = \frac{\omega_{\text{g},i}^2 + 2\text{i}\zeta_{\text{g},i}\omega_{\text{g},i}\omega}{\left(\omega_{\text{g},i}^2 - \omega^2\right) + 2\text{i}\zeta_{\text{g},i}\omega_{\text{g},i}\omega}, H_j(\omega) = \frac{\omega_{\text{g},j}^2 + 2\text{i}\zeta_{\text{g},j}\omega_{\text{g},j}\omega}{\left(\omega_{\text{g},j}^2 - \omega^2\right) + 2\text{i}\zeta_{\text{g},j}\omega_{\text{g},j}\omega} \tag{5.77}$$

where $\omega_{\text{g},\,i}$ and $\zeta_{\text{g},\,i}$ denote the predominant circular frequency and damping ratio of the site soil at Point i, respectively.

The other parameter values for implementing dimension-reduction simulation of the 1D-4V non-stationary ground motion stochastic vector process are listed in Table 5.2.

In this numerical example, the random orthogonal function defined in Eq. (5.37) with merely two elementary random variables is applied for simulation, and 233 sets of non-stationary ground motion representative time-histories are generated. Figure 5.2 describes the representative time-histories of Point B on Soil site I_1 and Point C on Soil site II. It is evident that the seismic acceleration time-histories display typical non-stationary features. Additionally, due to the consideration of

Table 5.2 Parameter values for dimension-reduction simulation of the 1D-4V non-stationary ground motion stochastic vector process

Parameter	Value
Upper cut-off frequency ω_{u} (rad/s)	240
Frequency interval $\Delta\omega$ (rad/s)	0.15
Number of discrete frequency terms N	1600
Duration T (s)	25
Time interval Δt (s)	0.01
Number of generated representative time-histories n_{sel}	233

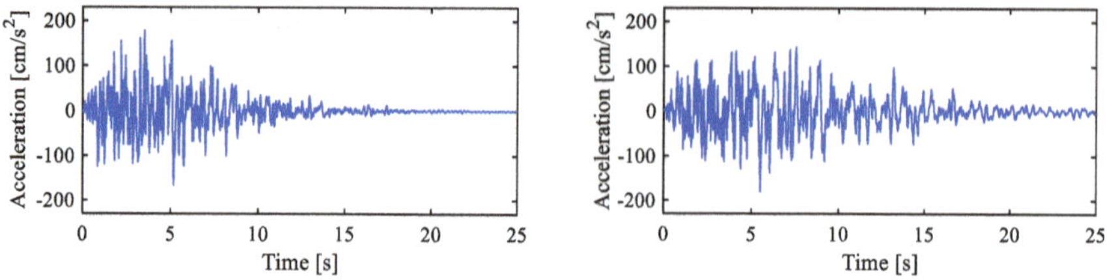

(a) Point B (Soil site I_1) (b) Point C (Soil site II)

Fig. 5.2 Representative acceleration time-histories of 1D-4V non-stationary ground motion stochastic vector process generated by the dimension-reduction method

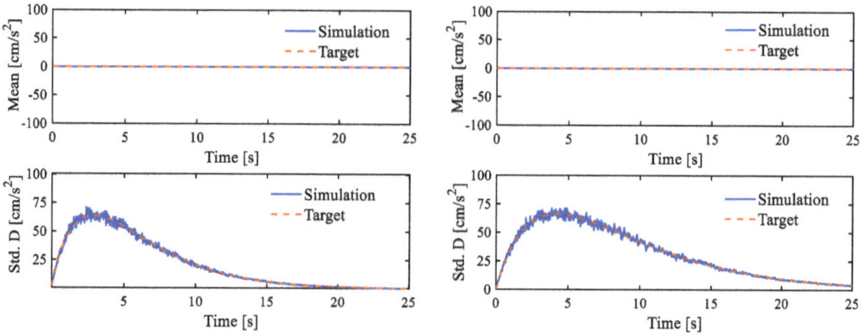

(a) Point B (Soil site I_1) (b) Point C (Soil site II)

Fig. 5.3 Comparison of mean and standard deviation between the simulated representative time-history sets with the corresponding target values

the spatial variability of seismic ground motion field, there are significant differences in the frequency variation and amplitude attenuation of the acceleration time-histories at the two different action points.

Figure 5.3 presents the comparison of the mean and standard deviation (Std. D) of the representative time-history sets at Points B and C and their corresponding target values. As evident from the figures, the simulation values of the representative time-history sets for the 1D-4V non-stationary seismic stochastic vector process generated through the dimension-reduction method closely align with the target values, thereby fully validating the effectiveness of the dimension-reduction method.

Figure 5.4 shows the comparison of the PSD between the simulated representative time-history sets at Points B and C and the corresponding target values at typical time instants. As can be observed from the figures, all the PSD values of the representative time-history sets for different soil sites fit well with the targets at different time instants.

Figure 5.5 depicts the comparison of the auto-correlation function at Point B and the cross-correlation function of Points B and C of the representative time-history sets with the corresponding target values, respectively. From the figures, it can be clearly seen that both the auto-correlation and cross-correlation functions have consistent fits with the target values.

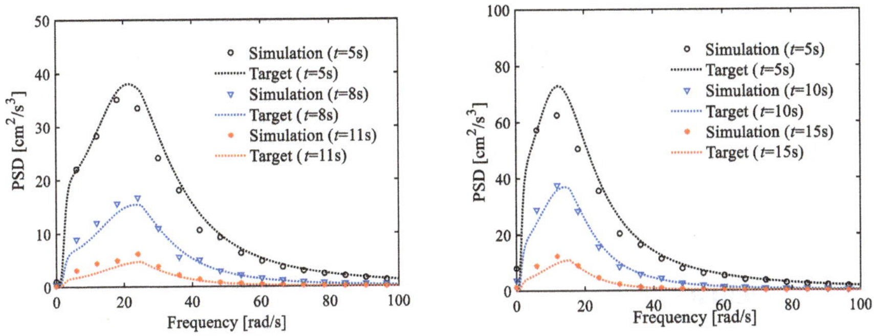

(a) Point B (Soil site I_1) (b) Point C (Soil site II)

Fig. 5.4 Comparison of PSD between the simulated representative time-history sets with the corresponding target values at typical time instants

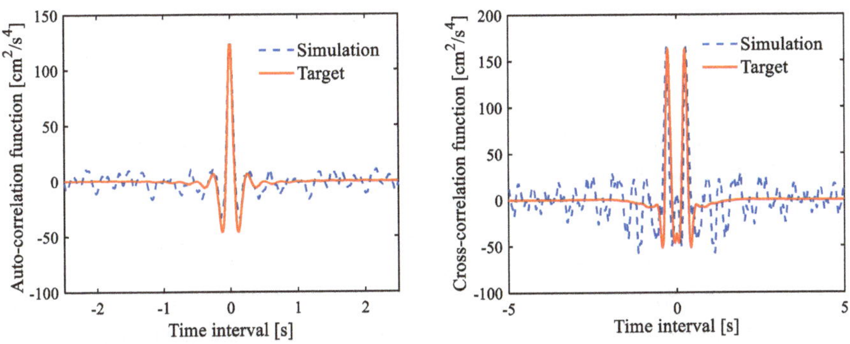

(a) Auto-correlation function (b) Cross-correlation function

Fig. 5.5 Comparison of auto–/cross-correlation function between the simulated representative time-history sets with the corresponding target values

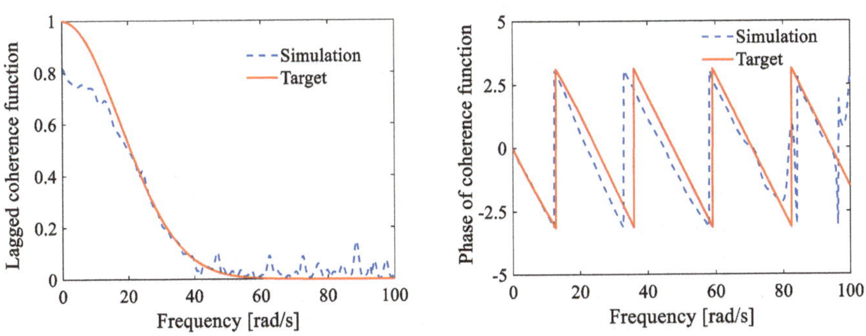

(a) lagged coherence function (b) phase of coherence function

Fig. 5.6 Comparison of the spatial coherence effect between the simulated representative time-history set and the corresponding target values

Figure 5.6 presents the comparison of spatial coherence effects between the simulated representative time-history sets and their corresponding target values. Specifically, Fig. 5.6a depicts the lagged coherence effect, and Fig. 5.6b displays the phase of the spatial coherence function. As evident from the figures, both the simulated values of the lagged coherence effect and the phase exhibit a satisfactory fit with the target values.

5.7 Conclusion Remarks

This chapter mainly deduces the simulation methods of multivariate non-stationary stochastic process (1D-nV non-stationary stochastic vector process), including the spectral decomposition methods based on frequency-domain analysis and the methods based on time-domain analysis. Further, through introducing the constraint form of the random function associated with the standard orthogonal random vector, the multivariate stochastic process can be precisely represented with merely 1–3 elementary random variables. The main conclusions drawn from this chapter are as follows:

1. For the frequency-domain analysis methods of multivariate stochastic process, this chapter elaborately derives the unified form of original spectral decomposition representation, which has a unified theoretical basis. Specifically, in the case that Cholesky decomposition is applied to the PSD or coherence function matrix, the conventional SRM can be obtained, and when eigen-decomposition is adopted, the conventional POD can be achieved. This implies that both the SRM and POD are special cases of the original spectral decomposition in essence. Therefore, if different spectral decomposition techniques are employed to the PSD or coherence function matrix, different orthogonal decomposition forms can be obtained.
2. The conventional Monte Carlo methods and the dimension-reduction methods for simulating multivariate stochastic process are both derived from the original spectral decomposition, implying that both the two approaches can be regarded as special cases of the original spectral decomposition. The differences and connections between the conventional Monte Carlo methods and the dimension-reduction methods are explained in detail. Both methods share identical necessary conditions for achieving simulation, but their sufficient conditions are different. In fact, the two methods are two parallel strategies.
3. The time-domain analysis methods can generally be extended from univariate stochastic process to multivariate stochastic process. This chapter mainly introduces the conventional time-domain representation method and the TVP-VAR model. Note that both time-domain analysis methods involve the solution of undetermined matrices, which commonly need to be solved through the correlation function matrix of the multivariate stochastic process. However, one advantage of the time-domain analysis method is that it can realize the dimension-

reduction simulation for multivariate stochastic process utilizing merely 1–2 elementary random variables. Obviously, the MA and ARMA models involved in the time-domain analysis methods are also effective methods that can be expected to achieve the dimension-reduction simulation.

References

Box GE, Jenkins GM, Reinsel GC et al (2015) Time series analysis: forecasting and control. Wiley, Hoboken, NJ

Deodatis G (1996) Non-stationary stochastic vector processes: seismic ground motion applications. Prob Eng Mech 11(3):149–167

Di Paola M, Zingales M (2000) Digital simulation of multivariate earthquake ground motions. Earthq Eng Struct Dyn 29(7):1011–1127

Li Y, Kareem A (1991) Simulation of multivariate nonstationary random processes by FFT. J Eng Mech 117(5):1037–1058

Liu ZX, Liu ZJ, Ruan XX et al (2018) Spectral representation-based dimension reduction for simulating multivariate non-stationary ground motions. Soil Dyn Earthq Eng 114:313–325

Liu ZJ, Liu ZX, Ruan XX et al (2019) POD-based dimension reduction representation of stochastic ground motion fields. Sci Sin Technol 49(5):589–601. (in Chinese)

Liu ZX, Liu ZJ, Ai QH et al (2023) Global reliability evaluation of a high-pier long-span continuous RC rigid frame bridge subjected to multi-point and multi-component stochastic ground motions. Soil Dyn Earthq Eng 164:107623

Liu ZX, Liu ZJ, Ruan XX et al (2025) Advances in dimension-reduction methods for simulating univariate and multivariate non-stationary stochastic processes via spectral representation. Prob Eng Mech 79:103720

Luco JE, Wong HL (1986) Response of a rigid foundation to a spatially random ground motion. Earthq Eng Struct Dyn 14(6):891–908

Matsushima Y (1976) Stochastic response of structures subjected to three directional seismic input. In: Collection of Papers and Reports of the Architectural Society of Japan. (in Japanese)

Priestley MB (1965) Evolutionary spectra and non-stationary processes. J R Stat Soc Ser B (Methodol) 27(2):204–229

Samaras E, Shinzuka M, Tsurui A (1985) ARMA representation of random processes. J Eng Mech 111(3):449–461

Chapter 6
Dimension-Reduction Representation of Stochastic Ground Motion Fields

Abstract This chapter systematically investigates the dimension-reduction representation for stochastic ground motion fields to address the substantial computational burden in simulating ground motions for large-scale engineering structures. The original frequency-wavenumber spectral representation (O-FWSR) for multi-dimensional, univariate (*m*D-1V) spatio-temporal stochastic fields is derived, revealing that the conventional Monte Carlo method is essentially a special case of the O-FWSR. By imposing random orthogonal functions on the standard orthogonal random variables, an efficient dimension-reduction simulation method is proposed, requiring merely two and three elementary random variables for 2D-1V and 3D-1V fields, respectively. Numerical examples demonstrate its high accuracy and efficiency.

Keywords Spatio-temporal stochastic field · Frequency-wavenumber spectral representation · Dimension-reduction simulation · FFT-embeded simulation

The previous chapter presents the dimension-reduction simulation approach for 1D-*n*V non-stationary stochastic vector process in detail. Nevertheless, when it comes to lifeline engineering structures with a large number of supports, like oil and gas pipelines and water supply networks, employing the simulation method of multivariate stochastic processes would cause enormous computational burdens, potentially even leading to numerical instability (Benowitz and Deodatis 2015; Peng et al. 2017). To address this issue, the *m*D-1V spatio-temporal stochastic field, also regarded as continuous spatio-temporal stochastic field, can be employed to generate a substantial number of non-uniform earthquake ground motion input time-histories at once. Chapter 2 has provided a brief introduction to the fundamental concepts of continuous spatio-temporal stochastic field. This chapter mainly deduces the original frequency-wavenumber spectral representation (O-FWSR) of spatio-temporal stochastic field, explains the simulation formula and its derivation for the conventional Monte Carlo method, and further elaborates on the latest advancements in dimension-reduction simulation method for spatio-temporal stochastic field.

Z. Liu et al., *Dimension-reduction Representation of Stochastic Ground Motion and Engineering Applications*, https://doi.org/10.1007/978-981-92-0168-6_6

6.1 Original Frequency-Wavenumber Spectral Representation of Spatio-temporal Stochastic Field

6.1.1 Basic Theory of Spatio-temporal Stochastic Field

Suppose $f_0(\boldsymbol{x}, t)$ is a real-valued non-stationary and non-homogeneous (r+1)D-1V spatio-temporal stochastic field with continuous mean-square value and zero-mean, which can be generally expressed as the following Fourier-Stieltjes integral form (Deodatis and Shinozuka 1989):

$$f_0(\boldsymbol{x}, t) = \int_{-\infty}^{\infty} \int_{-\infty}^{\infty} A(\boldsymbol{x}, t; \boldsymbol{\kappa}, \omega) \mathrm{e}^{\mathrm{i}\,(\boldsymbol{\kappa} \cdot \boldsymbol{x} + \omega t)} \mathrm{d}Z(\boldsymbol{\kappa}, \omega) \tag{6.1}$$

where $\boldsymbol{x} = (x_1, x_2, \cdots, x_r)^{\mathrm{T}}$ denotes an r-dimensional spatial vector. $\boldsymbol{\kappa} = (\kappa_1, \kappa_2, \cdots, \kappa_r)^{\mathrm{T}}$ denotes an r-dimensional wavenumber vector corresponding to the spatial vector $\boldsymbol{x}$. $A(\boldsymbol{x}, t; \boldsymbol{\kappa}, \omega)$ signifies a deterministic modulation function. $Z(\boldsymbol{\kappa}, \omega)$ is an orthogonal increment process in terms of frequency ω and wavenumber vector $\boldsymbol{\kappa}$, of which the increment $\mathrm{d}Z(\boldsymbol{\kappa}, \omega)$ satisfies the following conditions (Vanmarcke 1983; Deodatis and Shinozuka 1989):

$$E[\mathrm{d}Z(\boldsymbol{\kappa}, \omega)] = 0 \tag{6.2a}$$

$$E[\mathrm{d}Z(\boldsymbol{\kappa}, \omega)\mathrm{d}Z^*(\boldsymbol{\kappa}', \omega')] = \delta_{\kappa_1 \kappa_1'} \delta_{\kappa_2 \kappa_2'} \cdots \delta_{\kappa_r \kappa_r'} \delta_{\omega\omega'} \mathrm{d}\psi(\boldsymbol{\kappa}, \omega) \tag{6.2b}$$

where $\delta_{\kappa_i \kappa_i'} (i = 1, 2, \cdots, r)$ and $\delta_{\omega\omega'}$ are both the Kronecker-delta. $\psi(\boldsymbol{\kappa}, \omega)$ denotes a deterministic function.

Since $f_0(\boldsymbol{x}, t)$ is real-valued, the following symmetry conditions can be obtained:

$$A(\boldsymbol{x}, t; -\boldsymbol{\kappa}, -\omega) = A^*(\boldsymbol{x}, t; \boldsymbol{\kappa}, \omega), \quad \mathrm{d}Z(-\boldsymbol{\kappa}, -\omega) = \mathrm{d}Z^*(\boldsymbol{\kappa}, \omega) \tag{6.3}$$

Further, the auto-correlation function of $f_0(\boldsymbol{x}, t)$ can be written as follows:

$$\begin{aligned} R_{f_0}(\boldsymbol{x}, t; \boldsymbol{x} + \boldsymbol{\xi}, t + \tau) &= E[f_0{}^*(\boldsymbol{x}, t) f_0(\boldsymbol{x} + \boldsymbol{\xi}, t + \tau)] \\ &= \int_{-\infty}^{\infty} \int_{-\infty}^{\infty} A^*(\boldsymbol{x}, t; \boldsymbol{\kappa}, \omega) A(\boldsymbol{x} + \boldsymbol{\xi}, t + \tau; \boldsymbol{\kappa}, \omega) \mathrm{e}^{\mathrm{i}\,(\boldsymbol{\kappa} \cdot \boldsymbol{\xi} + \omega \tau)} \mathrm{d}\psi(\boldsymbol{\kappa}, \omega) \end{aligned} \tag{6.4}$$

In Eq. (6.4), let $\boldsymbol{\xi} = \boldsymbol{0}$ and $\tau = 0$, the mean-square function can be obtained as:

$$R_{f_0}(\boldsymbol{x}, t; \boldsymbol{x}, t) = E[f_0^2(\boldsymbol{x}, t)] = \int_{-\infty}^{\infty} \int_{-\infty}^{\infty} |A(\boldsymbol{x}, t; \boldsymbol{\kappa}, \omega)|^2 \mathrm{d}\psi(\boldsymbol{\kappa}, \omega) \tag{6.5}$$

If the deterministic function $\psi(\boldsymbol{\kappa}, \omega)$ is differentiable, we can define $\mathrm{d}\psi(\boldsymbol{\kappa}, \omega) = S(\boldsymbol{\kappa}, \omega)\mathrm{d}\boldsymbol{\kappa}\mathrm{d}\omega$, such that:

$$E\left[f_0^2(\boldsymbol{x},t)\right]=\int_{-\infty}^{\infty}\int_{-\infty}^{\infty}|A(\boldsymbol{x},t;\boldsymbol{\kappa},\omega)|^2 S(\boldsymbol{\kappa},\omega)\mathrm{d}\boldsymbol{\kappa}\mathrm{d}\omega \tag{6.6}$$

Then, the evolutionary frequency-wavenumber spectrum (EFWS) $S_{f_0}(\boldsymbol{x},t;\boldsymbol{\kappa},\omega)$ can be defined as follows:

$$S_{f_0}(\boldsymbol{x},t;\boldsymbol{\kappa},\omega)=|A(\boldsymbol{x},t;\boldsymbol{\kappa},\omega)|^2 S(\boldsymbol{\kappa},\omega) \tag{6.7}$$

where $S(\boldsymbol{\kappa},\omega)$ represents the two-sided frequency-wavenumber spectrum density function of the stationary and homogeneous spatio-temporal stochastic field corresponding to $f_0(\boldsymbol{x},t)$, referred to as the frequency-wavenumber spectrum (FWS). At this moment, $\psi(\boldsymbol{\kappa},\omega)$ is termed as the spectral distribution function of the spatio-temporal stochastic field corresponding to $f_0(\boldsymbol{x},t)$.

Thus, Eq. (6.1) is referred to as the frequency-wavenumber spectral representation (FWSR) of the spatio-temporal stochastic field, which demonstrates that any non-stationary spatio-temporal stochastic field with continuous mean-square can be represented as the superposition of an infinite number of harmonic waves with varying frequencies, jointly modulated by deterministic functions and orthogonal increments.

Furthermore, Eq. (6.1) can be approximately discretized as the following finite series form:

$$\begin{aligned}f_0(\boldsymbol{x},t)&\approx\sum_{m_1=-M_1}^{M_1}\sum_{m_2=-M_2}^{M_2}\cdots\sum_{m_r=-M_r}^{M_r}\sum_{n=-N}^{N}A(\boldsymbol{x},t;\boldsymbol{\kappa}_m,\omega_n)\Delta Z(\boldsymbol{\kappa}_m,\omega_n)\mathrm{e}^{\mathrm{i}(\boldsymbol{\kappa}_m\cdot\boldsymbol{x}+\omega_n t)}\\&=\sum_{m_1=-M_1}^{M_1}\sum_{m_2=-M_2}^{M_2}\cdots\sum_{m_r=-M_r}^{M_r}\sum_{n=-N}^{N}A(\boldsymbol{x},t;\boldsymbol{\kappa}_m,\omega_n)\Delta Z(\boldsymbol{\kappa}_m,\omega_n)\mathrm{e}^{\mathrm{i}\left(\kappa_{1m_1}x_1+\kappa_{2m_2}x_2+\cdots+\kappa_{rm_r}x_r+\omega_n t\right)}\end{aligned} \tag{6.8}$$

where $\boldsymbol{\kappa}_m=\left(\kappa_{1m_1},\kappa_{2m_2},\cdots,\kappa_{rm_r}\right)^{\mathrm{T}}$ denotes the discretized wavenumber vector, and ω_n is the discretized frequency, given by:

$$\kappa_{im_i}=\begin{cases}(m_i-0.5)\Delta\kappa_i, & m_i=1,2,\cdots,M_i\\(m_i+0.5)\Delta\kappa_i, & m_i=-1,-2,\cdots,-M_i\end{cases},\quad(i=1,2,\cdots,r) \tag{6.9a}$$

$$\omega_n=\begin{cases}(n-0.5)\Delta\omega, & n=1,2,\cdots,N\\(n+0.5)\Delta\omega, & n=-1,-2,\cdots,-N\end{cases} \tag{6.9b}$$

where $\Delta\kappa_i=\kappa_{iu}/M_i$ ($i=1, 2, \cdots, r$) denote the wavenumber steps; $\Delta\omega=\omega_u/N$ is the frequency step. κ_{iu} represents the upper cut-off wavenumber, and ω_u denotes the upper cut-off frequency, respectively. Note that the energy corresponding to the EFWS beyond the upper cut-off wavenumber κ_{iu} or the upper cut-off frequency ω_u can be neglected.

According to Eq. (6.2b) and considering $\mathrm{d}\psi(\boldsymbol{\kappa}, \omega) = S(\boldsymbol{\kappa}, \omega)\mathrm{d}\boldsymbol{\kappa}\mathrm{d}\omega$, the increment $\Delta Z(\boldsymbol{\kappa}_m, \omega_n)$ can be defined as follows:

$$\Delta Z(\boldsymbol{\kappa}_m, \omega_n) = \sqrt{S(\boldsymbol{\kappa}_m, \omega_n)\Delta\kappa_1\Delta\kappa_2\cdots\Delta\kappa_r\Delta\omega}P_{m_1m_2\cdots m_rn} \tag{6.10}$$

where $P_{m_1m_2\cdots m_rn}$ denotes the standard orthogonal complex random variables with zero-mean, satisfying the following conditions:

$$P_{(-m_1)(-m_2)\cdots(-m_r)(-n)} = P^*_{m_1m_2\cdots m_rn}, \quad E[P_{m_1m_2\cdots m_rn}] = 0 \tag{6.11a}$$

$$E\left[P_{m_1m_2\cdots m_rn}P^*_{m'_1m'_2\cdots m'_rn'}\right] = \delta_{m_1m'_1}\delta_{m_2m'_2}\cdots\delta_{m_rm'_r}\delta_{nn'} \tag{6.11b}$$

Furthermore, $P_{m_1m_2\cdots m_rn}$ can be written as the real-valued form, say $P_{m_1m_2\cdots m_rn} = (X_{m_1m_2\cdots m_rn} - \mathrm{i}Y_{m_1m_2\cdots m_rn})/\sqrt{2}$. According to the symmetry of $P_{m_1m_2\cdots m_rn}$, it can be achieved that $X_{(-m_1)(-m_2)\cdots(-m_r)(-n)} = X_{m_1m_2\cdots m_rn}$ and $Y_{(-m_1)(-m_2)\cdots(-m_r)(-n)} = -Y_{m_1m_2\cdots m_rn}$. Substituting Eq. (6.10) into Eq. (6.8), and utilizing the real form of $P_{m_1m_2\cdots m_rn}$ along with the Euler's formula, we can derive the real form of the spatio-temporal stochastic field $f_0(\boldsymbol{x}, t)$ as follows:

$$f(\boldsymbol{x}, t) = \sum_{\substack{I_i = \pm 1 \\ i = 1, 2, \cdots, r}} \sum_{m_1=1}^{M_1}\sum_{m_2=1}^{M_2}\cdots\sum_{m_r=1}^{M_r}\sum_{n=1}^{N} B^{I_1I_2\cdots I_r}_{m_1m_2\cdots m_rn}\left[X^{I_1I_2\cdots I_r}_{m_1m_2\cdots m_rn}\cos(\widetilde{\boldsymbol{\kappa}}_m\cdot\boldsymbol{x} + \omega_nt) + Y^{I_1I_2\cdots I_r}_{m_1m_2\cdots m_rn}\sin(\widetilde{\boldsymbol{\kappa}}_m\cdot\boldsymbol{x} + \omega_nt)\right] \tag{6.12a}$$

$$B^{I_1I_2\cdots I_r}_{m_1m_2\cdots m_rn} = \sqrt{2S_{f_0}(\boldsymbol{x}, t; \widetilde{\boldsymbol{\kappa}}_m, \omega_n)\Delta\kappa_1\Delta\kappa_2\cdots\Delta\kappa_r\Delta\omega} \tag{6.12b}$$

$$\widetilde{\boldsymbol{\kappa}}_m\cdot\boldsymbol{x} = I_1\kappa_{1m_1}x_1 + I_2\kappa_{2m_2}x_2 + \cdots + I_r\kappa_{rm_r}x_r \tag{6.12c}$$

$$\widetilde{\boldsymbol{\kappa}}_m = [I_1\kappa_{1m_1}, I_2\kappa_{2m_2}, \cdots, I_r\kappa_{rm_r}]^{\mathrm{T}} \tag{6.12d}$$

In Eq. (6.12a), $f(\boldsymbol{x},t)$ is the simulated spatio-temporal stochastic field. $\left\{X^{I_1I_2\cdots I_r}_{m_1m_2\cdots m_rn}, Y^{I_1I_2\cdots I_r}_{m_1m_2\cdots m_rn}\right\}$ is a set of standard orthogonal random variables, which should satisfy the following basic conditions:

$$E\left[X^{I_1I_2\cdots I_r}_{m_1m_2\cdots m_rn}\right] = E\left[Y^{I_1I_2\cdots I_r}_{m_1m_2\cdots m_rn}\right] = 0, \quad E\left[X^{I_1I_2\cdots I_r}_{m_1m_2\cdots m_rn}Y^{I'_1I'_2\cdots I'_r}_{m'_1m'_2\cdots m'_rn'}\right] = 0 \tag{6.13a}$$

$$E\left[X_{m_1m_2\cdots m_rn}^{I_1I_2\cdots I_r}X_{m'_1m'_2\cdots m'_rn'}^{I'_1I'_2\cdots I'_r}\right]=E\left[Y_{m_1m_2\cdots m_rn}^{I_1I_2\cdots I_r}Y_{m'_1m'_2\cdots m'_rn'}^{I'_1I'_2\cdots I'_r}\right]$$
$$=\delta_{I_1I'_1}\delta_{I_2I'_2}\cdots\delta_{I_rI'_r}\delta_{m_1m'_1}\delta_{m_2m'_2}\cdots\delta_{m_rm'_r}\delta_{nn'} \tag{6.13b}$$

where $I_i, I'_i = \pm 1$; $m_i, m'_i = 1, 2, \cdots, M_i(i = 1, 2, \cdots, r)$; n, $n' = 1, 2, \cdots, N$.

For Eq. (6.12a), the truncation error can be defined as:

$$\varepsilon(\boldsymbol{M},N)=1-\frac{\int_{\boldsymbol{0}}^{\boldsymbol{L}}\int_0^T\int_{-\boldsymbol{\kappa}_\mathrm{u}}^{\boldsymbol{\kappa}_\mathrm{u}}\int_0^{\omega_\mathrm{u}}S_{f_0}(\boldsymbol{x},t;\boldsymbol{\kappa},\omega)\,\mathrm{d}\boldsymbol{x}\,\mathrm{d}t\,\mathrm{d}\boldsymbol{\kappa}\,\mathrm{d}\omega}{\int_{\boldsymbol{0}}^{\boldsymbol{L}}\int_0^T\int_{-\infty}^{\infty}\int_0^{\infty}S_{f_0}(\boldsymbol{x},t;\boldsymbol{\kappa},\omega)\,\mathrm{d}\boldsymbol{x}\,\mathrm{d}t\,\mathrm{d}\boldsymbol{\kappa}\,\mathrm{d}\omega} \tag{6.14}$$

where $\boldsymbol{M} = (M_1, M_2, \cdots, M_r)^\mathrm{T}$; $\boldsymbol{\kappa}_\mathrm{u} = (\kappa_{1\mathrm{u}}, \kappa_{2\mathrm{u}}, \cdots, \kappa_{r\mathrm{u}})^\mathrm{T}$ denotes the upper cut-off wavenumber vector. T and $\boldsymbol{L}$ denote the duration and spatial scale of the simulated spatio-temporal stochastic field, respectively. Assuming that $\boldsymbol{\kappa}_\mathrm{u}$ and ω_u remain consistent, as $M_i \to \infty$ and $N \to \infty$, we have $\Delta\kappa_i \to 0$ and $\Delta\omega \to 0$. Generally, the truncation error is usually determined as $\varepsilon(\boldsymbol{M}, N) << 1$, which can be taken as $\varepsilon(\boldsymbol{M}, N) = 0.05$ in engineering applications. According to the sampling theorem, to avoid distortion, the spatial interval Δx_i and time interval Δt of $f(x, t)$ should satisfy the following conditions (Shinozuka and Deodatis 1996):

$$\Delta x_i \le \frac{\pi}{\kappa_{i,\mathrm{u}}},\ \Delta t \le \frac{\pi}{\omega_\mathrm{u}} \tag{6.15}$$

where Δx_i represents the space interval of the i-th dimension, and Δt denotes the time interval, respectively.

6.1.2 *Some Special Cases*

The spatial dimension of the (r+1)D-1V spatio-temporal stochastic field is r. As a matter of fact, the spatial dimension r generally has three cases, say $r = 1, 2, 3$, corresponding to one-, two- and three-dimensional space. Accordingly, the spatio-temporal stochastic fields are $f_0(x, t)$ (2D-1V), $f_0(x_1, x_2, t)$ (3D-1V) and $f_0(x_1, x_2, x_3, t)$ (4D-1V), respectively.

2D-1V Spatio-temporal Stochastic Field

For the case of $r = 1$, the spatial vector degenerates into a scalar, say $\boldsymbol{x} = x$, and correspondingly, the wavenumber vector also degenerates into a scalar, say $\boldsymbol{\kappa} = \kappa$. At this time, the EWFS can be expressed as follows:

$$S_{f_0}(x,t;\kappa,\omega) = |A(x,t;\kappa,\omega)|^2 S(\kappa,\omega) \tag{6.16}$$

The corresponding auto-correlation function is given by:

$$R_{f_0}(x,t;x+\xi,t+\tau) = \int_{-\infty}^{\infty}\int_{-\infty}^{\infty} A^*(x,t;\kappa,\omega)A(x+\xi,t+\tau;\kappa,\omega)\mathrm{e}^{\mathrm{i}(\kappa\xi+\omega\tau)}S(\kappa,\omega)\mathrm{d}\kappa\mathrm{d}\omega \tag{6.17}$$

If the spatio-temporal stochastic field $f_0(x, t)$ is homogeneous and non-stationary, the EWFS degenerates to the following:

$$S_{f_0}(t,\kappa,\omega) = |A(t,\omega)|^2 S(\kappa,\omega) \tag{6.18}$$

The corresponding auto-correlation function is as follows:

$$R_{f_0}(t,t+\tau,\xi) = \int_{-\infty}^{\infty}\int_{-\infty}^{\infty} A^*(t,\omega)A(t+\tau,\omega)\mathrm{e}^{\mathrm{i}(\kappa\xi+\omega\tau)}S(\kappa,\omega)\mathrm{d}\kappa\mathrm{d}\omega \tag{6.19}$$

In the case that the spatio-temporal stochastic field $f_0(x, t)$ is non-homogeneous and stationary, the EWFS degenerates to the following:

$$S_{f_0}(x,\kappa,\omega) = |A(x,\kappa)|^2 S(\kappa,\omega) \tag{6.20}$$

The corresponding auto-correlation function is as follows:

$$R_{f_0}(x,x+\xi,\tau) = \int_{-\infty}^{\infty}\int_{-\infty}^{\infty} A^*(x,\kappa)A(x+\xi,\kappa)\mathrm{e}^{\mathrm{i}(\kappa\xi+\omega\tau)}S(\kappa,\omega)\mathrm{d}\kappa\mathrm{d}\omega \tag{6.21}$$

Therefore, the real form of 2D-1V spatio-temporal stochastic field $f_0(x, t)$ can be presented as follows (Ruan et al. 2021):

$$f(x,t) = \sum_{i=1}^{2}\sum_{m=1}^{M}\sum_{n=1}^{N} B_{mn}^{(i)}\left[X_{mn}^{(i)}\cos(I_i\kappa_m x+\omega_n t)+Y_{mn}^{(i)}\sin(I_i\kappa_m x+\omega_n t)\right] \tag{6.22a}$$

$$B_{mn}^{(i)} = \sqrt{2S_{f_0}(x,t;I_i\kappa_m,\omega_n)\Delta\kappa\Delta\omega} \tag{6.22b}$$

$$\kappa_m = (m-0.5)\Delta\kappa, \quad \omega_n = (n-0.5)\Delta\omega \tag{6.22c}$$

where $\Delta\kappa = \kappa_u/M$, $\Delta\omega = \omega_u/N$ and $I_i = (-1)^{i+1}$. $f(x, t)$ denotes the simulated spatio-temporal stochastic field. $\{X_{mn}^{(i)}, Y_{mn}^{(i)}\}$ is a set of standard orthogonal random variables, and they should satisfy the following basic conditions:

$$E\left[X_{mn}^{(i)}\right]=E\left[Y_{mn}^{(i)}\right]=0,\ E\left[X_{mn}^{(i)}Y_{kl}^{(j)}\right]=0 \tag{6.23a}$$

$$E\left[X_{mn}^{(i)}X_{kl}^{(j)}\right]=E\left[Y_{mn}^{(i)}Y_{kl}^{(j)}\right]=\delta_{ij}\delta_{mk}\delta_{nl} \tag{6.23b}$$

where $i, j = 1, 2$; $m, k = 1, 2, \cdots, M$ and $n, l = 1, 2, \cdots, N$.

3D-1V Spatio-temporal Stochastic Field

For the case of $r = 2$, the spatial vector is $\boldsymbol{x} = (x_1, x_2)^{\mathrm{T}}$, and correspondingly, the wavenumber vector is $\boldsymbol{\kappa} = (\kappa_1, \kappa_2)^{\mathrm{T}}$. At this time, the EWFS can be written as follows:

$$S_{f_0}(x_1,x_2,t;\kappa_1,\kappa_2,\omega)=|A(x_1,x_2,t;\kappa_1,\kappa_2,\omega)|^2 S(\kappa_1,\kappa_2,\omega) \tag{6.24}$$

The corresponding auto-correlation function is as follows:

$$\begin{aligned}R_{f_0}(x_1,x_2,t;x_1+\xi_1,x_2+\xi_2,t+\tau)=&\int_{-\infty}^{\infty}\int_{-\infty}^{\infty}\int_{-\infty}^{\infty}A^*(x_1,x_2,t;\kappa_1,\kappa_2,\omega)A(x_1+\xi_1,x_2+\xi_2,t+\tau;\kappa_1,\kappa_2,\omega)\\&\cdot e^{\mathrm{i}(\kappa_1\xi_1+\kappa_2\xi_2+\omega\tau)}S(\kappa_1,\kappa_2,\omega)\mathrm{d}\kappa_1\mathrm{d}\kappa_2\mathrm{d}\omega\end{aligned} \tag{6.25}$$

The 3D-1V spatio-temporal stochastic field $f_0(x_1, x_2, t)$ can be presented as follows:

$$\begin{aligned}f(x_1,x_2,t)=&\sum_{\substack{I_i=\pm1\\i=1,2}}\sum_{m_1=1}^{M_1}\sum_{m_2=1}^{M_2}\sum_{n=1}^{N}B_{m_1m_2n}^{I_1I_2}\Big[X_{m_1m_2n}^{I_1I_2}\cos(I_1\kappa_{1m_1}x_1+I_2\kappa_{2m_2}x_2+\omega_nt)\\&+Y_{m_1m_2n}^{I_1I_2}\sin(I_1\kappa_{1m_1}x_1+I_2\kappa_{2m_2}x_2+\omega_nt)\Big]\end{aligned} \tag{6.26a}$$

$$B_{m_1m_2n}^{I_1I_2}=\sqrt{2S_{f_0}(x_1,x_2,t;I_1\kappa_{1m_1},I_2\kappa_{2m_2},\omega_n)\Delta\kappa_1\Delta\kappa_2\Delta\omega} \tag{6.26b}$$

$$\kappa_{im_i}=(m_i-0.5)\Delta\kappa_i,\quad i=1,2;\quad \omega_n=(n-0.5)\Delta\omega \tag{6.26c}$$

where $f(x_1, x_2, t)$ denotes the simulated spatio-temporal stochastic field. $\left\{X_{m_1m_2n}^{I_1I_2}, Y_{m_1m_2n}^{I_1I_2}\right\}$ is a set of standard orthogonal random variables, and they should satisfy the following basic conditions:

$$E\left[X_{m_1m_2n}^{I_1I_2}\right]=E\left[Y_{m_1m_2n}^{I_1I_2}\right]=0,\ E\left[X_{m_1m_2n}^{I_1I_2}Y_{m_1'm_2'n'}^{I_1'I_2'}\right]=0 \tag{6.27a}$$

$$E\left[X_{m_1m_2n}^{I_1I_2}X_{m_1'm_2'n'}^{I_1'I_2'}\right]=E\left[X_{m_1m_2n}^{I_1I_2}X_{m_1'm_2'n'}^{I_1'I_2'}\right]=\delta_{I_1I_1'}\delta_{I_2I_2'}\delta_{m_1m_1'}\delta_{m_2m_2'}\delta_{nn'} \tag{6.27b}$$

where $I_i, I_i' = \pm 1$; $m_i = 1, 2, \cdots, M_i (i = 1, 2)$ and $n, n' = 1, 2, \cdots, N$.

Obviously, as can be seen from Eqs. (6.13), (6.23) and (6.27), the specific probability distributions for the sets of standard orthogonal random variables $\left\{X_{m_1m_2\cdots m_rn}^{I_1I_2\cdots I_r}, Y_{m_1m_2\cdots m_rn}^{I_1I_2\cdots I_r}\right\}$, $\left\{X_{mn}^{(i)}, Y_{mn}^{(i)}\right\}$ and $\left\{X_{m_1m_2n}^{I_1I_2}, Y_{m_1m_2n}^{I_1I_2}\right\}$ have not been provided. Consequently, Eqs. (6.12a), (6.22a) and (6.26a) cannot be utilized directly for simulation, leading to them can be designated as the original frequency-wavenumber spectral representation (O-FWSR) of spatio-temporal stochastic field. In essence, the O-FWSR of spatio-temporal stochastic field can be regarded as an extension of the original spectral decomposition representation for stochastic processes. It also can be seen that the randomness degrees of Eqs. (6.12a), (6.22a) and (6.26a) are $2^{r+1}M_1M_2\cdots M_rN$, $4MN$ and $8M_1M_2N$, respectively. Next, we will delve into the conventional Monte Carlo simulation method and the updated dimension-reduction simulation approach for spatio-temporal stochastic field in detail.

6.2 Conventional Monte Carlo Methods of Spatio-temporal Stochastic Field

In the O-FWSR of spatio-temporal stochastic field, say Eqs. (6.12a), (6.22a) and (6.26a), the sets of standard orthogonal random variables $\left\{X_{m_1m_2\cdots m_rn}^{I_1I_2\cdots I_r}, Y_{m_1m_2\cdots m_rn}^{I_1I_2\cdots I_r}\right\}$, $\left\{X_{mn}^{(i)}, Y_{mn}^{(i)}\right\}$ and $\left\{X_{m_1m_2n}^{I_1I_2}, Y_{m_1m_2n}^{I_1I_2}\right\}$ just need to satisfy the basic conditions defined in Eqs. (6.13), (6.23) and (6.27), which can be regarded as a necessary condition for simulation. In order to implement the simulation, the probability distributions of the standard orthogonal random variables should be specified. Similar to the conventional Monte Carlo methods for univariate and multivariate stochastic processes described in Chaps. 3 and 5, the conventional simulation approaches of spatio-temporal stochastic field also include two alternatives, say the conventional random amplitude simulation formula (also regarded as the non-deterministic spectral amplitude method) and the conventional random phase simulation formula (also regarded as the deterministic spectral amplitude method). For simplicity, this chapter will just present the conventional random phase simulation formula for spatio-temporal stochastic field.

Taking $(r+1)$D-1Vspatio-temporal stochastic field, say Eq. (6.12a), as an example for illustrative purpose, the set of standard orthogonal random variables $\left\{X_{m_1m_2\cdots m_rn}^{I_1I_2\cdots I_r}, Y_{m_1m_2\cdots m_rn}^{I_1I_2\cdots I_r}\right\}$ can be defined as follows:

$$X_{m_1m_2\cdots m_rn}^{I_1I_2\cdots I_r}=\sqrt{2}\cos\varphi_{m_1m_2\cdots m_rn}^{I_1I_2\cdots I_r},\ Y_{m_1m_2\cdots m_rn}^{I_1I_2\cdots I_r}=-\sqrt{2}\sin\varphi_{m_1m_2\cdots m_rn}^{I_1I_2\cdots I_r} \tag{6.28}$$

where $I_i, I_i'=\pm 1$; $m_i, m_i'=1,2,\cdots,M_i(i=1,2,\cdots,r)$; $n,\ n'=1,\ 2,\ \cdots,\ N$. $\left\{\varphi_{m_1m_2\cdots m_rn}^{I_1I_2\cdots I_r}\right\}$ is a set of mutually independent random phases uniformly distributed over the interval $[0,2\pi)$. Obviously, the set of random variables defined in Eq. (6.28) can fully satisfy the basic conditions involved in Eq. (6.13).

Through substituting Eq. (6.28) into Eq. (6.12a), the conventional random phase simulation formula, also recognized as the deterministic spectral amplitude method, for the (r+1)D-1V spatio-temporal stochastic field can be obtained as follows (Deodatis and Shinozuka 1989):

$$f_1(\boldsymbol{x},t)=\sqrt{2}\sum_{\substack{I_i=\pm 1\\ i=1,2,\cdots,r}}\sum_{m_1=1}^{M_1}\sum_{m_2=1}^{M_2}\cdots\sum_{m_r=1}^{M_r}\sum_{n=1}^{N}B_{m_1m_2\cdots m_rn}^{I_1I_2\cdots I_r}\cos\left(\widetilde{\boldsymbol{\kappa}}_m\cdot\boldsymbol{x}+\omega_nt+\varphi_{m_1m_2\cdots m_rn}^{I_1I_2\cdots I_r}\right) \tag{6.29}$$

where $f_1(\boldsymbol{x},t)$ is the (r+1)D-1V spatio-temporal stochastic field simulated by conventional Monte Carlo method. It should be noted that the randomness degree of $f_1(\boldsymbol{x},t)$ is $2^{r+1}M_1M_2\cdots M_rN/2$, which is half of that in the O-FWSR, say Eq. (6.12a).

In the same way, for the O-FWSR of the 2D-1V spatio-temporal stochastic field, say Eq. (6.22a), the set of standard orthogonal random variables $\left\{X_{mn}^{(i)}, Y_{mn}^{(i)}\right\}$ can be defined as follows:

$$X_{mn}^{(i)}=\sqrt{2}\cos\varphi_{mn}^{(i)},\quad Y_{mn}^{(i)}=-\sqrt{2}\sin\varphi_{mn}^{(i)} \tag{6.30}$$

where $m=1,2,\cdots,M$; $n=1,2,\cdots N$; $i=1,2$. $\left\{\varphi_{mn}^{(i)}\right\}$ is a set of mutually independent random phases uniformly distributed over the interval $[0,2\pi)$. Obviously, the set of random variables defined in Eq. (6.30) can fully satisfy the basic conditions involved in Eq. (6.23).

Then, by means of substituting Eq. (6.30) into Eq. (6.22a), the conventional random phase simulation formula, also recognized as the deterministic spectral amplitude method, for the 2D-1V spatio-temporal stochastic field can be obtained as follows (Deodatis and Shinozuka 1989; Ruan et al. 2021):

$$f_1(x,t)=\sqrt{2}\sum_{i=1}^{2}\sum_{m=1}^{M}\sum_{n=1}^{N}B_{mn}^{(i)}\cos\left(I_i\kappa_mx+\omega_nt+\varphi_{mn}^{(i)}\right) \tag{6.31}$$

where $f_1(x,t)$ is the 2D-1V spatio-temporal stochastic field simulated by conventional Monte Carlo method. It should be noted that the randomness degree of $f_1(\boldsymbol{x},t)$ is $2MN$, which is half of that in the O-FWSR, say Eq. (6.22a).

From the derivation process of the conventional Monte Carlo method for spatio-temporal stochastic field described above, it is evident that, regardless of whether it pertains to the (r+1)D-1V or 2D-1V spatio-temporal stochastic field, the conventional random simulation formulas (Eqs. (6.29) and (6.31)) are both derived from the O-FWSR (Eqs. (6.12a) and (6.22a)). Consequently, the conventional Monte Carlo method essentially represents a special case of the O-FWSR. In this way, the relationship between the conventional Monte Carlo method and the O-FWSR of spatio-temporal stochastic field has been clarified.

When employing the conventional Monte Carlo method for simulating the spatio-temporal stochastic field, a substantial number of random variables are necessary in general. For instance, the conventional random phase simulation formula for the 2D-1V spatio-temporal stochastic field can exhibit an extremely high degree of randomness, reaching a certain magnitude of 10^6. Meanwhile, the conventional Monte Carlo method inherently relies on random sampling. Although the convergence rate of the Monte Carlo method is not affected by the number of random variables, its slow speed and stochastic feature of convergence would inevitably lead to the generation of a vast number of earthquake ground motion samples. Moreover, the probability information within the sample set remains incomplete, thereby limiting its application in refined nonlinear random dynamic response analysis and reliability evaluation of engineering structures. Consequently, there is an urgent demand for developing efficient dimension-reduction simulation method for spatio-temporal stochastic field.

6.3 Dimension-Reduction Methods of Spatio-temporal Stochastic Field

6.3.1 Dimension-Reduction Simulation Methods

As evident from the above derivation, defining the standard orthogonal random variables within the O-FWSR of the spatio-temporal stochastic field as random orthogonal functions (i.e., random phase) yields the conventional Monte Carlo method. Meanwhile, the number of random variables in the conventional random phase simulation formula is merely half of that in the O-FWSR. This demonstrates that by imposing appropriate constraints on the sets of standard orthogonal random variables, the number of random variables required for simulating spatio-temporal stochastic field can be effectively reduced. In the following, the 2D-1V spatio-temporal stochastic field is taken as an example to expound the dimension-reduction simulation approach in detail. The random orthogonal function for the 3D-1V spatio-temporal stochastic field is also provided.

As for the 2D-1V spatio-temporal stochastic field, say Eq. (6.22a), the set of standard orthogonal random variables $\{X_{mn}^{(i)}, Y_{mn}^{(i)}\}$ can be defined as the random orthogonal function form of the elementary random vector $\boldsymbol{\Theta}$, that is:

$$X_{mn}^{(i)} = g_{mn}^{(i)}(\boldsymbol{\Theta}), Y_{mn}^{(i)} = h_{mn}^{(i)}(\boldsymbol{\Theta}), \quad m = 1, 2, \cdots, \quad n = 1, 2, \cdots, N; \quad i = 1, 2 \quad (6.32)$$

where $g_{mn}^{(i)}(\cdot)$ and $h_{mn}^{(i)}(\cdot)$ denote the deterministic orthogonal functions. $\boldsymbol{\Theta} = (\Theta_1, \Theta_2, \cdots, \Theta_q)$ denotes the q-dimensional elementary random vector, of which the joint PDF, say $p_{\boldsymbol{\Theta}}(\boldsymbol{\theta})$, is known.

Apparently, the defined form of Eq. (6.32) should satisfy the basic conditions involved in Eq. (6.23), which can be regarded as a necessary condition for simulation. Consequently, Eq. (6.23) can be rewritten as the following:

$$E\left[X_{mn}^{(i)}\right] = E\left[g_{mn}^{(i)}(\boldsymbol{\Theta})\right] = \int_{\Omega} g_{mn}^{(i)}(\boldsymbol{\theta}) p_{\boldsymbol{\Theta}}(\boldsymbol{\theta}) \mathrm{d}\boldsymbol{\theta} = 0 \quad (6.33a)$$

$$E\left[Y_{mn}^{(i)}\right] = E\left[h_{mn}^{(i)}(\boldsymbol{\Theta})\right] = \int_{\Omega} h_{mn}^{(i)}(\boldsymbol{\theta}) p_{\boldsymbol{\Theta}}(\boldsymbol{\theta}) \mathrm{d}\boldsymbol{\theta} = 0 \quad (6.33b)$$

$$E\left[X_{mn}^{(i)} Y_{kl}^{(j)}\right] = E\left[g_{mn}^{(i)}(\boldsymbol{\Theta}) h_{kl}^{(j)}(\boldsymbol{\Theta})\right] = \int_{\Omega} g_{mn}^{(i)}(\boldsymbol{\theta}) h_{kl}^{(j)}(\boldsymbol{\theta}) p_{\boldsymbol{\Theta}}(\boldsymbol{\theta}) \mathrm{d}\boldsymbol{\theta} = 0 \quad (6.33c)$$

$$E\left[X_{mn}^{(i)} X_{kl}^{(j)}\right] = E\left[g_{mn}^{(i)}(\boldsymbol{\Theta}) g_{kl}^{(j)}(\boldsymbol{\Theta})\right] = \int_{\Omega} g_{mn}^{(i)}(\boldsymbol{\theta}) g_{kl}^{(j)}(\boldsymbol{\theta}) p_{\boldsymbol{\Theta}}(\boldsymbol{\theta}) \mathrm{d}\boldsymbol{\theta} = \delta_{ij}\delta_{mk}\delta_{nl} \quad (6.33d)$$

$$E\left[Y_{mn}^{(i)} Y_{kl}^{(j)}\right] = E\left[h_{mn}^{(i)}(\boldsymbol{\Theta}) h_{kl}^{(j)}(\boldsymbol{\Theta})\right] = \int_{\Omega} h_{mn}^{(i)}(\boldsymbol{\theta}) h_{kl}^{(j)}(\boldsymbol{\theta}) p_{\boldsymbol{\Theta}}(\boldsymbol{\theta}) \mathrm{d}\boldsymbol{\theta} = \delta_{ij}\delta_{mk}\delta_{nl} \quad (6.33e)$$

where $i, j = 1, 2$; $m, k = 1, 2, \cdots, M$; $n, l = 1, 2, \cdots, N$; Ω is the definition space domain of $\boldsymbol{\Theta}$.

Here, the form of random orthogonal functions involving merely two elementary random variables is addressed for convenience, such that (Ruan et al. 2021):

$$\begin{cases} \overline{X}_{kl}^{(i)} = \sqrt{2}\cos(I_i k\Theta_1 + l\Theta_2) \\ \overline{Y}_{kl}^{(i)} = -\sqrt{2}\sin(I_i k\Theta_1 + l\Theta_2) \end{cases}, \quad k = 1, 2, \cdots, M; \quad l = 1, 2, \cdots, N; \quad i = 1, 2 \quad (6.34)$$

where $I_i = (-1)^{i-1}(i = 1, 2)$. Θ_1 and Θ_2 are mutually independent elementary random variables uniformly distributed over the interval $[0, 2\pi)$.

It is easy to verify that the set of standard orthogonal random variables $\left\{\overline{X}_{kl}^{(i)}, \overline{Y}_{kl}^{(i)}\right\}$ defined in Eq. (6.34) fully satisfy the basic conditions included in Eq. (6.33), say the necessary conditions for simulation have been fulfilled. Furthermore, $\left\{\overline{X}_{kl}^{(i)}, \overline{Y}_{kl}^{(i)}\right\}$ defined in Eq. (6.34) are required to be transformed into the target set of standard orthogonal random variables $\{X_{mn}^{(i)}, Y_{mn}^{(i)}\}$ in Eq. (6.22a) through a deterministic one-to-one mapping relationship, say $\overline{X}_{kl}^{(i)} \to X_{mn}^{(i)}$ and $\overline{Y}_{kl}^{(i)} \to Y_{mn}^{(i)}$. Here, the transformation of $\overline{X}_{kl}^{(1)} \to X_{mn}^{(1)}$ is taken as an example to explain the procedure of the deterministic one-to-one mapping. Firstly, define the double-subscript random variables $X_{mn}^{(1)}$ and $\overline{X}_{kl}^{(1)}$ as the single-subscript random variables $X_p^{(1)}$ and $\overline{X}_q^{(1)}$, respectively, that is $X_p^{(1)}=X_{mn}^{(1)}$ and $\overline{X}_q^{(1)}=\overline{X}_{kl}^{(1)}$, where $p=m+(n-1)\times M$ and $q=k+(l-1)\times M$. In this way, the deterministic one-to-one mapping relationship can be conveniently established using the built-in functions rand('*state*',0) and randperm(·) from the MATLAB toolbox, namely by defining temp = randperm($M\times N$) and $X_p^{(1)}=\overline{X}_q^{(1)}$, while the one-to-one mapping relationship between p and q is determined by p = temp(q). In fact, this deterministic one-to-one mapping serves as a sufficient condition for the dimension-reduction simulation approach of 2D-1V non-stationary spatio-temporal stochastic field. Consequently, the target set of standard orthogonal random variables $\{X_{mn}^{(i)}, Y_{mn}^{(i)}\}$ can be uniquely identified.

Moreover, by substituting Eq. (6.34) into the O-FWSR of the 2D-1V non-stationary spatio-temporal stochastic field, say Eq. (6.22a), the dimension-reduction representation can be obtained as follows:

$$f_2(x,t)=\sqrt{2}\sum_{i=1}^{2}\sum_{m=1}^{M}\sum_{n=1}^{N}B_{mn}^{(i)}\cos\left[I_i\kappa_m x+\omega_n t+\alpha_{mn}^{(i)}(\boldsymbol{\Theta})\right] \tag{6.35a}$$

$$\alpha_{mn}^{(i)}(\boldsymbol{\Theta})=\overline{\alpha}_{kl}^{(i)}(\boldsymbol{\Theta})=I_ik\Theta_1+l\Theta_2 \tag{6.35b}$$

where $I_i=(-1)^{i+1}$. $f_2(x, t)$ is the 2D-1V spatio-temporal stochastic field simulated by dimension-reduction simulation method. The random variables Θ_1 and Θ_2 describe the randomness of phase. The deterministic one-to-one mapping relationship between $\alpha_{mn}^{(i)}$ and $\overline{\alpha}_{kl}^{(i)}$ is the same as the mapping relationship between $X_{mn}^{(i)}$ and $\overline{X}_{kl}^{(i)}$.

As for the 3D-1V spatio-temporal stochastic field, say Eq. (6.26a), the form of random orthogonal function involving merely three elementary random variables is presented as follows:

$$\begin{cases}\overline{X}_{k_1k_2l}^{I_1I_2}=\sqrt{2}\cos(I_1k_1\Theta_1+I_2k_2\Theta_2+l\Theta_3)\\ \overline{Y}_{k_1k_2l}^{I_1I_2}=-\sqrt{2}\sin(I_1k_1\Theta_1+I_2k_2\Theta_2+l\Theta_3)\end{cases},\quad k_1=1,2,\cdots,M_1;\ k_2=1,2,\cdots,M_2;\ l=1,2,\cdots,N \tag{6.36}$$

where Θ_1, Θ_2 and Θ_3 are mutually independent elementary random variables uniformly distributed over the interval $[0, 2\pi)$.

Similarly, the initial set of standard orthogonal random variables $\left\{\overline{X}_{k_1k_2l}^{I_1I_2}, \overline{Y}_{k_1k_2l}^{I_1I_2}\right\}$ defined in Eq. (6.36) is required to be transformed into the target set of standard orthogonal random variables $\left\{X_{m_1m_2n}^{I_1I_2}, Y_{m_1m_2n}^{I_1I_2}\right\}$ in Eq. (6.26a) through a deterministic one-to-one mapping relationship, say $\overline{X}_{k_1k_2l}^{I_1I_2}\rightarrow X_{m_1m_2n}^{I_1I_2}$ and $\overline{Y}_{k_1k_2l}^{I_1I_2}\rightarrow Y_{m_1m_2n}^{I_1I_2}$. Here, the transformation of $\overline{X}_{k_1k_2l}^{I_1I_2}\rightarrow X_{m_1m_2n}^{I_1I_2}$ is taken as an example to explain the procedure of the deterministic one-to-one mapping. Firstly, define the triple-subscript random variables $X_{m_1m_2n}^{I_1I_2}$ and $\overline{X}_{k_1k_2l}^{I_1I_2}$ as the single-subscript random variables $X_p^{I_1I_2}$ and $\overline{X}_q^{I_1I_2}$, respectively, that is $X_p^{I_1I_2}=X_{m_1m_2n}^{I_1I_2}$ and $\overline{X}_q^{I_1I_2}=\overline{X}_{k_1k_2l}^{I_1I_2}$, where

$p = (m_1-1)M_2N + (m_2-1)N + n$, $q = (k_1-1)M_2N + (k_2-1)N + l$; $m_1, k_1 = 1, 2, \cdots, M_1$; $m_2, k_2 = 1, 2, \cdots, M_2$; $n, l = 1, 2, \cdots, N$. In this way, the deterministic one-to-one mapping relationship can be conveniently established using the built-in functions rand('*state*',0) and randperm(·) from the MATLAB toolbox, namely by defining temp = randperm($M_1 \times M_2 \times N$) and $X_p^{I_1 I_2} = \overline{X}_q^{I_1 I_2}$, while the one-to-one mapping relationship between p and q is determined by p = temp(q). In fact, this deterministic one-to-one mapping serves as a sufficient condition for the dimension-reduction simulation approach of 3D-1V non-stationary spatio-temporal stochastic field. Consequently, the target set of standard orthogonal random variables $\left\{X_{m_1 m_2 n}^{I_1 I_2}, Y_{m_1 m_2 n}^{I_1 I_2}\right\}$ can be uniquely identified.

Then, by substituting Eq. (6.36) into the O-FWSR of the 3D-1V spatio-temporal stochastic field, say Eq. (6.26a), the dimension-reduction presentation can be obtained as follows:

$$f_2(x_1,x_2,t) = \sum_{\substack{I_i=\pm 1 \\ i=1,2}} \sum_{m_1=1}^{M_1} \sum_{m_2=1}^{M_2} \sum_{n=1}^{N} B_{m_1 m_2 n}^{I_1 I_2} \cos\left[I_1 \kappa_{1m_1} x_1 + I_2 \kappa_{2m_2} x_2 + \omega_n t + \alpha_{m_1 m_2 n}^{I_1 I_2}(\boldsymbol{\Theta})\right] \tag{6.37a}$$

$$\alpha_{m_1 m_2 n}^{I_1 I_2}(\boldsymbol{\Theta}) = \overline{\alpha}_{k_1 k_2 l}^{I_1 I_2}(\boldsymbol{\Theta}) = I_1 k_1 \Theta_1 + I_2 k_2 \Theta_2 + l\Theta_3 \tag{6.37b}$$

where $f_2(x_1, x_2, t)$ is the 3D-1V spatio-temporal stochastic field simulated by dimension-reduction simulation method. The deterministic one-to-one mapping relationship between $\alpha_{m_1 m_2 n}^{I_1 I_2}$ and $\overline{\alpha}_{k_1 k_2 l}^{I_1 I_2}$ is the same as the mapping relationship between $X_{m_1 m_2 n}^{I_1 I_2}$ and $\overline{X}_{k_1 k_2 l}^{I_1 I_2}$.

In the O-FWSR of spatio-temporal stochastic field, efficient dimension-reduction simulation can be accomplished by representing the standard orthogonal random variables as the constraint form of random orthogonal functions. Specifically, for a 2D-1V spatio-temporal stochastic field, the level of randomness decreases from the dimension of millions to just 2. And for a 3D-1V spatio-temporal stochastic field, the randomness degree drops from the dimension of billions to merely 3. Similarly, in the dimension-reduction methods, the orthogonality condition defined in Eq. (6.33) that the random orthogonal function must adhere to serves as a necessary condition for simulation. Meanwhile, the one-to-one mapping relationship between the initial set of standard orthogonal random variables and the target set of standard orthogonal random variables constitutes a sufficient condition for simulation. Besides, this section solely introduces the random orthogonal functions in terms of trigonometric form. Actually, akin to the previous chapters, various types of random orthogonal functions can be constructed, such as those corresponding to the conventional random amplitude method (also regarded as non-deterministic spectral amplitude method) (Liu et al. 2025) or orthogonal polynomial functions, provided that the necessary condition is satisfied. This showcases the flexibility and diversity of dimension-reduction simulation approaches.

6.3.2 Introduction of Fast Fourier Transform Technology

It is ready to find that the dimension-reduction representation of spatio-temporal stochastic field, say Eq. (6.35a), has the same expression form as the conventional FWSR, say Eq. (6.31), thus the Fast Fourier Transform (FFT) technology can be further introduced here to improve the simulation efficiency. In the following, we will elaborate on the specific implementation approaches for integrating the FFT technology into the simulation of 2D-1V and 3D-1V spatio-temporal stochastic fields based on the dimension-reduction method.

2D-1V Spatio-temporal Stochastic Field

In the case that $f_0(x, t)$ represents a non-stationary and non-homogeneous spatio-temporal stochastic field, the FFT algorithm cannot be directly utilized because its EFWS $S_{f_0}(x, t; \kappa, \omega)$ encompasses both spatial and temporal variables (Deodatis 1996), which poses a common challenge in the simulation of non-stationary stochastic process or spatio-temporal stochastic field. To address this issue, the variable separation method proposed by Peng et al. (2017) can be applied here to accomplish the decomposition of the square root upon the EFWS $S_{f_0}(x, t; \kappa, \omega)$ based on the Proper Orthogonal Decomposition (POD).

As stated in Sect. 6.1, the EFWS of the 2D-1V spatio-temporal stochastic field can be constructed using Eq. (6.16), which is obtained by multiplying the square of the deterministic modulation function, i.e., $A(x, t; \kappa, \omega)$, by the FWS, i.e., $S(\kappa, \omega)$, of the corresponding stationary and homogeneous spatio-temporal stochastic field. Among that, $S(\kappa, \omega)$ can be achieved by the following expression (Zerva 1992a):

$$S(\kappa, \omega) = \frac{1}{2\pi}\int_{-\infty}^{\infty} S(\omega)\gamma(\xi, \omega)\mathrm{e}^{-\mathrm{i}\kappa\xi}\mathrm{d}\xi = S(\omega)\chi(\kappa, \omega) \tag{6.38}$$

where $\gamma(\xi, \omega)$ represents the spatial coherence function, which is generally a complex function symmetric about the origin, say $\gamma(-\xi, -\omega) = \gamma^*(\xi, \omega)$. $S(\omega)$ denotes the two-sided power spectrum density (PSD). $\chi(\kappa, \omega)$ is the Fourier transform of the coherence function $\gamma(\xi, \omega)$.

For the ground motion spatio-temporal stochastic field, the modulation function can be considered a real-valued function that is independent of wavenumberκ, i.e., $A(x, t, \omega)$, for the sake of simplicity. Under such circumstances, the EFWS of the spatio-temporal stochastic field can be derived by multiplying the Fourier integral of the EPSD function with the coherence function, given by:

$$S_{f_0}(x,t;\kappa,\omega)=A^2(x,t,\omega)S(\kappa,\omega)=\widehat{S}(x,t,\omega)\chi(\kappa,\omega) \tag{6.39}$$

where $\widehat{S}(x,t,\omega)=A^2(x,t,\omega)S(\omega)$ denotes the evolutionary PSD related to location.

As indicated by Eq. (6.39), $\chi(\kappa,\omega)$ does not incorporate spatial or temporal variables. Consequently, the factorization of the square root in the EFWS can be realized through the factorization of $H(x,t,\omega)=\sqrt{\widehat{S}(x,t,\omega)}$. By employing the factorization method put forward by Peng et al. (2017), the EFWS of the spatio-temporal stochastic field can be represented as:

$$\sqrt{S_{f_0}(x,t;\kappa,\omega)}\approx\sum_{p=1}^{N_p}a_p(x,t)\eta_p(\kappa,\omega) \tag{6.40a}$$

in which

$$\eta_p(\kappa,\omega)=\psi_p(\omega)\sqrt{\chi(\kappa,\omega)} \tag{6.40b}$$

where $a_p(x,t)$ is a deterministic function of temporal and spatio variables. $\psi_p(\omega)$ denotes the even function of frequency. N_p represents the number of reconstructed truncation terms. Generally, the order of magnitude of N_p is 10^0.

Through substituting Eq. (6.40a) into Eq. (6.35a), we can obtain

$$f_2(x,t)\approx\sum_{p=1}^{N_p}a_p(x,t)\overline{f}_p(x,t) \tag{6.41}$$

where

$$\overline{f}_p(x,t)=2\sqrt{\Delta\kappa\Delta\omega}\sum_{i=1}^{2}\sum_{m=1}^{M}\sum_{n=1}^{N}\eta_p(I_i\kappa_m,\omega_n)\cos\left[I_i\kappa_m x+\omega_n t+\alpha_{mn}^{(i)}(\boldsymbol{\Theta})\right] \tag{6.42}$$

As can be observed from Eq. (6.41), a non-stationary and non-homogeneous spatio-temporal stochastic field can be approximately represented as the sum of a finite number (N_p) of stationary and homogeneous spatio-temporal stochastic fields $\overline{f}_p(x,t)$ modulated by the deterministic function of temporal and spatio variables, say $a_p(x,t)$. Therefore, the FFT algorithm can be utilized on $\overline{f}_p(x,t)$ in Eq. (6.42) to facilitate the rapid simulation of non-stationary and non-homogeneous spatio-temporal stochastic field.

For the stationary and homogeneous spatio-temporal stochastic field $f_0(x,t)$, its FWS is $S(\kappa,\omega)$. By means of applying the FFT algorithm, Eq. (6.35a) can be rewritten as follows (Ruan et al. 2021):

$$f_2(p_1\Delta x,p_2\Delta t)=\mathrm{Re}\Bigg\{\exp\left[\mathrm{i}\left(\frac{\pi p_1}{J}+\frac{\pi p_2}{K}\right)\right]\sum_{m=1}^{J}\sum_{n=1}^{K}C_{mn}^{(1)}\exp\left[\mathrm{i}\frac{2\pi(m-1)q_1}{J}+\mathrm{i}\frac{2\pi(n-1)q_2}{K}\right]$$
$$+\exp\left[\mathrm{i}\left(-\frac{\pi p_1}{J}+\frac{\pi p_2}{K}\right)\right]\sum_{m=1}^{J}\sum_{n=1}^{K}C_{mn}^{(2)}\exp\left[-\mathrm{i}\frac{2\pi(m-1)q_1}{J}+\mathrm{i}\frac{2\pi(n-1)q_2}{K}\right]\Bigg\} \tag{6.43}$$

in which

$$C_{mn}^{(i)}=\begin{cases}2\sqrt{S\left[(-1)^{i+1}\kappa_m,\omega_n\right]\Delta\kappa\Delta\omega}\exp\left[-\mathrm{i}\alpha_{mn}^{(i)}(\boldsymbol{\Theta})\right], & 1\le m\le M \text{ and } 1\le n\le N\\ 0, & M<m\le J \text{ or } N<n\le K\end{cases} \tag{6.44}$$

where $i = 1, 2$. Δx and Δt denote the space interval and time interval for simulation, respectively, which satisfy $\Delta x = 2\pi/(J\Delta\kappa)$ and $\Delta t = 2\pi/(K\Delta\omega)$. Re $[\cdot]$ denotes taking the real part. $p_1 = 0, 1, 2, \cdots, (2J-1)$, $p_2 = 0, 1, 2, \cdots, (2K-1)$; $q_1 = \mod(p_1, J)$, $q_2 = \mod(p_2, K)$, where mod$(\cdot)$ denotes taking the remainder. In order to avoid aliasing effects, the number of FFT transformation points J and K should satisfy the following conditions:

$$J\ge 2N,\quad K\ge 2M \tag{6.45}$$

3D-1V Spatio-temporal Stochastic Field

As stated in Sect. 6.1, the EFWS of the 3D-1V spatio-temporal stochastic field can be constructed using Eq. (6.24), which is obtained by multiplying the square of the deterministic modulation function, i.e., $A(x_1, x_2, t; \kappa_1, \kappa_2, \omega)$, by the FWS, i.e., $S(\kappa_1, \kappa_2, \omega)$, of the corresponding stationary and homogeneous spatio-temporal stochastic field. Among that, $S(\kappa_1, \kappa_2, \omega)$ can be achieved by the following expression:

$$S(\kappa_1,\kappa_2,\omega)=\frac{1}{4\pi^2}\int_{-\infty}^{\infty}\int_{-\infty}^{\infty}S(\omega)\gamma(\xi_1,\xi_2,\omega)\mathrm{e}^{-\mathrm{i}(\kappa_1\xi_1+\kappa_2\xi_2)}\mathrm{d}\xi_1\mathrm{d}\xi_2 \tag{6.46}$$

where $\gamma(\xi_1, \xi_2, \omega)$ denotes the two-dimensional spatial coherence function, which is related to the space interval ξ_1, ξ_2 and frequency ω.

In Eq. (6.46), it is required to calculate the double Fourier transform of the two-dimensional spatial coherence function. For simplicity, Deodatis and Shinozuka (1989) considered the nondispersive Rayleigh wave. In this case, there is a relationship between frequency and wavenumber, say $\omega/c = \kappa = \sqrt{\kappa_1^2+\kappa_2^2}$, where c is the phase velocity of the Rayleigh wave. Therefore, the FWS $S(\kappa_1, \kappa_2, \omega)$ can be transformed into the form of wavenumber spectrum $S(\kappa_1, \kappa_2)$, such that:

$$S(\kappa_1, \kappa_2, \omega) = S(\kappa_1, \kappa_2)\,\delta[\omega - g(\kappa_1, \kappa_2)] \tag{6.47}$$

where the wavenumber spectrum $S(\kappa_1, \kappa_2)$ can apply the model proposed by Shinozuka (1987).

Equation (6.47) posits the presence of a dominant large peak in the spectrum, enabling the wavefield to be characterized by a plane wave. Consequently, the simulation formula is simplified to merely necessitate summation within the wavenumber domain, thereby significantly reducing the computational burden of the simulation.

In fact, for certain special types of coherence functions, such as the model proposed by Loh and Lin (1990), the coherence function can be expressed as the product of the two one-dimensional coherence functions, namely $\gamma(\xi_1, \xi_2, \omega) = \gamma_1(\xi_1, \omega)\gamma_2(\xi_2, \omega)$. For the sake of simplicity, just considering the non-stationary and homogeneous case, in which the EFWS can be written as follows:

$$\begin{aligned} S_{f_0}(t; \kappa_1, \kappa_2, \omega) &= A^2(t, \omega) S(\kappa_1, \kappa_2, \omega) \\ &= A^2(t, \omega)\chi_1(\kappa_1, \omega)\chi_2(\kappa_2, \omega) S(\omega) \\ &= S(t, \omega)\chi_1(\kappa_1, \omega)\chi_2(\kappa_2, \omega) \end{aligned} \tag{6.48}$$

where $\chi_1(\kappa, \omega)$ and $\chi_2(\kappa, \omega)$ are the Fourier transforms of $\gamma_1(\xi, \omega)$ and $\gamma_2(\xi, \omega)$, respectively.

In Eq. (6.48), it can be seen that just the first term on the right-hand side of the third equality sign involves the time variable. Therefore, similar to the 2D-1V case, we can directly factorize the square root of the EFWS as follows:

$$\sqrt{S_{f_0}(t; \kappa_1, \kappa_2, \omega)} = \sum_{p=1}^{N_p} a_p(t)\eta_p(\kappa_1, \kappa_2, \omega) \tag{6.49a}$$

in which

$$\eta_p(\kappa_1, \kappa_2, \omega) = \sqrt{\chi_1(\kappa_1, \omega)\chi_2(\kappa_2, \omega)}\,\psi_p(\omega) \tag{6.49b}$$

where $a_p(t)$ denotes a deterministic function of temporal variable.

Through substituting Eq. (6.49a) into Eq. (6.37a), we have:

$$f_2(x_1, x_2, t) \approx \sum_{p=1}^{N_p} a_p(t)\bar{f}_p(x_1, x_2, t) \tag{6.50}$$

in which

$$\bar{f}_p(x_1,x_2,t)=2\sqrt{\Delta\kappa_1\Delta\kappa_2\Delta\omega}\sum_{\substack{I_i=\pm1\\ i=1,2}}\sum_{m_1=1}^{M_1}\sum_{m_2=1}^{M_2}\sum_{n=1}^{N}\eta_p(I_1\kappa_{1m_1},I_2\kappa_{2m_2},\omega_n)\cdot \cos\left[I_1\kappa_{1m_1}x_1+I_2\kappa_{2m_2}x_2+\omega_n t+\alpha_{m_1m_2n}^{I_1I_2}(\boldsymbol{\Theta})\right] \tag{6.51}$$

Similar to Eq. (6.42), by applying the FFT technology into Eq. (6.51) for the 3D-1V stationary and homogeneous spatio-temporal stochastic field, of which the rapid simulation can thus be realized.

6.4 Numerical Examples of Spatio-temporal Stochastic Field

This section validates the effectiveness of the proposed dimension-reduction method through numerical examples involving 2D-1V non-stationary ground motion acceleration field and 3D-1V non-stationary ground motion displacement field. Among these, for the 2D-1V non-stationary seismic field, both the dimension-reduction method and the conventional Monte Carlo method are utilized to generate the same sets of representative acceleration time-histories, and the accuracy and efficiency of these two methods are compared for investigation purpose.

6.4.1 Simulation of 2D-1V Non-stationary Ground Motion Acceleration Field

Regarding the spatial coherence function of seismic ground motion field, both the lagged coherence effect and the wave passage effect are taken into account. The lagged coherence effect is represented by the Luco-Wong model as specified in Eq. (2.108), and the wave passage effect is depicted using the models presented in Eqs. (2.110) and (2.111). Consequently, the coherence function can be expressed as follows:

$$\gamma(\xi,\omega)=\exp\left(-\alpha^2\omega^2\xi^2\right)\cdot\exp\left(-\frac{\mathrm{i}\omega\xi}{v_{\mathrm{app}}}\right),\quad \omega>0 \tag{6.52}$$

where the first term on the right-hand side of Eq. (6.52) denotes the lagged coherence effect. For simplicity, define $\alpha=\frac{\eta}{v_s}$, which describes the coherence decay and is valued by $\alpha = 2\times10^{-4}$ s/m herein. The second term on the right-hand side of

Eq. (6.52) denotes the wave passage effect, where the apparent propagation velocity is taken as $v_{\mathrm{app}} = 1000$ m/s.

Through substituting Eq. (6.52) into Eq. (6.38) and carrying out the integration, the FWS of the stationary and homogeneous stochastic seismic ground motion field be derived as follows (Zerva 1992a, b):

$$S(\kappa,\omega) = \frac{S(\omega)}{2\alpha\omega\sqrt{\pi}} \exp\left(-\frac{\left(\kappa + \omega/v_{\mathrm{app}}\right)^2}{4\alpha^2\omega^2}\right), \quad \omega > 0 \tag{6.53}$$

where the PSD of ground motion acceleration process, say $S(\omega)$, takes the Clough-Penzien (C-P) model presented in Eq. (2.10). In this numerical example, consider the situation that the soil site category is II and the design earthquake group is the second group, thus the parameter values for the C-P model are determined as $\omega_{\mathrm{g}} = 5\pi$ rad/s, $\omega_{\mathrm{f}} = 0.5\pi$ rad/s, $\zeta_{\mathrm{g}} = \zeta_{\mathrm{f}} = 0.72$, $a_{\max} = 200\mathrm{cm/s}^2$ and $r = 3.0$.

For the time-frequency modulation function of the fully non-stationary ground motion, the improved Deodatis & Shinozuka model proposed by Liu et al. (2017), presented in Eq. (2.67), is applied here. The parameter of the modulation function is taken as $a = 0.2$. As a result, according to Eq. (6.18), the EFWS of the 2D-1V non-stationary ground motion field can be obtained.

It is important to note that the afore-mentioned EFWS model for the 2D-1V non-stationary and homogeneous ground motion spatio-temporal stochastic field can simultaneously account for the full time-frequency non-stationarity, lagged coherence effect and wave passage effect of the seismic ground motion field. However, since this EFWS is derived from the stationary and homogeneous FWS, it cannot incorporate local site effect in Eq. (6.52). If it is necessary to consider the non-homogeneous characteristic of the seismic ground motion field, the soil site conditions and non-homogeneous modulation function for various regions within the simulated seismic ground motion field should be specified to develop an EFWS model that can also consider the local site effect for generating 2D-1V non-stationary and non-homogeneous seismic spatio-temporal stochastic field.

To simulate the 2D-1V non-stationary and homogeneous ground motion spatio-temporal stochastic field, 1024 discrete points along the x-direction on ground surface are selected, with a distance of 10 m between two adjacent points. The simulation parameters for this case are detailed in Table 6.1.

In this numerical example, $n_{\mathrm{sel}} = 233, 377, 610, 987$ sets of representative time-histories of 2D-1V ground motion acceleration field are generated, respectively. To thoroughly investigate the accuracy and efficiency of the updated dimension-reduction simulation method, an equivalent number of samples using the conventional Monte Carlo method are also generated for comparison purpose. Then, a detailed comparison is conducted between the two strategies in terms of mean error, standard deviation (Std. D) error, and computational time cost, as detailed shown in Table 6.2. It should be clarified that the mean error and Std. D error listed in the table denote the average relative error across all the simulated action points on the ground surface, and the computational time cost indicates the

Table 6.1 Parameter values for simulating 2D-1V non-stationary ground motion acceleration field

Parameter	Value
Upper cut-off wavenumber (rad/m)	$\kappa_u = \pi/10$
Total number of wavenumber terms	$M = 1024$
Upper cut-off frequency (rad/s)	$\omega_u = 50\pi$
Total number of frequency terms	$N = 1024$
Simulated spatial scale (m)	$L = 20480$
Space interval (m)	$\Delta x = 10.0$
Simulated duration (s)	$T = 40.96$
Time interval (s)	$\Delta t = 0.02$

Table 6.2 Comparison of simulation accuracy and efficiency between dimension-reduction method and conventional Monte Carlo method

		Number of samples (n_{sel})			
Terms	Strategy	233	377	610	987
Mean error	Dimension-reduction	4.32%	3.30%	2.50%	1.44%
	Monte Carlo	5.28%	4.07%	3.22%	2.44%
Std. D error	Dimension-reduction	3.67%	2.85%	2.23%	1.73%
	Monte Carlo	3.80%	2.94%	2.29%	1.86%
Computational time cost	Dimension-reduction	490 s	775 s	1254 s	1995 s
	Monte Carlo	426 s	690 s	1150 s	1889 s

total time expense required to generate all samples. As shown in the table, when generating an equal number of seismic acceleration time-histories, the dimension-reduction method exhibits smaller mean error and Std. D error compared to the conventional Monte Carlo method. This demonstrates that the dimension-reduction method offers higher accuracy, while the computational time cost required by both methods is similar. Further analysis reveals that when generating merely 233 sets of representative time-histories, both the mean error and Std. D error of the dimension-reduction method remain below 5%. This indicates that merely over 200 sets of representative time-histories are required to fulfill the engineering accuracy requirements using the dimension-reduction method. As a result, compared to the conventional Monte Carlo method, which requires generating a larger number of samples, using the representative time-history set of ground motion acceleration process generated by the dimension-reduction method as the input for structural dynamic response analysis can significantly enhance the computational efficiency.

Figure 6.1 presents the whole representative samples of 2D-1V non-stationary seismic acceleration field generated by the dimension-reduction method. Among these, Fig. 6.1a presents the acceleration representative time-histories at all action points within the generated ground motion field. As evident from the figure, the simulated seismic field displays the typical characteristics of a pre-defined spatio-temporal stochastic field, which is homogeneous in the space domain and non-stationary in the time domain. For better observation, Fig. 6.1b displays the representative time-histories of ground motion acceleration process at three typical locations, i.e., $x_1 = 100$m, $x_2 = 140$m and $x_3 = 200$m. As can be seen from Fig. 6.1b,

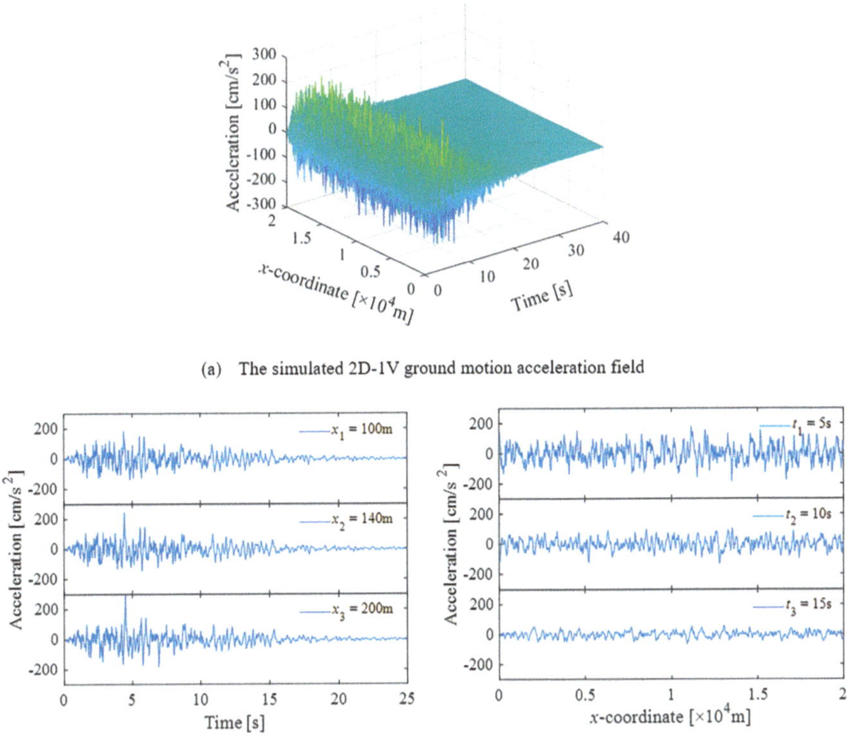

(a) The simulated 2D-1V ground motion acceleration field

(b) Three typical locations

(c) Three typical time instants

Fig. 6.1 Representative time-histories generated by the dimension-reduction method. (**a**) The simulated 2D-1V ground motion acceleration field (**b**) Three typical locations (**c**) Three typical time instants

the seismic acceleration representative time-histories exhibit typical non-stationarity. Additionally, the representative time-histories at the three locations display similar curve patterns, which can reflect the spatial correlation effect. The ground motion acceleration wave fronts at three typical time instants, say $t_1 = 5\text{s}$, $t_2 = 10\text{s}$ and $t_3 = 15\text{s}$ are described in Fig. 6.1c. Evidently, the ground motion acceleration wavefronts exhibit typical spatial non-homogeneous feature. Simultaneously, as time goes by, the ground motion intensity gradually decreases, aligning with the characteristics of the given modulation function.

Figure 6.2 presents a comparison between the mean and Std. D of representative time-history sets of ground motions at three typical locations, say $x_1 = 100\text{m}$, $x_2 = 140\text{m}$ and $x_3 = 200\text{m}$, generated by the dimension-reduction methods, and their corresponding target values. The number of generated representative time-histories is $n_{\text{sel}} = 377$. It is evident from the figure that both the mean and Std. D for these three locations closely match their respective target values.

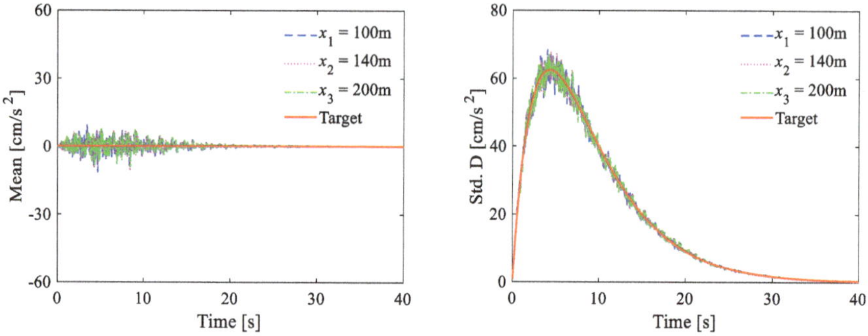

(a) Mean (b) Std. D

Fig. 6.2 Comparison of mean and standard deviation between the simulated representative time-history sets with the corresponding target values. (**a**) Mean (**b**) Std. D

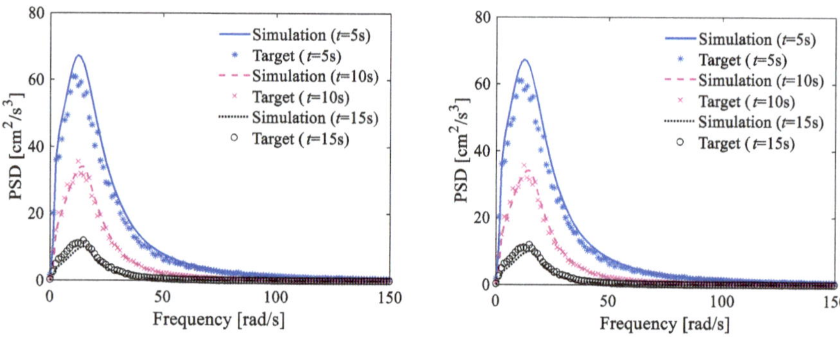

(a) $x_1 = 100\text{m}$ (b) $x_2 = 140\text{m}$

Fig. 6.3 Comparison of PSD between the simulated representative time-history sets with the corresponding target values at typical time instants. (**a**) $x_1 = 100\text{m}$ (**b**) $x_2 = 140\text{m}$

Figure 6.3 presents a comparison of the PSD between the representative time-history sets at two typical locations, say $x_1 = 100\text{m}$ and $x_2 = 140\text{m}$, and their corresponding target values at three typical time instants, say $t_1 = 5\text{s}$, $t_2 = 10\text{s}$ and $t_3 = 15\text{s}$. As evident from the figure, the simulated PSD at different locations and time instants closely align with the target values, demonstrating the effectiveness of the dimension-reduction method in simulating non-stationary ground motion stochastic spatial-temporal field.

Figure 6.4 illustrates the comparison of the spatial coherence effect between the simulated representative time-history set and the corresponding target values. As depicted in Fig. 6.4a, the simulated lagged coherence function closely matches the target value and exhibits the characteristic of rapid attenuation of coherence effects with increasing distance. In Fig. 6.4b, it is evident that, apart from the weaker coherence and significant noise interference in the high-frequency range (above

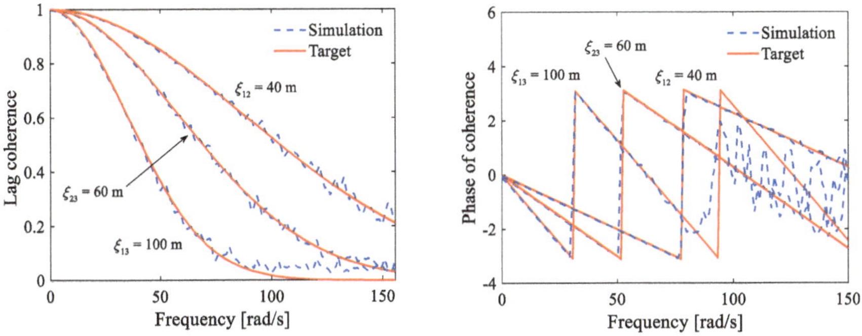

(a) Lagged coherence function (b) Phase of coherence function

Fig. 6.4 Comparison of the spatial coherence effect between the simulated representative time-history set and the corresponding target values. (**a**) Lagged coherence function (**b**) Phase of coherence function

80 rad/s), the simulated phase of the coherence function also aligns well with the target values. The aforementioned simulation results further confirm the effectiveness of the dimension-reduction method.

6.4.2 Simulation of 3D-1V Non-stationary Ground Motion Displacement Field

The representative time-histories of 3D-1V non-stationary ground motion displacement field are generated utilizing the dimension-reduction method in this numerical example. At first, the FWS can be constructed according to Eq. (6.46). For the two-dimensional spatial coherence function, it is essential to extend the Luco-Wong model for the one-dimensional lagged coherence effect to the two-dimensional case. Accounting for the wave passage effect at the meantime and assuming that seismic ground motion propagates along the main direction of the x_1-axis, the two-dimensional spatial coherence function can be expressed as follows:

$$\gamma(\xi_1, \xi_2, \omega) = \exp\left(-\alpha_2^2\omega^2\xi_2^2\right)\cdot\exp\left(-\alpha_1^2\omega^2\xi_1^2\right)\cdot\exp\left(-\frac{i\omega\xi_1}{v_{\text{app}}}\right) \tag{6.54}$$

where the model parameters are valued by $\alpha_1 = \alpha_2 = 2\times10^{-4}$s/m, say the decay rate of coherence effect in both x_1- and x_2- directions is the same. The apparent propagation velocity is taken as $v_{\text{app}} = 500$ m/s.

Through substituting Eq. (6.54) into Eq. (6.46), the FWS for the 3D-1V stationary and homogeneous seismic field can be constructed as follows:

$$S(\kappa_1,\kappa_2,\omega)=\frac{S_u(\omega)}{4\pi\alpha_1\alpha_2\omega^2}\exp\left[-\frac{\left(\kappa_1+\omega/v_{\mathrm{app}}\right)^2}{4\alpha_1^2\omega^2}\right]\exp\left(-\frac{\kappa_2^2}{4\alpha_2^2\omega^2}\right) \tag{6.55}$$

where $S_u(\omega)$ denotes the PSD of the stationary ground motion displacement process, which employs the Clough-Penzien displacement model, such that:

$$S_u(\omega)=\frac{\omega_{\mathrm{g}}^4+4\zeta_{\mathrm{g}}^2\omega_{\mathrm{g}}^2\omega^2}{\left(\omega^2-\omega_{\mathrm{g}}^2\right)^2+4\zeta_{\mathrm{g}}^2\omega_{\mathrm{g}}^2\omega^2}\cdot\frac{1}{\left(\omega^2-\omega_{\mathrm{f}}^2\right)^2+4\zeta_{\mathrm{f}}^2\omega_{\mathrm{f}}^2\omega^2}S_0 \tag{6.56}$$

In Eq. (6.56), the values of model parameters are taken as the same as Sect. 6.4.1.

For the time-frequency modulation function, the model proposed by Deodatis & Shinozuka presented in Eq. (2.67) is applied here, where the parameter value is taken as $a=0.2$. It should be clarified that since this numerical example simulates the seismic ground motion displacement field, the improved Deodatis & Shinozuka model proposed by Liu et al. (Eq. (2.69)) is not recommended herein. The reason is that the cut-off frequency of seismic displacement commonly does not exceed ω_{g}, while the frequency decay rate of the model in Eq. (2.69) reaches its peak at ω_{g}.

To simulate the 3D-1V seismic ground motion displacement field, it is necessary to discretize the two-dimensional space. As illustrated in Fig. 6.5, a certain number of 512×512 for discrete action points are selected on the plane x_1-x_2, with the assumption that the distance between adjacent discrete points is equal in both the x_1- and x_2-directions.

Other simulation parameter values can be found in Table 6.3.

In this numerical example, $n_{\mathrm{sel}}=266$ sets of representative time-histories for the 3D-1V non-stationary ground motion displacement field are generated by means of the dimension-reduction simulation method for investigation purpose. Figure 6.6

Fig. 6.5 Schematic diagram of discrete action points distributed on the plane of the simulated 3D-1V seismic ground motion field

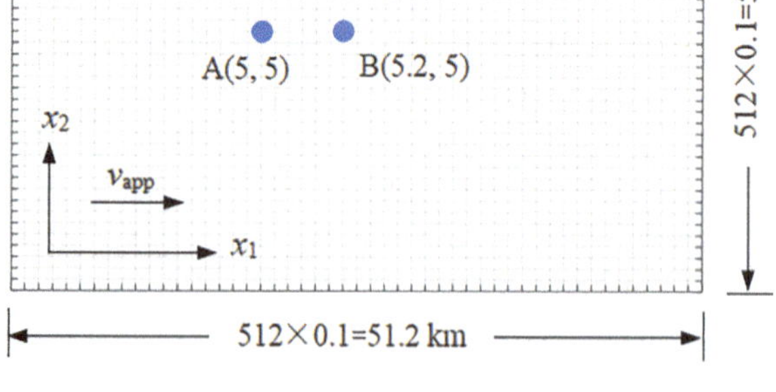

Table 6.3 Parameter values for simulating 3D-1V ground motion displacement field

Parameter	Value
Upper cut-off wavenumber along x_1-direction (rad/m)	$\kappa_{1u} = \pi/100$
Upper cut-off wavenumber along x_2 -direction (rad/m)	$\kappa_{2u} = \pi/100$
Total number of wavenumber terms along x_1-direction	$M_1 = 256$
Total number of wavenumber terms along x_2-direction	$M_2 = 256$
Upper cut-off frequency (rad/s)	$\omega_u = 5\pi$
Total number of frequency terms	$N = 128$
Simulated spatial scale along x_1-direction (m)	$L_1 = 51200$
Simulated spatial scale along x_2-direction (m)	$L_2 = 51200$
Space interval along x_1-direction (m)	$\Delta x_1 = 100$
Space interval along x_2-direction (m)	$\Delta x_2 = 100$
Simulated duration (s)	$T = 51.2$
Time interval (s)	$\Delta t = 0.2$

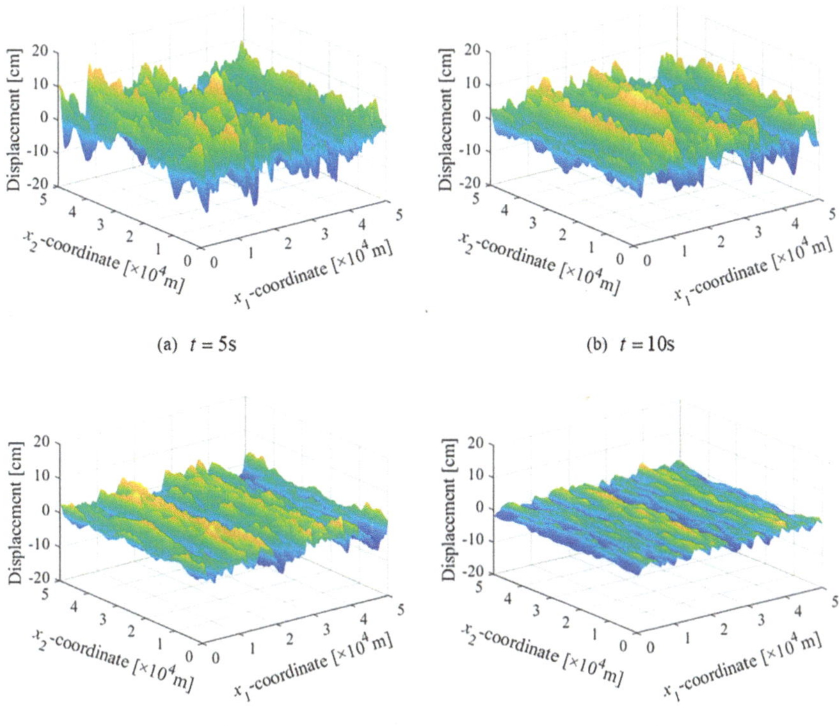

(a) $t = 5$s (b) $t = 10$s (c) $t = 15$s (d) $t = 20$s

Fig. 6.6 Wavefront of the seismic ground motion displacement field at typical time instants. (**a**) $t = 5$s (**b**) $t = 10$s (**c**) $t = 15$s (**d**) $t = 20$s depicts the wavefronts of seismic displacement field at four typical time instants, i.e., $t = 5$s, 10s, 15s, 20s, for the representative time-histories of the 100th set. As can be seen from the figures, the simulated ground motion displacement time-histories are homogeneous in both the x_1- and x_2-directions, with notable fluctuations observed in the x_1-direction. Meanwhile, it can be observed that the amplitude of the seismic displacement wavefront gradually decreases over time, reflecting the non-stationary characteristics of the seismic displacement field. The above analytical results align with the predefined characteristics of the simulated 3D-1V non-stationary seismic displacement field.

Figure 6.7 presents the displacement time-histories (the 100th set and 200th set of representative samples) of Points A, B, and C (as depicted in Fig. 6.5) within the 3D-1V non-stationary ground motion displacement field. For the convenience of observation, the displacement time-histories of the first 30 s have been captured. As shown in the figures, the displacement time-histories between Points A and B, as well as between Points A and C, display a noticeable waveform lag. In contrast, the displacement time-histories between Points B and C exhibit no lag and are almost entirely coincident. The main reason is that Points B and C share identical coordinates of x_1-direction, whereas Point A has a different coordinate of x_1-direction from these two points. Given that the wave passage effect along the x_1-direction is taken into account in this case, it leads to waveform lag between Points A and B, as well as between Points A and C. In contrast, although the coordinates of Points B and C in x_2-direction are different, due to the neglect of the wave passage effect in this direction, there is no noticeable waveform delay. Furthermore, it can be observed from Fig. 6.7b that the maximum amplitude of the displacement time-history at Point A occurs at 5.8 s, whereas at Point B, it occurs at 6.2 s. Given that the distance between these two points is 200 m, it can be calculated that the wave passage velocity is 500 m/s, which aligns precisely with the preset value of the apparent wave velocity $v_{\text{app}} = 500$ m/s in Eq. (6.54). In addition, it can also be observed from the figures that the displacement time-histories at different action points within the

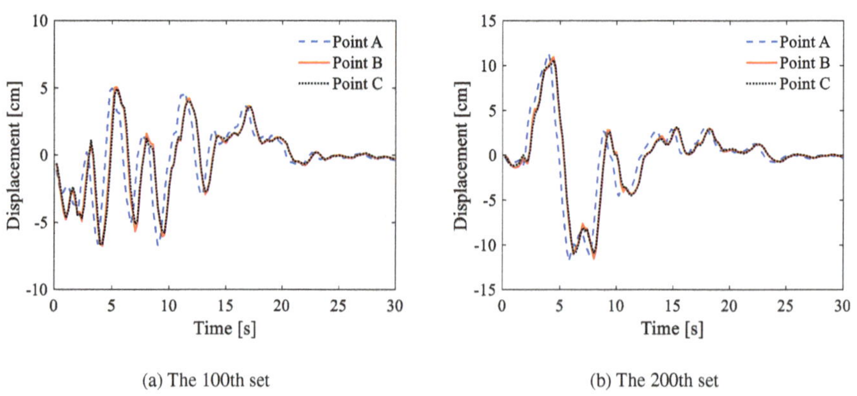

Fig. 6.7 Representative time-histories of ground motion displacement at typical action points. (**a**) The 100th set (**b**) The 200th set

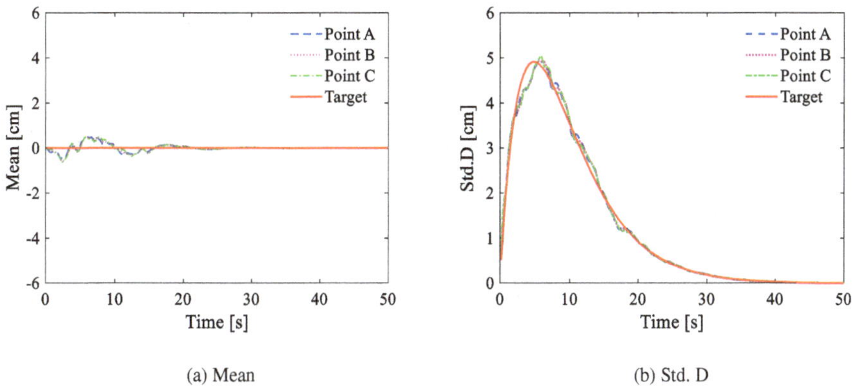

Fig. 6.8 Comparison of mean and standard deviation between the simulated representative time-history sets with the corresponding target values. (**a**) Mean (**b**) Std. D

same representative sample set exhibit similarity, indicating the lagged coherence effect between them.

Figure 6.8 presents the comparison of the mean and Std. D between the representative displacement time-history sets at Points A, B, and C in the 3D-1V non-stationary seismic displacement field generated applying the dimension-reduction method and their corresponding target values. It is worth noting that since the simulated displacement field is homogeneous, the target values of both the mean and Std. D at different action points are identical. As can be seen from the figures, both the mean and Std. D of the simulated values at different locations are well fitted with the target values, verifying the effectiveness of the dimension-reduction method for simulating the 3D-1V ground motion spatio-temporal stochastic field.

6.5 Conclusion Remarks

This chapter systematically deduces the O-FWRS of the *m*D-1V spatio-temporal stochastic field and explores the origins of the conventional Monte Carlo method. Furthermore, the dimension-reduction approach is expounded, and the effectiveness of which has been validated through numerical examples. The key conclusion remarks drawn from this chapter are as follows:

1. For the *m*D-1V spatio-temporal stochastic field, a detailed derivation of the O-FWSR and the conventional Monte Carlo method is performed. Thus, the relationship between the two methods has been clarified. Specifically, the conventional Monte Carlo method is essentially derived from the O-FWRS, indicating that the conventional Monte Carlo method is a special case of the O-FWSR in essence.

2. By constructing the constraint form of random orthogonal functions for the set of standard orthogonal random variables in the O-FWSR, efficient dimension-reduction simulation of *m*D-1V spatio-temporal stochastic field is achieved. Specifically, for 2D-1V spatio-temporal stochastic field, merely 2 elementary random variables are required, and for 3D-1V spatio-temporal stochastic field, just 3 elementary random variables are needed, which effectively reduces the randomness from the degree of thousands or millions to an extremely low level.
3. Compared with the conventional Monte Carlo method, the dimension-reduction method offers higher simulation accuracy. Moreover, fewer generated representative samples are needed to satisfy the required engineering accuracy using the dimension-reduction method, which significantly enhances the computational efficiency of structural dynamic response analysis. Additionally, the probabilistic information of the representative sample set produced by the dimension-reduction method is complete, allowing for integration with probability density evolution method to conduct refined stochastic seismic response analysis and dynamic reliability assessment of complex engineering structures.
4. The FWSR of *m*D-1V spatio-temporal stochastic field is essentially a continuous approach, enabling efficient simulation of large-scale ground motion field. It is particularly suitable for seismic analysis and design of large-scale and multi-support lifeline engineering projects, such as underground pipeline networks. Additionally, the FWSR-based dimension-reduction method for the *m*D-1V spatio-temporal stochastic field can be further extended to the case of *m*D-*n*V spatio-temporal stochastic field according to engineering practices.

References

Benowitz BA, Deodatis G (2015) Simulation of wind velocities on long span structures: a novel stochastic wave based model. J Wind Eng Ind Aerodyn 147:154–163

Deodatis G (1996) Non-stationary stochastic vector processes: seismic ground motion applications. Probabilistic Eng Mech 11(3):149–168

Deodatis G, Shinozuka M (1989) Simulation of seismic ground motion using stochastic waves. J Eng Mech 115(12):2723–2737

Liu ZJ, Liu ZH, Liu W (2017) Probability model of fully non-stationary ground motion with the target response spectrum compatible. J Vibr Shock 36(2):32–38. (in Chinese)

Liu ZX, Liu ZJ, Ruan XX et al (2025) Advances in dimension-reduction methods for simulating univariate and multivariate non-stationary stochastic processes via spectral representation. Probabilistic Eng Mech 79:103720

Loh CH, Lin SG (1990) Directionality and simulation in spatial variation of seismic waves. Eng Struct 12:134–143

Peng LL, Huang GQ, Chen XZ et al (2017) Simulation of multivariate nonstationary random processes: hybrid stochastic wave and proper orthogonal decomposition approach. J Eng Mech 143(9):04017064

Ruan XX, Liu ZJ, Liu ZX et al (2021) Dimension-reduction representation of stochastic ground motion fields based on wavenumber-frequency spectrum for engineering purposes. Soil Dyn Earthq Eng 143:106604

Shinozuka M (1987) Stochastic fields and their digital simulation. In: Schueller GI, Shinozuka M (eds) Stochastic methods in structural dynamics. M. Nijhoff Publishers, Dordrecht, pp 93–133

Shinozuka M, Deodatis G (1996) Simulation of multi-dimensional Gaussian stochastic fields by spectral representation. Appl Mech Rev 49(1):29–53
Vanmarcke EH (1983) Random fields: analysis and synthesis. MIT Press, Cambridge, MA
Zerva A (1992a) Seismic ground motion simulations from a class of spatial variability models. Earthq Eng Struct Dyn 21(4):351–361
Zerva A (1992b) Spatial incoherence effects on seismic ground strains. Probabilistic Eng Mech 7(4):217–226

Chapter 7
Dimension-Reduction Modeling of Stochastic Ground Motion Via Strong Motion Records

Abstract To overcome the limitations of empirical parameters in traditional seismic ground motion models, which inadequately capture the natural variability observed in real earthquakes, this chapter presents a dimension-reduction modeling approach for non-stationary ground motions based on extensive strong motion records. Model parameters are identified through both frequency-domain and time-domain analysis methods, followed by statistical characterization of their probability distributions and correlation structures. By incorporating a small number of elementary random variables, an efficient dimension-reduction simulation scheme is established that explicitly accounts for parameter randomness. Numerical examples demonstrate that the proposed model effectively reproduces the key features and inherent variability of real ground motions, providing a robust foundation for stochastic response analysis and reliability assessment of complex engineering structures.

Keywords Strong motion records · Natural variability · Dimension-reduction modeling · Parameter identification · Statistical model of ground motion parameters

In the previous chapters, the model parameters for simulating stochastic ground motions, such as the evolutionary power spectrum density (EPSD), impulse response functions and so forth, are commonly determined based on empirical values or by fitting code response spectra. This treatment might result in small variations among the generated ground motion time-histories, thereby underestimating the natural variability of actual earthquake events. To sufficiently and reasonably account for the inherent randomness and natural variability of ground motions, this chapter conducts a detailed investigation into the statistical properties of the parameters in stochastic ground motion models according to a large number of measured strong motion records, and provides recommended parameter values. The numerical case study on the dimension-reduction simulation of univariate stochastic ground motions considering the randomness, correlation and variability of model parameters is addressed utilizing both the frequency-domain and time-domain analysis methods.

Z. Liu et al., *Dimension-reduction Representation of Stochastic Ground Motion and Engineering Applications*, https://doi.org/10.1007/978-981-92-0168-6_7

7.1 Statistical Modeling of Stochastic Ground Motion Parameters in Frequency Domain

7.1.1 Selection of Strong Motion Records

The measured strong motion records serve as the most authentic representation of the engineering characteristics for earthquake ground motions, providing an indispensable foundation for the establishment, validation and refinement of stochastic ground motion simulation methods. In engineering practice, selecting representative ground motion records relies on systematic and high-quality strong earthquake databases. Among these, the Next Generation Attenuation (NGA) series of databases maintained by the Pacific Earthquake Engineering Research Center (PEER) in the United States have become a primary data source for international seismic analysis due to the global representativeness of their records, standardized processing, and widespread use in both academia and engineering. Other important national and regional observation systems, such as the China National Earthquake Data Center (CNEDC) and Japan Strong-motion Seismograph Networks (K-NET, Kik-NET), have also accumulated a large amount of valuable record data.

The rational selection of ground motion records should carefully consider several key parameters that control spectral characteristics: magnitude, which governs the energy and low-frequency content of the motion; source-to-site distance, which influences its attenuation process and spectral shape; and site shear-wave velocity, which directly reflects the filtering and amplification effects of local soil conditions. Based on these principles and considering international comparability and engineering practicality, the analyses in this section utilize the PEER NGA-West2 database (PEER (n.d.); Power et al. 2008). The specific screening criteria for the measured strong earthquake records are outlined as follows:

(a) The fault distance should be greater than 10 km to mitigate the influence of near-fault effect.
(b) The moment magnitude of the measured seismic records should be greater than 5 to exclude ground motions that have little impact on the structures.

In accordance with the above selection criteria, a total of 2453 strong motion records from 101 earthquakes are screened. The selected suite of ground motion records encompasses a broad range of parameters: the Joyner-Boore distance ranging from 0 km to 540 km, the average shear wave velocity at 30 m underground $V_{S,30}$ ranging from 180 m/s to 1526 m/s, and the moment magnitude M_w ranging from 4.7 to 7.9. According to the $V_{S,30}$, the site categories are categorized into three groups. Table 7.1 shows the distribution of $V_{S,30}$ and the quantities of measured strong motion records across different site conditions.

It should be noted that soil site I corresponds to hard rock or soft rock conditions, and is broadly equivalent to soil site B (hard soil) in European Code 8 (Eurocode 8 2001) and soil site I in the "Code for Seismic Design of Buildings" (GB50011-2010

Table 7.1 Classification of soil sites for strong motion records

Classification of soil sites	I	II	III
Number of strong motion records	722	1061	670
$V_{S,30}$ (m/s)	>500	300~500	180~300

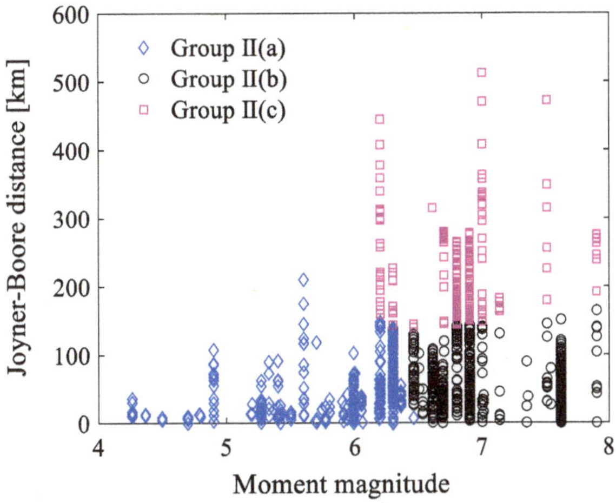

Fig. 7.1 Grouping results by K-means clustering method

2016) of China. Soil site II represents medium-stiff soil and generally corresponds to the softer soil of soil site B and harder soil of soil site C in Eurocode 8, as well as soil site II in GB50011-2010. Soil site III refers to medium-soft soil and is roughly comparable to soil site C in Eurocode 8, and to soil site III and softer soil of soil site II in GB50011-2010. In addition, the acceleration amplitudes of all measured ground motions are adjusted to 200 cm/s^2 by fourth-order Butterworth filter to avoid the influence of peak ground acceleration (PGA) on the statistical properties of the spectra.

To further consider the impact of magnitude and propagation distance on ground motions, the K-means clustering method (Jain 2010) is adopted to group the ground motion records, in which the moment magnitude and Joyner-Boore distance are used as cluster parameters. Since moment magnitude and Joyner-Boore distance differ in order of magnitude, normalization is required prior to cluster analysis, that is, their ranges are linearly transformed to [0,1]. Here, the ground motion acceleration records of soil site II are taken as an example, which are clustered into three groups through the above steps. As shown in Fig. 7.1, group a, b, and c refer to small magnitude and short Joyner-Boore distance, large magnitude and short Joyner-Boore distance, and large magnitude and long Joyner-Boore distance, respectively. The mean of response spectra of ground motion acceleration records belonging to different groups are shown in Fig. 7.2. As clearly shown in Fig. 7.2, significant differences among the three groups indicate that the long period components of the ground motions grow with increasing magnitude and Joyner-Boore distance. This finding highlights the need to categorize the selected ground motions on a site-specific basis.

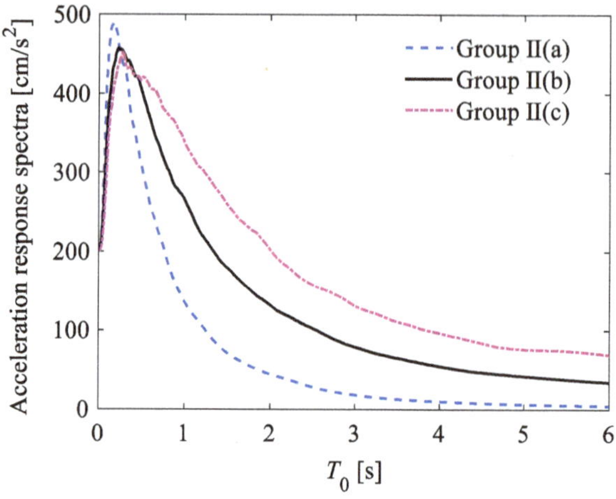

Fig. 7.2 Mean response spectra of different cluster groups

7.1.2 *Parameter Identification of EPSD Model*

Accurate identification of uncertain parameters in the EPSD model from strong motion records is fundamental to properly accounting for their inherent stochasticity and correlations. To this end, this section first clarifies the physical meaning and functional roles of the key parameters embedded in the EPSD model. Subsequently, an identification criterion is established to determine these parameters in a consistent and robust manner via strong motion records.

Parameters of EPSD Model

In earthquake engineering, the EPSD model is an effective tool to describe seismic non-stationary and statistical properties. According to Priestley's evolutionary spectrum theory (Priestley 1965), the EPSD function of a non-stationary ground motion in parametric form can be expressed as:

$$S_U(\omega, t; \boldsymbol{\lambda}_U) = A^2(t, \omega; \boldsymbol{\lambda}_A) S(\omega; \boldsymbol{\lambda}_S) \tag{7.1}$$

where $S_U(\omega, t; \boldsymbol{\lambda}_U)$ is the one-sided EPSD function of the non-stationary seismic ground motion acceleration process $U(t)$, and $\boldsymbol{\lambda}_U$ denotes the parameter vector of the EPSD, respectively. $A(t, \omega; \boldsymbol{\lambda}_A)$ stands for the real-valued time-frequency modulation function, and $\boldsymbol{\lambda}_A$ denotes its parameter vector, respectively. $S(\omega; \boldsymbol{\lambda}_S)$ represents the one-sided PSD function of the corresponding stationary seismic acceleration process, and $\boldsymbol{\lambda}_S$ refers to its parameter vector, respectively.

In this study, the Clough-Penzien spectrum is adopted for the one-sided PSD function $S(\omega; \boldsymbol{\lambda}_S)$, given by (Clough and Penzien 1975):

$$S(\omega; \boldsymbol{\lambda}_S) = \frac{\omega_g^4 + 4\zeta_g^2\omega_g^2\omega^2}{\left(\omega^2 - \omega_g^2\right)^2 + 4\zeta_g^2\omega_g^2\omega^2} \cdot \frac{\omega^4}{\left(\omega^2 - \omega_f^2\right)^2 + 4\zeta_f^2\omega_f^2\omega^2} \cdot S_0 \tag{7.2}$$

where ω_g and ζ_g denote the dominant frequency and damping ratio of the site soil, respectively. ω_f and ζ_f represent the dominant frequency and damping ratio of the bedrock, respectively. Generally, it is assumed that the following relationship holds, i.e., $\omega_f = 0.1\omega_g$ and $\zeta_f = \zeta_g$. S_0 is the spectral intensity of ground motion acceleration process, which is expressed as follows (Liu et al. 2016; Deodatis 1996):

$$S_0 = 2 \times \frac{a_{\max}^2}{r^2 \pi \omega_g \left(2\zeta_g + \frac{1}{2\zeta_g}\right)} \tag{7.3}$$

where $a_{\max}$ denotes the mean of the PGA, and r is the peak factor, respectively.

Since the PGAs of all the strong motion records are scaled to 200 cm/s^2, the PGA and peak factor r are not included in the parameter vector. Thus, the parameter vector of the one-sided PSD function can be presented as $\boldsymbol{\lambda}_S = (\omega_g, \zeta_g)$.

For the time-frequency modulation function $A(t, \omega; \boldsymbol{\lambda}_A)$, the model proposed by Deodatis and Shinozuka (1989) is applied:

$$A(t, \omega; \boldsymbol{\lambda}_A) = \frac{\exp(-a \times t) - \exp[-(c \times \omega + b) \times t]}{\exp(-a \times t^*) - \exp[-(c \times \omega + b) \times t^*]}, \quad \omega > 0 \tag{7.4a}$$

$$t^* = \frac{\ln(c \times \omega + b) - \ln a}{c \times \omega + (b - a)} \tag{7.4b}$$

where the parameters a, b and c control the peak arrival time and energy decay rate of ground motions, and they satisfy $b = a + 0.001$ and $c = 0.005$, respectively. Accordingly, only one parameter a in the time-frequency modulation function $A(t, \omega; \boldsymbol{\lambda}_A)$ needs to be determined, i.e., $\boldsymbol{\lambda}_A = a$.

In conclusion, the parameter vector for the EPSD model of seismic ground motion using the frequency-domain analysis method can be rewritten as $\boldsymbol{\lambda}_U = (\boldsymbol{\lambda}_S, \boldsymbol{\lambda}_A) = (\omega_g, \zeta_g, a)$.

Identification Criterion of Parameters

Since the EPSD is a function about both time and frequency, the energy distribution function (EDF) of seismic ground motion can be defined as follows (Wang et al. 2017):

$$E(\omega;\boldsymbol{\lambda}_U)=\int_0^\infty S_{U_g}(\omega,t;\boldsymbol{\lambda}_U)\mathrm{d}t \tag{7.5}$$

Substituting Eqs. (7.1) and (7.4a) into Eq. (7.5), it yields (Ruan 2023):

$$E(\omega;\boldsymbol{\lambda}_U)=\frac{\frac{1}{2b+2c\omega}+\frac{1}{2a}-\frac{2}{a+b+c\omega}}{\{\exp(-a\times t^*)-\exp[-(c\times\omega+b)\times t^*]\}^2}\cdot S(\omega;\boldsymbol{\lambda}_S) \tag{7.6}$$

For any measured seismic record, its PSD and EPSD functions can be estimated. Specifically, for the i-th non-stationary seismic acceleration record, the PSD function of the corresponding stationary seismic acceleration record is obtained by averaging the EPSD function over time, effectively relaxing the constraint of its time-varying energy distribution, given by (Liang et al. 2007):

$$S_i(\omega)=\frac{1}{T_{0,i}}\int_0^{T_i}S_i(\omega,t)\mathrm{d}t \tag{7.7}$$

By deforming Eq. (7.7), the EDF of the i-th strong motion record can be obtained as:

$$E_i(\omega)=\int_0^{T_i}S_i(\omega,t)\mathrm{d}t=T_{0,i}\cdot S_i(\omega) \tag{7.8}$$

where $S_i(\omega)$ and $S_i(\omega,t)$ denote the PSD function and EPSD function of the i-th strong motion record, respectively. T_i and $T_{0,\ i}$ represent the total duration and strong-motion duration, respectively.

The strong-motion duration $T_{0,\ i}$ can be determined by (Vanmarcke and Lai 1980):

$$T_{0,i}=\frac{I_{0,i}}{\sigma_{0,i}^2} \tag{7.9}$$

where $\sigma_{0,i}^2$ and $I_{0,\ i}$ are the mean square acceleration and total motion energy, respectively, which can be expressed as:

$$\sigma_{0,i}^2=\int_0^\infty S_i(\omega)\mathrm{d}\omega \tag{7.10}$$

$$I_{0,i}=\int_0^{T_i}U_i^2(t)\mathrm{d}t \tag{7.11}$$

where $U_i(t)$ is the i-th strong motion record.

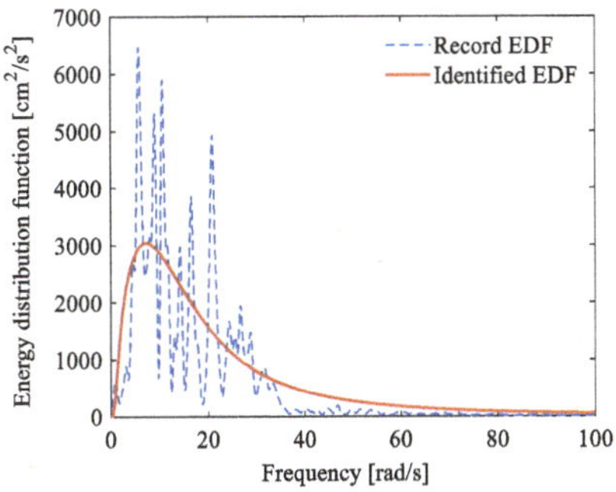

(a) Typical EDF

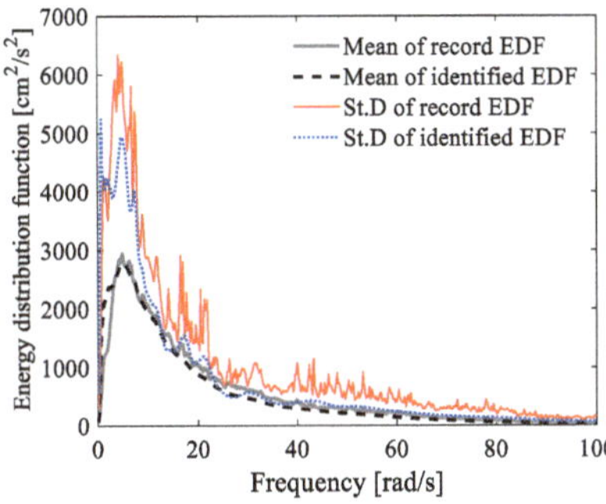

(b) Mean and Std. D of EDFs of soil site II(b)

Fig. 7.3 The identified result of EDF. (**a**) Typical EDF (**b**) Mean and Std. D of EDFs of soil site II (b)

The sample values of parameter vector can be identified from the EDF of records. Upon the rule of the least square approximation, the identified sample values $\boldsymbol{\lambda}_{U,\,i} = (\omega_{g,\,i}, \zeta_{g,\,i}, a_i)$ satisfy (Ruan 2023):

$$\int_0^{\infty} |E(\omega; \boldsymbol{\lambda}_{U,i}) - E_i(\omega)|^2 \mathrm{d}\omega \rightarrow \min \tag{7.12}$$

To circumvent the difficulties associated with time-frequency domain parameter identification and to reduce computational cost, the equal energy principle is employed to identify the EDF of the strong motion record, rather than the EPSD function.

Figure 7.3a shows the EDF comparison between the strong motion record of the Chi-Chi earthquake recorded by the ILA062 station and the identified value. It can be observed that the identified EDF can be regarded as the "mean" in the sense of the best square approximation of the recorded EDF. Meanwhile, Fig. 7.3b presents comparison results of the mean and standard deviation (Std. D) between the recorded EDF of soil site II(b) and identified EDF. Both the mean and Std. D of the recorded and identified EDFs show excellent agreement, demonstrating the effectiveness of the identified results.

7.1.3 Statistical Modeling of Parameters

Following the identification of random parameters in the EPSD model, it is essential to investigate their probabilistic distributions to represent the natural variability of ground motions. The parameters derived from recorded ground motions generally exhibit significant dispersion and correlation, reflecting the complex coupling effects of seismic source properties, path attenuation and site amplification. Ignoring these statistical dependencies may lead to unrealistic or biased stochastic simulations. Therefore, this section develops probabilistic models for the EPSD parameters by statistically analyzing their distributions and correlation structures based on a large set of identified samples. These probabilistic descriptions serve as the stochastic foundation for subsequent ground motion simulations, ensuring that the generated ground motions faithfully reproduce the natural variability observed in real earthquake records (Douglas and Aochi 2008; Vlachos et al. 2016).

Optimal Probability Distribution of Parameters

The maximum likelihood estimation is used to estimate the marginal distribution of random variables, and five probability distributions, including lognormal, gamma, Weibull, Gumbel and generalized extreme value (GEV) distributions, are utilized to fit the marginal distribution of parametric variables. Figure 7.4 presents the distribution histograms of parametric variables and the corresponding five probability density functions (PDFs). According to Kolmogorov-Smirnov (K-S) test and Akaike

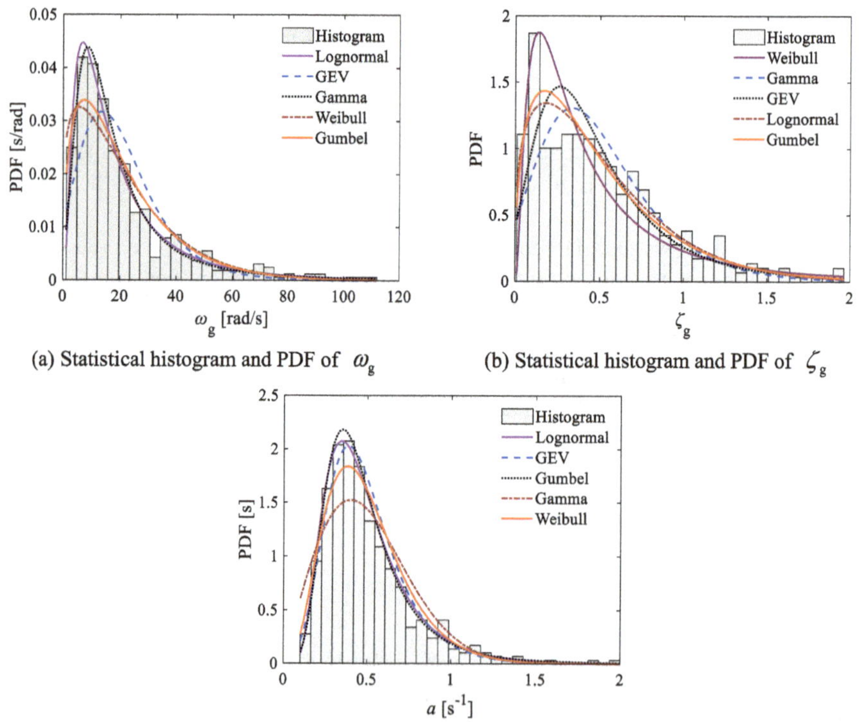

Fig. 7.4 Statistical histograms and PDF of parametric variables. (**a**) Statistical histogram and PDF of ω_g (**b**) Statistical histogram and PDF of ζ_g (**c**) Statistical histogram and PDF of a

Table 7.2 Optimal marginal distributions of parametric variables

Variable	Probabilistic distribution	PDF $f(x)$	Distribution parameters	K-S test results
ω_g	Lognormal	$\frac{1}{\sqrt{2\pi}(\sigma x)}\exp\left[-\frac{(\ln x-\mu)^2}{2\sigma^2}\right]$	$\mu = 2.702$ $\sigma = 0.884$	Pass
ζ_g	Weibull	$\frac{k}{\lambda^k}x^{k-1}\exp\left[-\left(\frac{x}{\lambda}\right)^k\right]$	$\lambda = 0.548$ $k = 1.300$	Pass
a	Lognormal	$\frac{1}{\sqrt{2\pi}(\sigma x)}\exp\left[-\frac{(\ln x-\mu)^2}{2\sigma^2}\right]$	$\mu = -0.819$ $\sigma = 0.493$	Pass

information criterion (AIC) (Akaike 1974), the most suitable model is selected, as shown in Table 7.2.

Correlation Structure of Parameters Via the Vine Copula Method

In general, both the randomness and correlation for the EPSD model have significant effects on the natural variability of ground motions. In the previous section, the marginal distributions of model parameters are fitted to reflect their randomness. However, if the correlation of the model parameters is not adequately considered, the natural variability of generated ground motion samples would be underestimated compared with the actual cases (Rezaeian and Sun 2014). In this section, the Vine Copula method is adopted to characterize the correlation structure of the model parameters.

The Copula theory, first proposed by Sklar (1959), can express any multivariable joint distribution as the product of a Copula function and several marginal distributions, of which the most attractive feature is that the selection of Copula function and the estimation of the marginal distribution function of variables can be carried out independently. The multi-dimensional Copula function can be directly used to construct the correlation structure between multi-dimensional variables, but the correlation structures between any two variables of the Copula function constructed are identical. For this reason, Joe (1996) proposed a more flexible Vine Copula structure, which establishes the correlation structure of multi-dimensional variables by constructing the correlation structure between pairwise variables. Subsequently, it was systematically developed and improved by Bedford and Cooke (2002). In the Vine Copula method, the regular Vine (R-Vine) is utilized to organize and construct multi-dimensional Copula functions, with drawable Vine (D-Vine) and canonical Vine (C-Vine) being the two common configurations.

For a three-dimensional (3D) random vector $\boldsymbol{X} = (X_1, X_2, X_3)$, the joint PDF expressed in form of both the C-Vine and D-Vine Copula configuration is given by (Aas et al. 2009):

$$f(\boldsymbol{x})=c_{12}(F_1,F_2)c_{13}(F_1,F_3)c_{23|1}\left(F_{2|1},F_{3|1}\right)f_1f_2f_3 \tag{7.13}$$

where F_i and f_i ($i = 1,2,3$) are defined as the marginal cumulative distribution function (CDF) and the marginal PDF of X_i, respectively. $F_{i|1}(i = 2,3)$ is the conditional CDF of X_i given X_1. $c_{1i}(\cdot\ ,\ \cdot)$ $(i = 2,3)$ represents the bivariate Copula density functions between X_1 and X_i. $c_{23|1}(\cdot\ ,\ \cdot)$ denotes the bivariate Copula density function between the conditional random variables $X_2|X_1$ and $X_3|X_1$.

In Eq. (7.13), the conditional distribution function $F_{i|1}(x_i|x_1)(i = 2,3)$ can be given as (Aas et al. 2009):

$$F_{i|1}(x_i|x_1)=\frac{\partial C_{1i}[F_1(x_1),F_i(x_i)]}{\partial F_1(x_1)},\qquad i=2,3 \tag{7.14}$$

where C_{1i} $(\cdot\ ,\ \cdot)$ $(i = 2,3)$ denotes the bivariate Copula distribution functions between X_i and X_1.

According to the work of Dißmann et al. (2013), the process of fitting R-Vine correlation structure from a given data set can be divided into the following three steps:

(a) Selection of R-Vine structure. In Vine Copula models, high-dimensional dependence is built by coupling bivariate Copulas, which leads to a combinatorial complexity in the number of possible structures and a consequent difficulty in identifying the optimal R-Vine. Dißmann et al. (2013) proposed a sequential method for selecting R-Vine structure, in which the variable pairs with the largest correlation and satisfying the R-Vine connection condition are selected as the blocks of each layer of the tree. Since just three random variables are required herein, R-Vine has just one structure, say C-Vine and D-Vine are equivalent, and the structure has three combinations.
(b) Selection of Copula family. In this study, seven Copula functions are employed to fit the correlation between each variable pair, say Gaussian, Clayton, Clayton 180-asym, Plackett, Frank, Gumbel, and Gumbel 180-asym Copulas.
(c) Copula parameter estimation. According to the empirical distribution function of the EPSD parameters in this study, the semiparametric method (Kim et al. 2007) is adopted to estimate the parameters of the Copula function, and the AIC is utilized to judge the optimal Copula.

The C-Vine structure of the three parameters in the EPSD function for soil site II (b) used in this example is shown in Fig. 7.5. The AIC values of all combinations for the C-Vine structure are listed in Table 7.3. It can be seen from the table that when the AIC value is the minimized, the optimal Vine Copula structure corresponds to the parameter ordering ω_g, ζ_g, and a. This ordering is consistent with the sequential method that satisfies the R-Vine connection condition and exhibits the strongest correlations, indicating that the sequential method is feasible for selecting the R-Vine structure. The optimal Copula family corresponding to the optimal Vine Copula structure and its parameters are shown in Table 7.4.

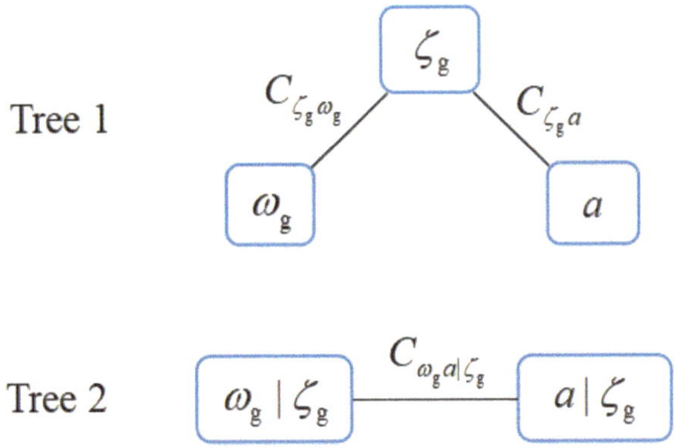

Fig. 7.5 Vine structure of parametric variables

Table 7.3 AIC values of Vine Copula

Sequence of variables	AIC
$\omega_g - \zeta_g - a$	−172.02
$a - \omega_g - \zeta_g$	−172.74
$\zeta_g - \omega_g - a$	−174.85

Table 7.4 Parameters of optimal Copula

Tree	Tree 1		Tree 2
Copula	$C_{\zeta_g \omega_g}$	$C_{\zeta_g a}$	$C_{\omega_g a \mid \zeta_g}$
Optimal copula	Plackett	Plackett	Plackett
Parameter	θ	θ	θ
Value	0.3298	0.2675	0.5476

Simulation of Parameters Considering Randomness and Correlation

Based on the obtained optimal Copula function and optimal marginal distribution function among the above three parameters, the representative points with corresponding correlation structure can be generated by using conditional sampling method (Aas et al. 2009). The procedure for generating d-dimensional m representative points are as follows.

(a) Generate the representative point set $W_m = \{w_{i,j}; i = 1, 2, \cdots, m; j = 1, 2, \cdots, d\}$ for the d-dimensional random variables that are uniformly distributed on the interval [0,1) and are independent from each other.
(b) According to the conditional marginal distribution function and its inverse function of Copula function, the point set $U_m = \{u_{i,j}; i = 1, 2, \cdots, m; j = 1, 2, \cdots, d\}$ of d-dimensional random variables with correlation structure between [0,1) can be obtained, given as:

$$
\begin{aligned}
u_{i,1} &= w_{i,1} \\
u_{i,2} &= F_{2|1}^{-1}(w_{i,2}|u_{i,1}) \\
\cdots &= \cdots \\
u_{i,d} &= F_{d|12\cdots d-1}^{-1}(w_{i,d}|u_{i,1}, u_{i,2}, \cdots, u_{i,d-1})
\end{aligned}
\tag{7.15}
$$

(c) Generate the representative point set $X_m = \{x_{i,j};\ i = 1, 2, \cdots, m;\ j = 1, 2, \cdots, d\}$ based on the marginal CDF of the random variables, such that:

$$
x_{i,j} = F_j^{-1}(u_{i,j}), \quad (i = 1, 2, \cdots, m;\ j = 1, 2, \cdots, d) \tag{7.16}
$$

Taking soil site II(b) as an example, 300 sample points are generated through conditional sampling method, as shown in Fig. 7.6. It can be seen that the distribution and correlation characteristics between the generated samples and the original data are consistent. This indicates that the joint distribution of the seismic parametric variables can be commendably represented by the Vine Copula method.

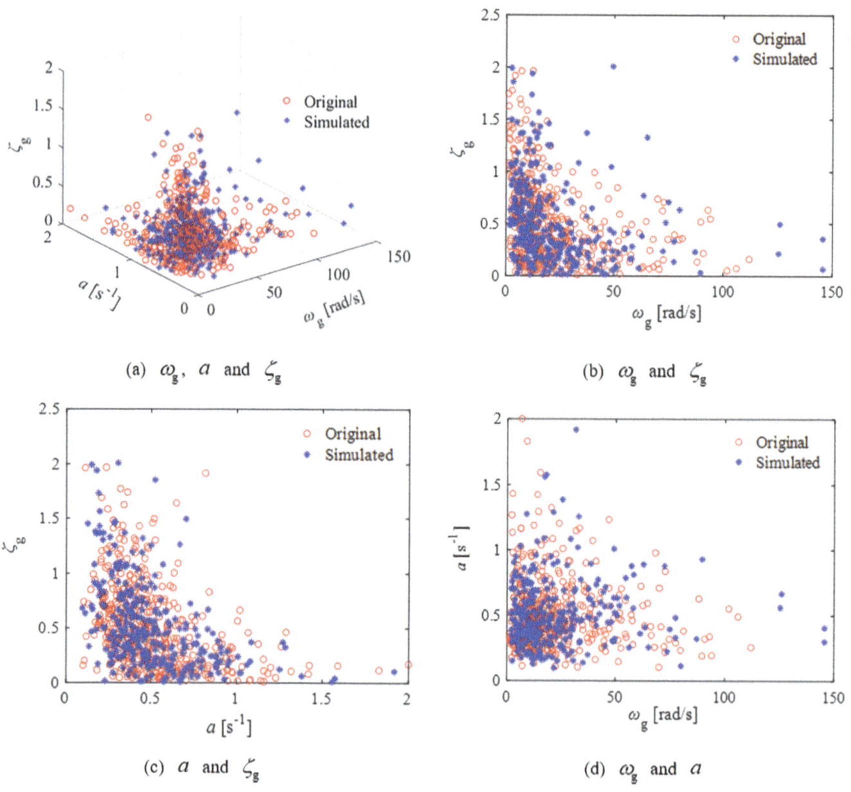

Fig. 7.6 Comparisons between simulated samples and original data. (**a**) ω_g, a and ζ_g (**b**) ω_g and ζ_g (**c**) a and ζ_g (**d**) ω_g and a

7.2 Statistical Modeling of Stochastic Ground Motion Parameters in Time Domain

7.2.1 Selection of Strong Motion Records

The measured strong motion records are selected from the PEER NGA-West2 ground motion database according to the following criteria:

1. The fault distance is restricted to the range of 10–100 km to reduce near-fault effects and exclude low-intensity ground motions.
2. The moment magnitude of the recorded strong ground motion is required to be greater than M_w 6.0, thereby excluding ground motions with minimal structural impact.
3. The equivalent shear wave velocity in the upper 30 m beneath the site, say $V_{S,30}$, should range from 265 to 550 m/s (Guo et al. 2011), corresponding to soil site II as defined in the "Seismic Ground Motion Parameter Zonation Map of China" (GB18306-2015 2015).

Finally, 454 measured strong motion records from 31 earthquake events are selected. The basic information, including the earthquake names, number of stations, magnitudes, fault distance ranges, and $V_{S,30}$ ranges of the selected records, is presented in Table 7.5. Subsequently, the fourth-order Butterworth filter is applied to the selected strong ground motion records, and the time-histories are truncated to retain only the segment containing 1%–99% of the total cumulative energy, thereby ensuring the reliability of the results. Furthermore, to facilitate parameter identification, the PGA of each record is also normalized to 200 cm/s^2. In the following section, the method for identifying time-domain model parameters will be presented in detail.

7.2.2 Parameter Identification of Time-Domain Model

Parameterization of the Time-Domain Model

To depict the full non-stationarities of seismic ground motion in both time and frequency domains, the intensity modulation function and time-varying site parameters are employed in the time-domain ground motion model, namely Eq. (4.52) or Eq. (4.79).

In this section, the three-segment intensity modulation function defined in Eq. (2.64) is used to represent the standard deviation (Std. D) function of non-stationary ground motion processes, such that:

Table 7.5 Basic information of the measured strong motion records

Earthquake name	Year	Magnitude	Fault distance (km)	$V_{S,30}$ (m/s)	Number of records
Big Bear-01	1992	6.46	26.47 ~ 82.98	282.14 ~ 537.95	25
Cape Mendocino	1992	7.01	20.41 ~ 31.46	355.18 ~ 492.74	2
Chalfant Valley-02	1986	6.19	30.11 ~ 51.98	359.23 ~ 529.39	3
Chi-Chi_ China Taiwan	1999	7.62	10.97 ~ 99.24	265.45 ~ 549.43	84
Chi-Chi_ China Taiwan-03	1999	6.2	16.63	549.43	1
Chi-Chi_ China Taiwan-04	1999	6.2	25.79	544.74	1
Chi-Chi_ China Taiwan-06	1999	6.3	10.05 ~ 98.9	270.22 ~ 543.06	48
Christchurch_ New Zealand	2011	6.2	14.41 ~ 91.34	280.26 ~ 485.04	12
Chuetsu-oki_ Japan	2007	6.8	14.81 ~ 97.68	277.29 ~ 528.19	16
Coalinga-01	1983	6.36	26.38 ~ 55.77	267.21 ~ 541.73	26
Darfield_ New Zealand	2010	7	11.86 ~ 95.85	280.26 ~ 511.16	13
Denali_ Alaska	2002	7.9	43 ~ 54.78	341.56 ~ 399.35	3
Duzce_ Turkey	1999	7.14	34.3 ~ 97.53	297 ~ 537.24	3
Friuli_ Italy-01	1976	6.5	15.82 ~ 80.41	352.05 ~ 507.23	2
Imperial Valley-06	1979	6.53	15.19 ~ 30.33	316.64 ~ 471.53	3
Irpinia_ Italy-01	1980	6.9	17.64 ~ 53.16	356.39 ~ 496.46	3
Irpinia_ Italy-02	1980	6.2	20.39 ~ 64.37	356.39 ~ 496.46	5
Iwate_ Japan	2008	6.9	11.12 ~ 94.82	275.07 ~ 540.42	63
Kern County	1952	7.36	38.89	385.43	1
Kobe_ Japan	1995	6.9	22.5 ~ 31.69	312	2
Kocaeli_ Turkey	1999	7.51	15.37 ~ 60.43	281.86 ~ 476.62	5
LAquila_ Italy	2009	6.3	11.15 ~ 96.14	356.39 ~ 547	9
Mammoth Lakes-01	1980	6.06	15.46	537.16	1
Northridge-01	1994	6.69	12.44 ~ 82.32	266.9 ~ 549.75	68
Northridge-02	1994	6.05	27.19 ~ 40.24	297.07 ~ 452.15	3
Parkfield	1966	6.19	63.34	493.5	1
San Fernando	1971	6.61	22.63 ~ 96.81	298.68 ~ 487.23	9
Southern calif	1952	6	73.41	493.5	1
St Elias_ Alaska	1979	7.54	26.46	306.37	1
Tabas_ Iran	1978	7.35	13.94 ~ 91.14	302.64 ~ 471.53	3
Tottori_ Japan	2000	6.61	11.29 ~ 98.02	271.29 ~ 538.73	37

$$\sigma_U(t;\boldsymbol{\lambda}_\sigma)=\sigma_{\max}f(t;\boldsymbol{\lambda}_f)=\sigma_{\max}\begin{cases} t^2/t_1^2, & 0\le t<t_1 \\ 1, & t_1\le t\le t_2 \\ \mathrm{e}^{-c(t-t_2)}, & t_2<t\le T \end{cases} \tag{7.17}$$

where $\boldsymbol{\lambda}_f = (t_1, t_2, c, T)$ and $\boldsymbol{\lambda}_\sigma = (\sigma_{\max}, \boldsymbol{\lambda}_f) = (\sigma_{\max}, t_1, t_2, c, T)$. $\sigma_U(t;\boldsymbol{\lambda}_\sigma)$ denotes the Std. D function of the non-stationary ground motion process $U(t)$. $\sigma_{\max}$ is the maximum value of $\sigma_U(t;\boldsymbol{\lambda}_\sigma)$. t_1 and t_2 are the start and end times of the stationary seismic ground motion segment, respectively. c is the parameter determining the decay rate of the descending segment. T is the ground motion duration.

For the power spectrum of the stationary seismic ground motion, the Hu-Zhou model (Hu and Zhou 1962) presented in Eq. (2.8) is adopted, given by:

$$S(\omega)=\frac{\omega_{\mathrm{g}}^4+4\zeta_{\mathrm{g}}^2\omega_{\mathrm{g}}^2\omega^2}{\left(\omega^2-\omega_{\mathrm{g}}^2\right)^2+4\zeta_{\mathrm{g}}^2\omega_{\mathrm{g}}^2\omega^2}\cdot\frac{\omega^2}{\omega^2+\omega_{\mathrm{c}}^2}S_0 \tag{7.18}$$

where ω_{c} is the parameter determining the low-frequency content of the ground motion, and can be taken as $\omega_{\mathrm{c}} = 2\pi/T$. S_0 is adopted as $S_0 = 1$ owing to the standardization in Eq. (4.52) or Eq. (4.79).

Based on Eqs. (4.58) and (7.18), the complex-valued function $H(\omega)$ can be derived as:

$$H(\omega)=-\frac{\omega_{\mathrm{g}}^2+2\mathrm{i}\zeta_{\mathrm{g}}\omega_{\mathrm{g}}\omega}{\left(\omega_{\mathrm{g}}^2-\omega^2\right)+2\mathrm{i}\zeta_{\mathrm{g}}\omega_{\mathrm{g}}\omega}\cdot\frac{\omega}{\omega-\mathrm{i}\omega_{\mathrm{c}}} \tag{7.19}$$

Substituting Eq. (7.19) into the second formula of Eq. (4.59), the impulse response function $h(t)$ can be determined as (Ruan et al. 2024):

$$h(t)=\frac{\omega_{\mathrm{g}}\mathrm{e}^{-\zeta_{\mathrm{g}}\omega_{\mathrm{g}}t}}{A\sqrt{1-\zeta_{\mathrm{g}}^2}}(B_1+B_2)\sin\omega_{\mathrm{D}}t+C\omega_{\mathrm{c}}\mathrm{e}^{-\omega_{\mathrm{c}}t} \tag{7.20}$$

where $\omega_{\mathrm{D}}=\omega_{\mathrm{g}}\sqrt{1-\zeta_{\mathrm{g}}^2}$. A, B_1, B_2 and C can be expressed as follows:

$$\begin{cases} A=\alpha^2-2\alpha\zeta_{\mathrm{g}}+1 \\ B_1=4\alpha\zeta_{\mathrm{g}}^3-2\zeta_{\mathrm{g}}^2-3\alpha\zeta_{\mathrm{g}}+1 \\ B_2=\left(-4\alpha\zeta_{\mathrm{g}}^2+2\zeta_{\mathrm{g}}+\alpha\right)\sqrt{1-\zeta_{\mathrm{g}}^2} \\ C=\dfrac{1-2\alpha\zeta_{\mathrm{g}}}{2\alpha\zeta_{\mathrm{g}}-\alpha^2-1} \end{cases} \tag{7.21}$$

where $\alpha = \omega_{\mathrm{c}}/\omega_{\mathrm{g}}$.

For Eq. (7.20), the time-varying parameter $\boldsymbol{\lambda}_h(\tau)$ of the impulse response function $h(t-\tau)$ can be determined as $\boldsymbol{\lambda}_h(\tau) = [\omega_g(\tau), \zeta_g(\tau)]$, in which $\omega_g(\tau)$ is an exponential function of time and $\zeta_g(\tau)$ is adopted as a constant value, that is:

$$\begin{cases} \omega_g(\tau; \boldsymbol{\lambda}_\omega) = 2\pi\eta_0 e^{-\gamma\tau} \\ \zeta_g(\tau) = \zeta_g \end{cases} \tag{7.22}$$

where $\boldsymbol{\lambda}_\omega = (\eta_0, \gamma)$ is the undetermined parameter of the predominant circular frequency for site soil.

In conclusion, the parameter vector for the time-domain model of seismic ground motion can be rewritten as $\boldsymbol{\lambda}_U = (\boldsymbol{\lambda}_\sigma, \boldsymbol{\lambda}_h) = (\sigma_{\max}, t_1, t_2, c, T, \eta_0, \gamma, \zeta_g)$.

Identification Criterion of Parameter

Regarding the time-varying Std. D (or intensity modulation function) governing the amplitude and energy of the seismic ground motion process, the parameter $I(t; \boldsymbol{\lambda}_\sigma) = E\left[\int_0^t U^2(t)\mathrm{d}t\right] = \int_0^t f^2(t; \boldsymbol{\lambda}_\sigma)\mathrm{d}t$ can be determined using the cumulative energy of the ground motion. For the i-th strong motion record, the cumulative energy can be expressed as follows (Vanmarcke and Lai 1980):

$$I_i(t) = \int_0^t U_i^2(\tau)\mathrm{d}\tau, \quad 0 \le t \le T_i \tag{7.23}$$

where T_i is the duration of the i-th strong motion record.

For the non-stationary ground motion process $U(t)$, the expectation of its cumulative energy $I(t; \boldsymbol{\lambda}_\sigma)$ is given by:

$$I(t; \boldsymbol{\lambda}_\sigma) = E\left[\int_0^t U^2(\tau)\mathrm{d}\tau\right] = \int_0^t \sigma^2(\tau; \boldsymbol{\lambda}_\sigma)\mathrm{d}\tau \tag{7.24}$$

Using Eqs. (7.23) and (7.24) and the least squares criterion, the parameter vector $\boldsymbol{\lambda}_{\sigma,\,i}$ can be identified as:

$$\int_0^{T_i} |I(t; \boldsymbol{\lambda}_{\sigma,i}) - I_i(t)|^2 \mathrm{d}t \to \min \tag{7.25}$$

Regarding the predominant circular frequency function of the site soil, which governs the central frequency of the time-domain model, the frequency at which the ground motion time-history crosses zero in the upward direction can generally be used as an approximation of the predominant frequency of the record. Consequently, the cumulative number of zero-up-crossings from measured ground motion acceleration time-histories can be utilized to fit the predominant circular frequency function (Rezaeian and Der Kiureghian 2010). Denoting $N_i^+(t)$ as the number of zero-up-crossings in the strong motion record and $N^+(t;\boldsymbol{\lambda}_\omega)$ as the expected number of zero-up-crossings, which can be calculated by the following expression:

$$N^+(t;\boldsymbol{\lambda}_\omega) = \frac{1}{2\pi}\int_0^t \omega_{\mathrm{g}}(\tau;\boldsymbol{\lambda}_\omega)\mathrm{d}\tau = \frac{\eta_0}{\gamma}\left(1-\mathrm{e}^{-\gamma t}\right) \tag{7.26}$$

The parameter vector $\boldsymbol{\lambda}_{\omega,\ i}$ can be identified as:

$$\int_0^{T_i} \left|N^+(t;\boldsymbol{\lambda}_{\omega,i}) - N_i^+(t)\right|^2 \mathrm{d}t \rightarrow \min \tag{7.27}$$

With the parameter vectors $\boldsymbol{\lambda}_{f,\ i}$ and $\boldsymbol{\lambda}_{\omega,\ i}$ determined, fitting the mean response spectrum of a small sample set (10 samples in this book) to the measured response spectrum $S_{U,\ i}(T_0)$ enables the achievement of the damping ratio $\zeta_{\mathrm{g},\ i}$:

$$\int_0^{T_{\mathrm{u}}} \left|S_U(T_0;\zeta_{\mathrm{g},i}) - S_{U,i}(T_0)\right|^2 \mathrm{d}T_0 \rightarrow \min \tag{7.28}$$

where $S_U(T_0;\zeta_{\mathrm{g},\ i})$ denotes the model value of ground motion response spectrum. T_{u} is the upper bound of the response spectrum, and it is adopted as 6s in this book.

Taking the Kern County earthquake recorded from the Taft Lincoln School station as an example, the aforementioned time-domain model and parameter identification method are employed to identify the parameters of the Std. D function and the impulse response function of the seismic ground motion. The parameter identification results are shown in Fig. 7.7. It can be observed that the model predictions agree well with the record values in terms of both the cumulative energy curve and the cumulative number of zero-up-crossings. Furthermore, for the acceleration response spectrum, the measured spectrum is largely encompassed within the range of the simulated response spectrum. These results confirm the effectiveness of the parameter identification method proposed in this book. The parameter identification results of this representative case are summarized in Table 7.6.

7.2.3 *Statistical Modeling of Parameters*

As aforementioned, the time-domain model of ground motion contains a total of eight parameters, i.e., $\boldsymbol{\lambda}_U = (\boldsymbol{\lambda}_\sigma, \boldsymbol{\lambda}_h) = (\sigma_{\max}, t_1, t_2, c, T, \eta_0, \gamma, \zeta_{\mathrm{g}})$. For the 454 selected strong motion records, the statistical results of these eight parameters are presented in Table 7.7, where $t_{\mathrm{s}} = t_2 - t_1$ is used in place of t_2 to facilitate analysis. The results indicate that the variability of the parameters $\sigma_{\max}$ and ζ_{g} is relatively small, whereas the remaining parameters exhibit considerably larger variability. In addition, since the duration T can be directly specified during simulation, only the five parameters t_1,

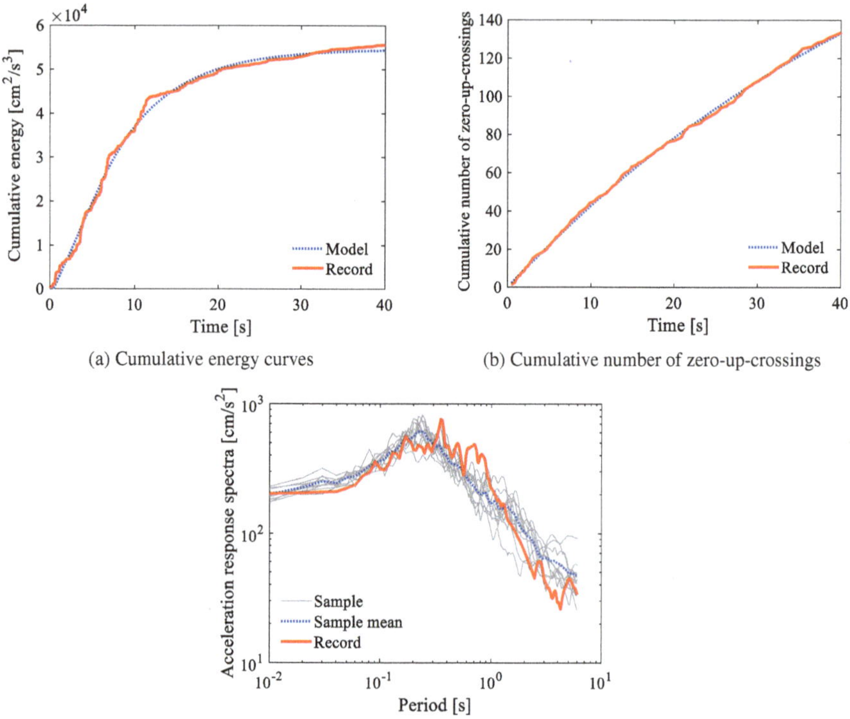

(a) Cumulative energy curves
(b) Cumulative number of zero-up-crossings
(c) Acceleration response spectra

Fig. 7.7 Parameter identification of typical strong motion records. (**a**) Cumulative energy curves (**b**) Cumulative number of zero-up-crossings (**c**) Acceleration response spectra

Table 7.6 Parameter identification results of typical strong motion records

Ground motion parameters		Fitting value
Std. D function	$\sigma_{\max}$ (cm/s^2)	65.856
	t_1 (s)	0.531
	t_2 (s)	5.718
	c	0.0685
	T (s)	40.67
Impulse response function	η_0 (Hz)	4.660
	γ	0.0179
	ζ_g	0.477

t_s, c, η_0, and γ are subjected to probabilistic density fitting in the following part. Furthermore, the minimum value of the decay exponent γ of the predominant circular frequency function is negative, indicating that the predominant frequency of the recorded ground motion does not always decay over time but may increase in certain cases.

Table 7.7 Statistical results of ground motion parameters

Ground motion parameters	Mean	Std. D	Minimum value	Maximum value
$\sigma_{\max}$ (cm/s^2)	68.980	10.990	37.290	107.540
t_1 (s)	5.400	3.430	0.044	15.810
t_s (s)	6.210	6.400	0.011	42.800
c	0.120	0.083	0.016	0.740
T (s)	42.740	21.890	8.730	176.210
η_0 (Hz)	6.220	3.170	1.400	18.570
γ	0.025	0.020	−0.032	0.097
ζ_g	0.380	0.130	0.025	0.730

Table 7.8 Fitting results of optimal marginal distribution of ground motion parameters

Ground motion parameter	Probabilistic distribution	Probability density function (PDF)	Distribution parameter
t_1	Weibull	$f(x)=\frac{k}{\lambda^k}x^{k-1}e^{-\left(\frac{x}{\lambda}\right)^k}$	$\lambda=5.986$ $k=1.553$
t_s	Exponential	$f(x)=\frac{1}{\mu}e^{-\frac{x}{\mu}}$	$\mu=6.207$
c	Lognormal	$f(x)=\frac{1}{\sqrt{2\pi}(\sigma x)}e^{-\frac{(\ln x-\mu)^2}{2\sigma^2}}$	$\mu=-2.297$ $\sigma=0.610$
η_0	Lognormal	$f(x)=\frac{1}{\sqrt{2\pi}(\sigma x)}e^{-\frac{(\ln x-\mu)^2}{2\sigma^2}}$	$\mu=1.704$ $\sigma=0.501$
γ	Generalized extreme value	$f(x)=\frac{1}{\lambda}e^{-\psi^{-1/k}}\cdot\psi^{-1-1/k}$ $\psi=1+k(x-\mu)/\lambda$	$k=-0.175$ $\lambda=0.0186$ $\mu=0.0169$

In the following section, seven candidate probabilistic distributions, namely the normal, lognormal, gamma, Weibull, Gumbel, generalized extreme value, and exponential distributions, are employed to determine the aforementioned five parameters. Every distribution of each parameter is estimated using the maximum likelihood method, and the best-fitting model is selected by minimizing the value of AIC. Subsequently, the K-S test (Ang and Tang 2007) is conducted to evaluate whether the selected model adequately represents the marginal distribution of the ground motion parameters, with significance level of 5%.

The optimal probability distribution fitting results for the parameters t_1, t_s, c, η_0 and γ are summarized in Table 7.8. The corresponding fitted optimal PDF and CDF of the above five parameters are depicted in Fig. 7.8. The results demonstrate that the fitted probability model effectively captures the distribution characteristics of these parameters.

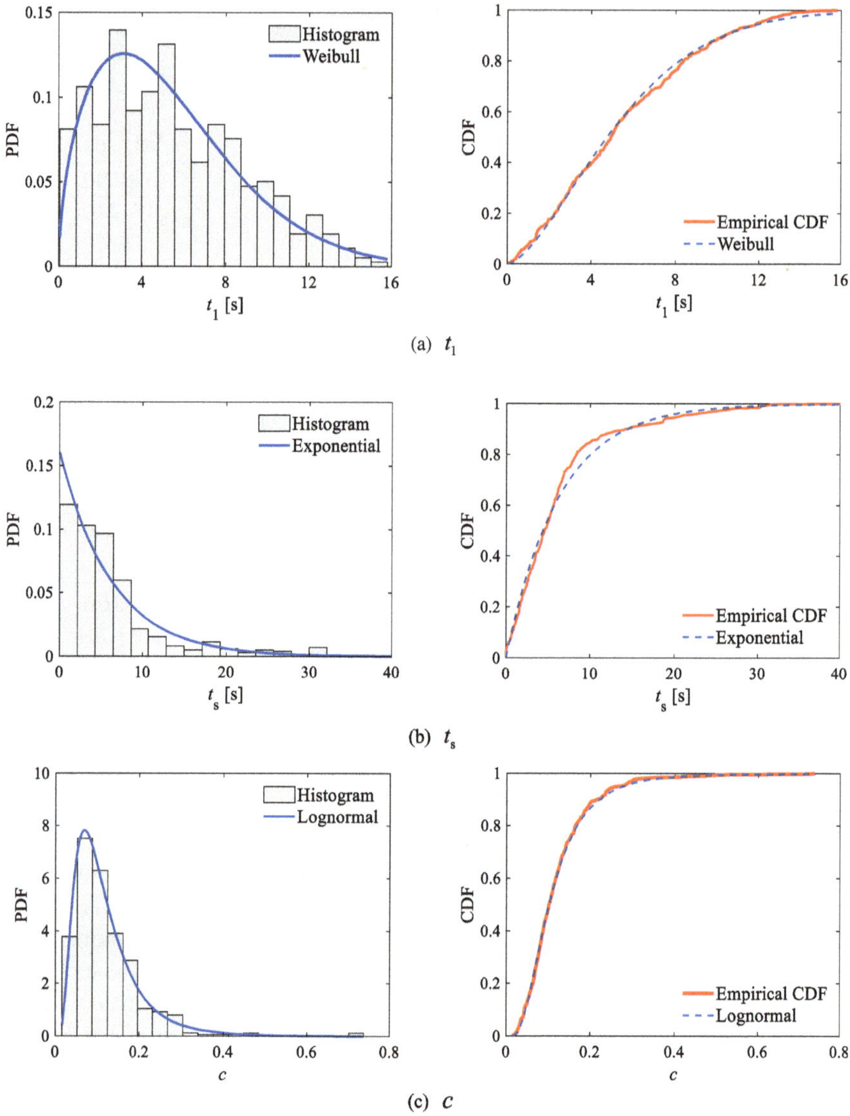

Fig. 7.8 Optimal PDF and CDF of parameters t_1, t_s, c, η_0, and γ. **(a)** t_1 **(b)** t_s **(c)** c **(d)** η_0 **(e)** γ

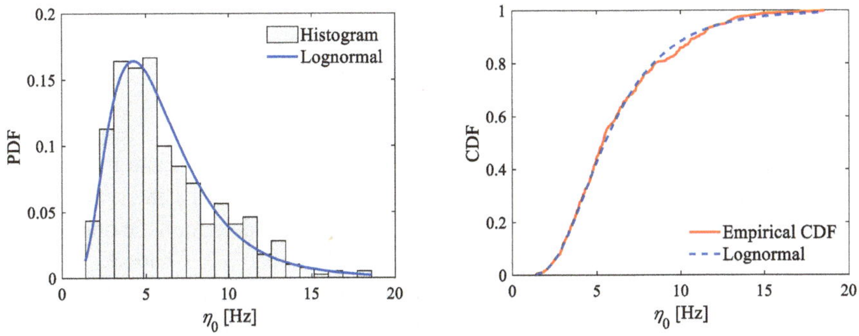

(d) η_0

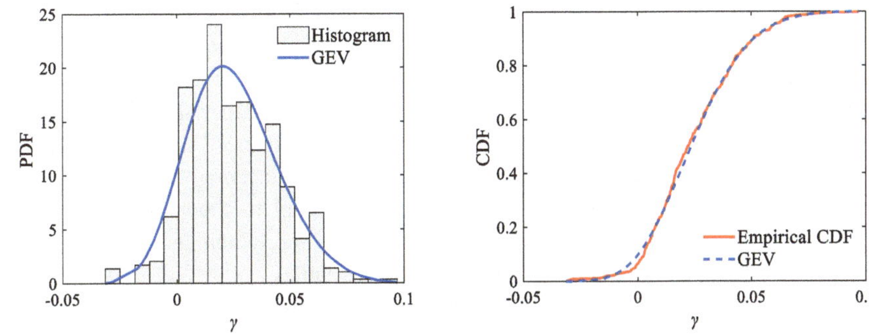

(e) γ

Fig. 7.8 (continued)

7.3 Dimension-Reduction Simulation of Stochastic Ground Motion with Random Parameters

7.3.1 Dimension-Reduction Simulation of Stochastic Ground Motion in Frequency Domain

Dimension-Reduction Simulation Method

Building upon the probabilistic characterization of the EPSD parameters, this section integrates the dimension-reduction representation based on frequency-domain analysis method proposed in Chap. 3 to efficiently simulate non-stationary stochastic ground motions.

For a univariate real-valued non-stationary process with one-sided EPSD $S_U(\omega, t; \boldsymbol{\lambda}_U)$ characterized by random parameters $\boldsymbol{\lambda}_U$, the process can be represented using the dimension-reduction representation on the basis of random phase shown in Eq. (3.27) as follows:

$$U(t) = \sum_{n=1}^{N} \sqrt{2S_U(\omega, t; \lambda_U)\Delta\omega} \cos(\omega_n t - \bar{n}\Theta - \varphi) \tag{7.29}$$

where $\omega_n = (n-0.5)\Delta\omega$, and $\Delta\omega$ denotes the frequency interval, respectively. N indicates the number of frequency intervals. Θ represents a random variable uniformly distributed on the interval $[0, 2\pi)$. φ is a constant in the interval $[0, 2\pi)$, and is valued by $\varphi = \pi/4$ herein. The implementation of the deterministic one-to-one mapping relationship between n and $\bar{n}$ can be found in Sect. 3.3, Chap. 3.

From the above, it can be seen that in the dimension-reduction representation of non-stationary ground motion process, there are total four elementary random variables, say ω_g, ζ_g, a and Θ. Specifically, ω_g and ζ_g describe the randomness of soil site conditions and affect the amplitude and frequency components of ground motion, a represents the randomness of the strong-motion duration, and Θ represents the randomness of the phase of ground motion, respectively. Since the number of random variables involved in Eq. (7.29) is merely four, rather than hundreds or thousands, say ($2N$+3), it is straightforward to apply the advanced sampling methods that are applicable to lower dimensions and offer higher accuracy than the Monte Carlo method. Examples include the number-theoretical method (NTM) (Li and Chen 2007), the rotational quasi-symmetric point method (Chen and Zhang 2013), among others, to sample the elementary random variables.

To generate the representative samples of the non-stationary ground motion processes, the following procedures need to be implemented.

(a) Determine the representative point set of the elementary random variables. For the four-dimensional elementary random variables that are independently and uniformly distributed over the interval $[0, 1)^4$, the initial representative point set $\{\vartheta_{1i}, \vartheta_{2i}, \vartheta_{3i}, \vartheta_{4i}\}_{i=1}^{n_{\text{sel}}}$ can be selected using the NTM. Simultaneously, the assigned probability P_i of each representative point $\{\vartheta_{1i}, \vartheta_{2i}, \vartheta_{3i}, \vartheta_{4i}\}$ can be determined as $P_i = 1/n_{\text{sel}}$, and we have $\sum_{i=1}^{n_{\text{sel}}} P_i = 1$.
(b) Generate the target representative point set for the random variables (ω_g, ζ_g, a, Θ). For the parametric variables of EPSD, i.e., (ω_g, ζ_g, a), the target representative point set can be obtained by transforming the initial representative point set $\{\vartheta_{1i}, \vartheta_{2i}, \vartheta_{3i}\}_{i=1}^{n_{\text{sel}}}$ using the conditional sampling method described in Sect. 7.1.3. For the random variable Θ, its representative point set can be obtained through a linear transformation of the initial representative point set $\{\vartheta_{4i}\}_{i=1}^{n_{\text{sel}}}$. In this manner, the target representative point set $\{\theta_{1i}, \theta_{2i}, \theta_{3i}, \theta_{4i}\}_{i=1}^{n_{\text{sel}}}$ of the random variables (ω_g, ζ_g, a, Θ) can be generated.
(c) Generate the representative time-histories of the non-stationary ground motion process with random parameters. By substituting the n_{sel} target representative points $\{\theta_{1i}, \theta_{2i}, \theta_{3i}, \theta_{4i}\}_{i=1}^{n_{\text{sel}}}$ into Eq. (7.29), the representative time-histories of non-stationary ground motions can be obtained. The assigned probability of each generated representative time-history is also P_i, and the total probability of the simulated ground motion representative time-history set equals one, say $\sum_{i=1}^{n_{\text{sel}}} P_i = 1$.

By following the above steps, the dimension-reduction simulation of stochastic seismic ground motion with uncertain parameters can be achieved. Compared with conventional simulation methods, this method substantially improves computational efficiency while maintaining high fidelity to the natural variability inherent in seismic excitations.

Numerical Examples

To investigate the influence of parameter correlations on the generated ground motion acceleration representative time-histories, simulations are performed for cases in which the model parameters are dependent and independent, respectively. Here, soil site II(b) is taken as an illustrative example. Table 7.9 shows the specific cases investigated in this section. It should be noted that, in the case where the parametric variables are independent, the parametric variables should be obtained using the equal-probability inverse transformation method (Ang and Tang 2007), rather than the conditional sampling method in step (b) of Sect. 7.1.3. Other common parameters for simulating ground motion acceleration are given as follows: the number of frequency intervals is $N = 1600$, the frequency interval is $\Delta\omega = 0.15$ rad/s, the upper cut-off frequency is $\omega_u = 240$ rad/s, the PGA is $a_{max} = 200$ cm/s^2, the seismic duration is $T = 30$ s, the time interval is $\Delta t = 0.01$ s and the number of representative time-histories is $n_{sel} = 307$, respectively.

Figure 7.9 presents the representative samples of ground motion with different parameter attributes, i.e., dependent parameters and independent parameters. It can

Table 7.9 Cases of model parameters for simulating ground motion

Case	Description
Dependent	Both randomness and correlation of model parameters are taken into account.
Independent	Only randomness of model parameters is taken into account.

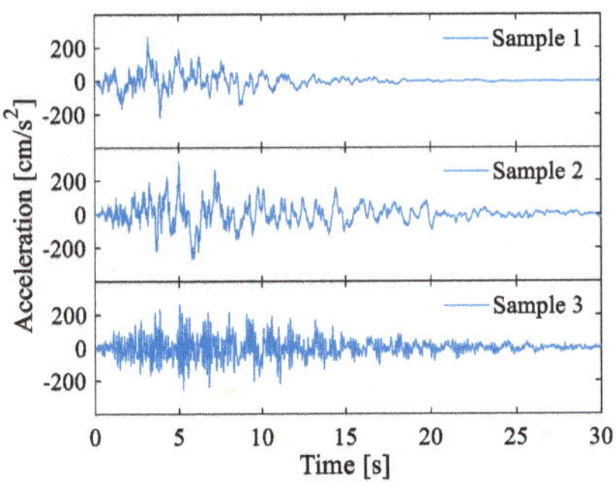

(a) Dependent parameters

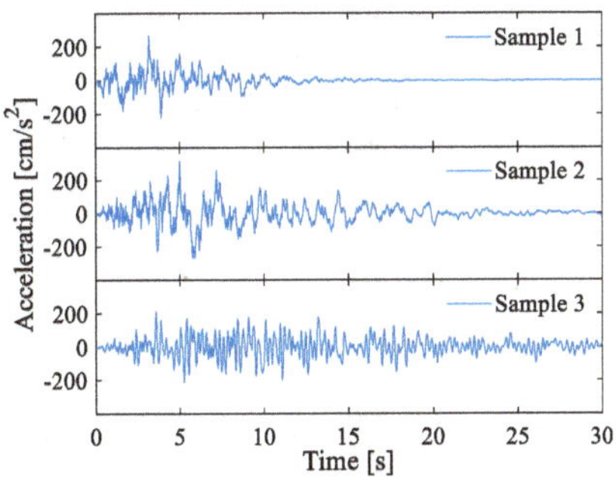

(b) Independent parameters

Fig. 7.9 Representative samples of ground motion with different parameter attributes at soil site II (b). (**a**) Dependent parameters (**b**) Independent parameters

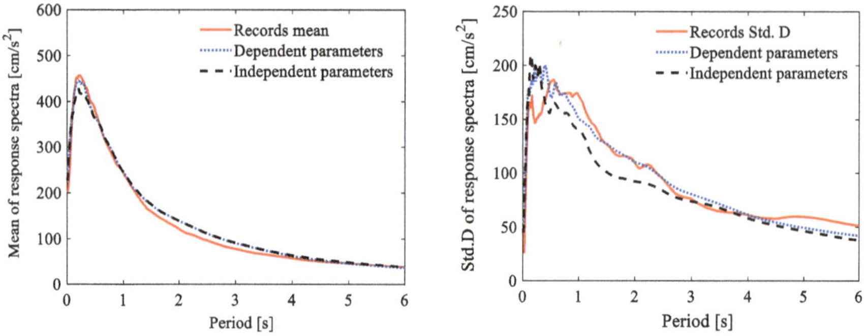

(a) Mean of response spectra (b) Std. D of response spectra

Fig. 7.10 Comparisons of the mean and Std. D between simulated acceleration response spectra with records. (**a**) Mean of response spectra (**b**) Std. D of response spectra

Table 7.10 Fitting errors of mean and Std. D of response spectra in different cases

	Mean of response spectra		Std. D of response spectra	
Case	Average RE	Maximum RE	Average RE	Maximum RE
Dependent	8.31%	17.56%	8.30%	35.54%
Independent	10.58%	19.14%	13.85%	37.53%

be observed that ground motion representative samples generated with dependent parameters and independent parameters show significant differences in frequency contents, waveforms and durations, due to the natural variability of ground motion processes.

To further illustrate the effectiveness and engineering applicability of the proposed model, the comparison between the generated representative samples of non-stationary ground motions and the measured strong motion records is conducted. Figure 7.10 compares the mean and Std. D of the simulated acceleration response spectra with those from the recorded data. As seen in Fig. 7.10a, the mean acceleration response spectra for the case of dependent parameters shows a better fit in the peak region, while the two cases are similar elsewhere. For the Std. D (Fig. 7.10b), the fitting degree of dependent parameters is better than that of independent parameters. To quantify the fitting degree of the mean and Std. D between the simulated representative sample set and the record set, the average relative error (RE) and the maximum RE are calculated respectively, as specified in Table 7.10. All REs for the dependent-parameters case are smaller than those for the independent-parameters case, particularly the average RE for the Std. D. This analysis demonstrates that the proposed model, using random parameters, can effectively capture the variability of ground motions. Furthermore, accounting for parametric correlation yields more accurate result.

7.3.2 *Dimension-Reduction Simulation of Stochastic Ground Motion in Time Domain*

Dimension-Reduction Simulation Method

For a fully non-stationary seismic ground motion acceleration process $U(t)$, considering the intensity modulation function $f(t;\boldsymbol{\lambda}_f)$ and time-varying impulse response function $h[t;\boldsymbol{\lambda}_h(\tau)]$ with uncertain parameters, Eq. (4.52) or Eq. (4.79) can be used to express the process as follows:

$$U(t)=\sigma_{\max}f(t;\boldsymbol{\lambda}_f)\sum_{i=1}^{n}\widehat{u}_i(t)V_i=\sigma_U(t;\boldsymbol{\lambda}_\sigma)\sum_{i=1}^{n}\widehat{u}_i(t)V_i \tag{7.30}$$

where $\widehat{u}_i(t)$ denotes the standardized linear filter:

$$\widehat{u}_i(t)=\frac{h[t-t_i;\boldsymbol{\lambda}_h(t_i)]}{\sqrt{\sum_{j=1}^{n}h^2\left[t-t_j;\boldsymbol{\lambda}_h(t_j)\right]}} \tag{7.31}$$

In Eq. (7.30), $\{V_i\}_{i=1}^{n}$ is a set of standard orthogonal random variables. To implement the dimension-reduction simulation of seismic ground motion process in time domain, the random orthogonal function based on the trigonometric orthogonal bases is adopted herein, followed by:

$$V_{2i-1}=\sqrt{2}\,\sin(j\Theta+\alpha),\quad V_{2i}=\sqrt{2}\,\cos(j\Theta+\alpha) \tag{7.32}$$

where $i,j=1,2,\cdots,N/2$. The elementary random variable Θ is uniformly distributed over the interval $[0,2\pi)$. α is an arbitrary constant value over the interval $[0,2\pi)$, and is valued by $\alpha=\pi/3$ herein. The implementation of the deterministic one-to-one mapping relationship between i and j can be found in Sect. 3.3, Chap. 3.

Therefore, the dimension-reduction simulation based on time-domain analysis method involves six elementary random variables, i.e., $\boldsymbol{\xi}=(t_1,t_s,c,\eta_0,\gamma,\Theta)$. The Stratified-NTM (Xu and Dang 2019) is employed to generate the initial six-dimensional representative point set. The assigned probability P_l for each representative point is also calculated as $P_i=1/n_{\text{sel}}$, and we have $\sum_{i=1}^{n_{\text{sel}}}P_i=1$. Subsequently, the target representative point set for $\boldsymbol{\xi}=(t_1,t_s,c,\eta_0,\gamma,\Theta)$ can be obtained through the equal-probability inverse transformation method (Ang and Tang 2007). Finally, by substituting the target representative points into Eqs. (7.30)–(7.32), the representative time-histories of the non-stationary ground motion process can be obtained.

In this numerical example, eight parameters are required to simulate the non-stationary ground motion process. The parameters t_1, t_s, c, η_0, and γ are treated as random variables with their probability distributions given in Table 7.8. The correlations between the parameters are neglected for simplicity. t_2 is taken as t_1+t_s. The mean values of the parameters σ_{max} and ζ_g are adopted herein, say $\sigma_{max} = 68.98$ cm/s^2 and $\zeta_g = 0.38$. The simulation duration is taken as the mean value plus one Std. D, approximately set as $T = 60$ s. Other parameters in this example are: the time interval is $\Delta t = 0.01$ s and the number of representative samples is $n_{sel} = 610$.

Numerical Examples

The ground motion acceleration representative samples for soil site II generated by the proposed method are shown in Fig. 7.11. Those three simulated representative samples exhibit significant randomness across both time and frequency domains. Key time-domain differences include the PGA, the start and end times of the strong-motion stationary stage, and the decay rate. Similarly, their frequency components and evolution also differ. These features collectively reflect the natural variability of earthquake ground motions.

Figure 7.12 compares the simulated mean value and Std. D of the 610 representative samples with their corresponding target values. It can be seen that both the simulated mean value and Std. D agree well with the target values, thereby validating the effectiveness of the dimension-reduction simulation based on the time-domain analysis method. It is worth noting that, because the three parameters t_1, t_2 and c are treated as random variables, the Std. D curve of the ground motion process remains a smooth function, even though the modulation function in Eq. (7.17) is a three-segment model.

To further evaluate the simulation accuracy of the proposed method, Table 7.11 presents the average REs for the simulated representative sample sets in the cases of $n_{sel} = 377$, 610, and 987, respectively. The definition of average REs upon the mean value and Std. D can be found in the reference (Liu et al. 2017). As shown in

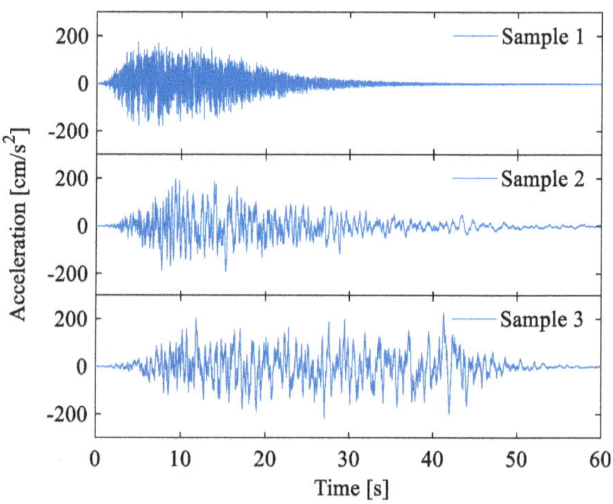

Fig. 7.11 Representative samples of ground motion acceleration

Fig. 7.12 Comparison of the mean and Std. D between the simulated representative samples and targets

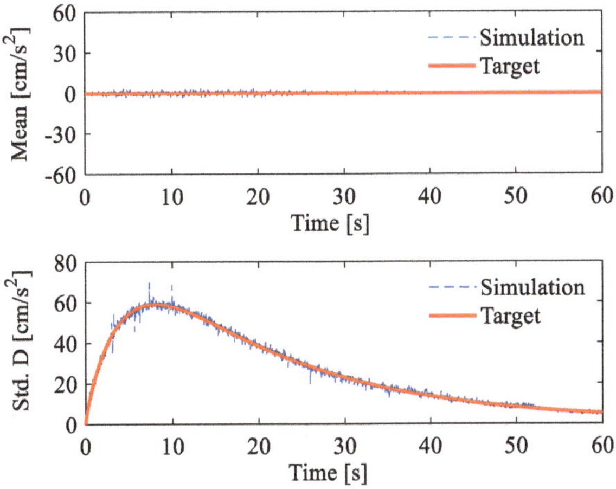

Table 7.11 Simulation error of ground motion process

Error type	Number of the representative samples (n_{sel})		
	377	610	987
Average RE of the mean (%)	3.66	2.66	2.14
Average RE of the Std. D (%)	6.31	4.37	3.81

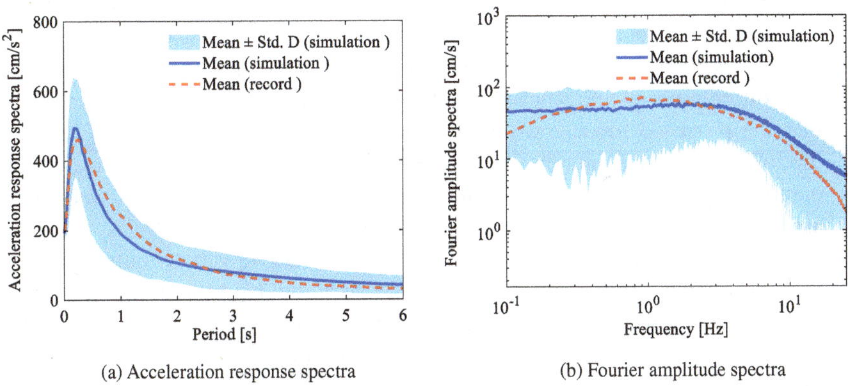

Fig. 7.13 Comparison of acceleration response spectra and Fourier amplitude spectra between simulated representative time-histories and measured records. (**a**) Acceleration response spectra (**b**) Fourier amplitude spectra

Table 7.11, the average REs of both the mean and Std. D gradually decrease as the number of representative samples increases, demonstrating the convergence of the dimension-reduction method. When the number of representative samples reaches 610, all the average REs are below 5%, satisfying the required accuracy criteria.

Figure 7.13 presents the comparison of the response spectrum and Fourier amplitude spectrum between the generated ground motion representative time-histories using the proposed method with those of the strong motion records. It can be seen that the mean curve of the records lies within the range of the simulated mean value plus and minus one Std. D. The results show a good agreement between the simulated value and the measured data, thereby validating the sound engineering characteristics of the proposed method.

7.4 Conclusion Remarks

This chapter conducts dimension-reduction modeling of univariate non-stationary ground motion processes applying both the frequency-domain and time-domain analysis methods based on measured strong motion records. The statistical properties of ground motion parameters, including the probability distribution and correlation structure are investigated in detail. Numerical results verify the effectiveness of the proposed approach. The key conclusion remarks are as follows:

1. In ground motion modeling, the frequency-domain approach typically involves fewer parameters, which can lead to strong correlations that should be accounted for. In contrast, the time-domain approach generally includes a larger number of parameters, allowing those with low variability to be treated as deterministic values. These findings highlight that the selection of core models, such as the power spectrum model and modulation function, directly influences parameter handling strategies and therefore warrants further in-depth research.
2. The representative samples generated using the proposed model show considerable differences in frequency content, duration, and waveform, demonstrating that the proposed model can effectively capture natural variability of seismic motion with a small number of elementary random variables. This capability greatly facilitates random dynamic response analysis and seismic reliability assessment of complex engineering structures.
3. Notably, this chapter focuses on one-dimensional univariate non-stationary ground motions processes. Future research should extend the proposed framework to more complex scenarios, including multi-support and multi-component ground motions, near-fault pulse-like ground motions, long-period ground motions, mainshock-aftershock sequences and so forth, to enhance its applicability.

References

Aas K, Czado C, Frigessi A et al (2009) Pair-copula constructions of multiple dependence. Insur. Math. Econ. 44(2):182–198

Akaike H (1974) A new look at the statistical model identification. Springer, Berlin

Ang AHS, Tang WH (2007) Probability concepts in engineering: emphasis on applications to civil and environmental engineering. Wiley, New York

Bedford T, Cooke RM (2002) Vines: a new graphical model for dependent random variables. Ann Stat 30(4):1031–1068

Chen JB, Zhang SH (2013) Improving point selection in cubature by a new discrepancy. SIAM J Sci Comput 35(5):A2121–A2149

Clough RW, Penzien J (1975) Dynamics of structures. McGraw-Hill, New York

Deodatis G (1996) Non-stationary stochastic vector processes: seismic ground motion applications. Probabilistic Eng Mech 11(3):149–167

Deodatis G, Shinozuka M (1989) Simulation of seismic ground motion using stochastic waves. J Eng Mech 115(12):2723–2737

Dißmann J, Brechmann EC, Czado C et al (2013) Selecting and estimating regular vine copulae and application to financial returns. Comput Stat Data Anal 59:52–69

Douglas J, Aochi H (2008) A survey of techniques for predicting earthquake ground motions for engineering purposes. Surv Geophys 29:187–220

Eurocode 8 (2001) Design of structures for earthquake resistance. European Committee for Standardization

GB18306-2015 (2015) Seismic ground motion parameters zonation map of China. Standards Press of China, Beijing. (in Chinese)

GB50011-2010 (2016) Code for seismic design of buildings. China Architecture & Building press, Beijing. (in Chinese)

Guo F, Wu DM, Xu GF et al (2011) Site classification corresponding relationship between Chinese and the overseas seismic design codes. J Civ Eng Manag 28(2):63–66

Hu YX, Zhou XY (1962) The response of the elastic system under the stationary and nonstationary ground motions [M]. Earthquake engineering research report (no.1) by the Institute of Civil Engineering, Chinese Academy of Sciences. Science Press, Beijing. (in Chinese)

Jain AK (2010) Data clustering: 50 years beyond K-means. Pattern Recogn Lett 31(8):651–666

Joe H (1996) Families of m-variate distributions with given margins and m(m-1)/2 bivariate dependence parameters. Institute of Mathematical Statistics 28:120–141

Kim G, Silvapulle MJ, Silvapulle P (2007) Comparison of semiparametric and parametric methods for estimating copulas. Comput Stat Data Analy 51(6):2836–2850

Li J, Chen JB (2007) The number theoretical method in response analysis of nonlinear stochastic structures. Comput Mech 39(6):693–708

Liang J, Chaudhuri SR, Shinozuka M (2007) Simulation of nonstationary stochastic processes by spectral representation. J Eng Mech 133(6):616–627

Liu ZJ, Liu W, Peng YB (2016) Random function based spectral representation of stationary and non-stationary stochastic processes. Probabilistic Eng Mech 45:115–126

Liu ZJ, Liu ZX, Peng YB (2017) Dimension reduction of Karhunen-Love expansion for simulation of stochastic processes. J Sound Vib 408:168–189

Pacific Earthquake Engineering Research (PEER) (n.d.) NGA-West strong ground motion database. http://peer.berkeley.edu/ngawest/nga_models.html

Power M, Chiou B, Abrahamson N et al (2008) An overview of the NGA project. Earthquake Spectra 24(1):3–21

Priestley MB (1965) Evolutionary spectra and non-stationary processes. J R Stat Soc 27(2): 204–237

Rezaeian S, Der Kiureghian A (2010) Simulation of synthetic ground motions for specified earthquake and site characteristics. Earthq Eng Struct Dyn 39(10):1155–1180

Rezaeian S, Sun X (2014) Stochastic ground motion simulation. Springer, Berlin

Ruan XX (2023) Study on time-varying reliability of existing buildings considering material performance degradation and random earthquake action. Doctoral Dissertation of Wuhan Institute of Engineering, Wuhan. https://doi.org/10.27727/d.cnki.gwhxc.2023.000004. (in Chinese)

Ruan XX, Liu ZJ, Jiang YM (2024) Time-domain dimension-reduction modeling of ground motion processes. J Vib Shock 43(6):311–319. (in Chinese)

Sklar A (1959) Fonctions de répartition à n dimensions et leurs marges. Publications de l'Institut de Statistique de L'Universite de Paris:229–231

Vanmarcke EH, Lai SSP (1980) Strong-motion duration and RMS amplitude of earthquake records. Bull Seismol Soc Am 70(4):1293–1307

Vlachos C, Papakonstantinou KG, Deodatis G (2016) A multi-modal analytical non-stationary spectral model for characterization and stochastic simulation of earthquake ground motions. Soil Dyn Earthq Eng 80:177–191

Wang D, Fan ZL, Hao SW et al (2017) An evolutionary power spectrum model of fully nonstationary seismic ground motion. Soil Dyn Earthq Eng 105:1–10

Xu J, Dang C (2019) A novel fractional moments-based maximum entropy method for high-dimensional reliability analysis. Appl Math Model 75:749–768

Chapter 8
Engineering Applications

Abstract This chapter integrates the dimension-reduction simulation method of stochastic ground motions with the probability density evolution method (PDEM) to systematically address nonlinear dynamic response analysis and global reliability assessment of complex engineering structures. Three representative case studies, including a concrete gravity dam, a high-rise frame-core tube structure, and a high-pier long-span rigid-frame bridge, demonstrate the effectiveness of the proposed framework in accommodating multi-component and/or multi-point seismic inputs, refining nonlinear response analysis and structural global reliability assessment, and identifying vulnerable components. The work offers a robust scientific basis for the seismic retrofitting of existing structures and the design of new ones.

Keywords Engineering application · Probability density evolution method · Nonlinear dynamic analysis · Seismic reliability · Weak component identification

This chapter mainly illustrates the practical engineering application of the proposed dimension-reduction representation of seismic ground motions in the accurate dynamic response analysis and reliability assessment of randomly-excited nonlinear structural systems. To address this, the probability density evolution method (PDEM) for general stochastic dynamic systems is firstly introduced in Sect. 8.1, detailing its fundamental principles and the standard procedures for probability density evolution analysis and structural reliability assessment. Section 8.2 presents the random seismic response and reliability analysis of a nonlinear concrete gravity dam subjected to stochastic seismic ground motions. Section 8.3 addresses the generation of three-dimensional ground motions compatible with design response spectrum, and the performance-based global seismic reliability assessment for a high-rise frame core-tube structure subjected to such three-dimensional ground motions. Section 8.4 details the global reliability assessment of a high-pier long-span continuous rigid-frame bridge subjected to multi-point and multi-component stochastic ground motions.

Z. Liu et al., *Dimension-reduction Representation of Stochastic Ground Motion and Engineering Applications*, https://doi.org/10.1007/978-981-92-0168-6_8

8.1 Introduction to Probability Density Evolution Method

In general, the dynamic response of structures with deterministic parameters subjected to stochastic seismic ground motions belongs to the analytical category of random vibration. In the case that the uncertainty of structural parameters is further considered, it gives rise to a more complex issue of composite random vibration or random structural analysis. However, traditional random vibration analysis strategies, which yield just second-order statistics, are often encounter tremendous challenges in conducting refined reliability evaluation of complex structures. To solve the problem, Li and Chen (2009) have systematically proposed and developed the PDEM in recent years. This effectively constructs a probability density evolution mechanism driven by physical states, thereby revealing the propagation laws of randomness within dynamical systems (Li 2016). Therefore, PDEM is still applicable to the quantification of simultaneous aleatory and epistemic uncertainty of basic parameters of structures (Chen and Wan 2019). Moreover, a unified analytical approach (Li et al. 2007; Chen and Li 2007) has been proposed for assessing the accurate reliability of structures against disasters within the PDEM. In fact, the PDEM has become one of the pivotal approaches in the third-generation structural design theory, which will be introduced briefly in the following.

8.1.1 Generalized Probability Density Evolution Equation

Generally, the motion equation of an n_f-degree-of-freedom structural system can be written as:

$$\boldsymbol{M}\ddot{\boldsymbol{X}} + \boldsymbol{C}\dot{\boldsymbol{X}} + \boldsymbol{G}(\boldsymbol{X}) = \boldsymbol{\Gamma}\boldsymbol{F}(t) \tag{8.1}$$

where $\ddot{\boldsymbol{X}}$, $\dot{\boldsymbol{X}}$ and $\boldsymbol{X}$ denote the n_f-dimensional acceleration, velocity, and displacement vectors of the system, respectively. $\boldsymbol{M}$ and $\boldsymbol{C}$ are the $n_f \times n_f$-order mass and damping matrices, respectively. $\boldsymbol{G}(\cdot)$ is an n_f-dimensional linear or nonlinear restoring force vector. For a linear system, we have $\boldsymbol{G} = \boldsymbol{K}\boldsymbol{X}$, where $\boldsymbol{K}$ is the $n_f \times n_f$-order stiffness matrix. $\boldsymbol{\Gamma}$ is the $n_f \times r$-order influence matrix, and $\boldsymbol{F}(t)$ is the r-dimensional excitation vector. Under uniform seismic excitation $\ddot{X}_g(t)$, we have $\boldsymbol{\Gamma} = -\boldsymbol{M}\boldsymbol{I}$, where $\boldsymbol{I}$ is an n_f-dimensional unit vector. In this case, $\boldsymbol{F}(t)$ reduces to a scalar, say $F(t) = \ddot{X}_g(t)$.

In Eq. (8.1), the fundamental parameters of $\boldsymbol{M}$, $\boldsymbol{X}$ and $\boldsymbol{G}(\cdot)$ may all exhibit inherent randomness, reflecting the variability in the basic properties of the engineering structure itself, such as the strength, elastic modulus and density of concrete materials. Meanwhile, the excitation process $\boldsymbol{F}(t)$ may also be a stochastic process or a vector stochastic process. Let the random vectors $\boldsymbol{\xi}_1$ and $\boldsymbol{\xi}_2$ represent the randomness associated with the structural parameters and external excitation, respectively. For notational consistency, define the random vector as $\boldsymbol{\Theta} = (\boldsymbol{\xi}_1, \boldsymbol{\xi}_2) = (\Theta_1, \Theta_2, \cdots, \Theta_s)$, where s denotes the total number of basic random variables in the entire system. Consequently, Eq. (8.1) can be rewritten as:

$$\boldsymbol{M}(\boldsymbol{\Theta})\ddot{\boldsymbol{X}}+\boldsymbol{C}(\boldsymbol{\Theta})\dot{\boldsymbol{X}}+\boldsymbol{G}(\boldsymbol{\Theta},\boldsymbol{X})=\boldsymbol{\Gamma}\boldsymbol{F}(\boldsymbol{\Theta},t) \tag{8.2}$$

The derivation of the motion equation for stochastic dynamic systems demonstrates that the physical solution to Eq. (8.2) not only exists, but also unique, and it continuously depends on the random vector $\boldsymbol{\Theta}$ of basic parameters. For convenience, the solution to Eq. (8.2) is expressed as follows:

$$\boldsymbol{X}=\boldsymbol{H}(\boldsymbol{\Theta},t) \tag{8.3}$$

Furthermore, its velocity process can be expressed as:

$$\dot{\boldsymbol{X}}=\boldsymbol{h}(\boldsymbol{\Theta},t) \tag{8.4}$$

where $\boldsymbol{H}=(H_1,H_2,\cdots,H_{n_t})^{\mathrm{T}}$ and $\boldsymbol{h}=(h_1,h_2,\cdots,h_{n_t})^{\mathrm{T}}$, respectively.

As a matter of fact, any physical quantity governed by the structural system in Eq. (8.2), including displacement, velocity, acceleration, deformation, stress, strain and so forth, is well-posed. These solutions collectively exist, possess uniqueness, and exhibit continuous dependence on the random vector $\boldsymbol{\Theta}$. Thus, by denoting the physical quantity of interest as $\boldsymbol{Z}=(Z_1,Z_2,\cdots,Z_m)^{\mathrm{T}}$, the following relation holds:

$$\boldsymbol{Z}=\boldsymbol{H}_Z(\boldsymbol{\Theta},t) \tag{8.5}$$

Correspondingly, its velocity process can be expressed as:

$$\dot{\boldsymbol{Z}}=\boldsymbol{h}_Z(\boldsymbol{\Theta},t) \tag{8.6}$$

where $\boldsymbol{H}_Z=(H_{Z,1},H_{Z,2},\cdots,H_{Z,m})^{\mathrm{T}}$ and $\boldsymbol{h}_Z=(h_{Z,1},h_{Z,2},\cdots,h_{Z,m})^{\mathrm{T}}$, respectively. Clearly, if the displacement response is the physical quantity of interest, then Eqs. (8.5) and (8.6) become Eqs. (8.3) and (8.4), respectively.

For the augmented system $(\boldsymbol{Z},\boldsymbol{\Theta})$ comprising $\boldsymbol{Z}$ and $\boldsymbol{\Theta}$, the stochastic source is entirely governed by the random vector $\boldsymbol{\Theta}$. Consequently, $(\boldsymbol{Z},\boldsymbol{\Theta})$ constitutes a probability-preserved system. Denoting the joint probability density function (PDF) of $(\boldsymbol{Z},\boldsymbol{\Theta})$ as $p_{Z\boldsymbol{\Theta}}(z,\boldsymbol{\theta},t)$, the generalized probability density evolution equation (GPDEE) can be derived as follows (Li and Chen 2009):

$$\frac{\partial p_{Z\boldsymbol{\Theta}}(z,\boldsymbol{\theta},t)}{\partial t}+\sum_{l=1}^{m}\dot{Z}_l(\boldsymbol{\theta},t)\frac{\partial p_{Z\boldsymbol{\Theta}}(z,\boldsymbol{\theta},t)}{\partial z_l}=0 \tag{8.7}$$

In the case that just a single physical quantity is considered, i.e., $m=1$, Eq. (8.7) reduces to:

$$\frac{\partial p_{Z\Theta}(z,\boldsymbol{\theta},t)}{\partial t}+\dot{Z}(\boldsymbol{\theta},t)\frac{\partial p_{Z\Theta}(z,\boldsymbol{\theta},t)}{\partial z}=0 \tag{8.8}$$

Obviously, the dimension m in both Eqs. (8.7) and (8.8) is independent of the number of freedom degrees in the original physical system described by Eq. (8.2). This independence effectively decouples the GPDEE from the physical equation, significantly facilitating subsequent numerical solutions and thereby addressing the computational challenge of high-dimensional probability density evolution.

Furthermore, the joint PDF $p_{\mathbf{Z}}(z, t)$ of the random vector $\mathbf{Z}(t)$ is given by:

$$p_{\mathbf{Z}}(z,t)=\int_{\Omega_{\Theta}}p_{Z\Theta}(z,\boldsymbol{\theta},t)\mathrm{d}\boldsymbol{\theta} \tag{8.9}$$

where Ω_{Θ} represents the distribution space of $\boldsymbol{\Theta}$.

For Eq. (8.7), the boundary conditions may generally be specified as:

$$p_{Z\Theta}(z,\boldsymbol{\theta},t)|_{z_l\to\pm\infty}=0,\quad l=1,2,\cdots,m \tag{8.10}$$

And the initial condition can be given by:

$$p_{Z\Theta}(z,\boldsymbol{\theta},t)|_{t=t_0}=\delta(z-z_0)p_{\Theta}(\boldsymbol{\theta}) \tag{8.11}$$

where z_0 is the deterministic initial value.

The general procedures for obtaining the numerical solution of GPDEE are as follows:

1. Discretize the probability space. Select a set of representative points for the random vector $\boldsymbol{\Theta}$ in the domain Ω_{Θ}, denoted by $\{\boldsymbol{\theta}_q\}_{q=1}^{n_{\text{sel}}}$, where n_{sel} is the number of representative points, $\boldsymbol{\theta}_q=(\theta_{1,\ q},\theta_{2,\ q},\cdots,\theta_{s,\ q})^{\mathrm{T}}$. Simultaneously, compute the assigned probability P_q for each representative point, given by:

$$P_q=\int_{V_q}p_{\Theta}(\boldsymbol{\theta})\mathrm{d}\boldsymbol{\theta},\quad q=1,2,\cdots,n_{\text{sel}} \tag{8.12}$$

   where V_q is the region of representative points $\boldsymbol{\theta}_q$, which can be taken as the Voronoi region (Liu et al. 2022a):

$$V_q=\{\boldsymbol{x}\in R^s:\|\boldsymbol{x}-\boldsymbol{\theta}_q\|\le\|\boldsymbol{x}-\boldsymbol{\theta}_j\|,\text{for all }j\} \tag{8.13}$$

2. Solve the deterministic dynamic systems. By sequentially substituting the assigned values of the random vector with each representative point, i.e., $\boldsymbol{\Theta}=\boldsymbol{\theta}_q(q=1,2,\cdots,n_{\text{sel}})$, the physical Eq. (8.2) can be solved using the dynamic time-history analysis method to obtain the corresponding physical quantity $\dot{Z}_l(\boldsymbol{\theta}_q,t)$ $(l=1,2,\cdots,m)$.

3. Solve the GPDEE. Similarly, by substituting the assigned values of the random vector with each representative point, that is $\boldsymbol{\Theta} = \boldsymbol{\theta}_q (q = 1, 2, \cdots, n_{\text{sel}})$, Eq. (8.7) can be correspondingly rewritten as:

$$\frac{\partial p_{Z\boldsymbol{\Theta}}(z, \boldsymbol{\theta}_q, t)}{\partial t} + \sum_{l=1}^{m} \dot{Z}_l(\boldsymbol{\theta}_q, t) \frac{\partial p_{Z\boldsymbol{\Theta}}(z, \boldsymbol{\theta}_q, t)}{\partial z_l} = 0, \quad l = 1, 2, \cdots, m \tag{8.14}$$

Here the corresponding initial conditions are given by:

$$p_{Z\boldsymbol{\Theta}}(z, \boldsymbol{\theta}_q, t)\big|_{t=t_0} = \delta(z - z_0) P_q \tag{8.15}$$

Equation (8.14) can be solved using the finite difference method, and the specific procedures can be found in the reference (Li and Chen 2009).

4. Compute the PDF $p_Z(z, t)$. The PDF $p_Z(z, t)$ can then be obtained by summing $p_{Z\boldsymbol{\Theta}}(z, \boldsymbol{\theta}_q, t)$ over all representative points, that is:

$$p_Z(z, t) = \sum_{q=1}^{n_{\text{sel}}} p_{Z\boldsymbol{\Theta}}(z, \boldsymbol{\theta}_q, t) \tag{8.16}$$

It can be seen that the numerical procedures for the probability density evolution process described above combine the solutions of a series of deterministic dynamic systems with the solution of the PDEE. This effectively reflects the underlying philosophy that the evolution of the PDF is governed by the evolution of the physical state associated with the dynamic systems.

8.1.2 Reliability Analysis Based on Extreme Value Distribution

The dynamic reliability of engineering structures generally encompasses two types of problems: first-passage failure and cumulative damage failure. The latter typically refers to failure induced by fatigue. In this book, the discussion is confined to the former, say the problem of first-passage failure reliability.

Based on the structural failure criterion, the dynamic reliability can be generally defined as:

$$R(t) = \Pr\{X(\tau) \in \Omega_s, 0 \le \tau \le t\} \tag{8.17}$$

where $X(\tau)$ denotes the physical quantity that causes structural failure. Ω_s is the safety region. t is the concerned time period (or structural life). $R(t)$ can be called the reliability function.

Furthermore, Eq. (8.17) can be rewritten as the reliability of a series system:

$$R(t)=\Pr\{X(\tau)\in\Omega_s,0\leq\tau\leq t\}=\Pr\{\cap_{0\leq\tau\leq t}[X(\tau)\in\Omega_s]\} \tag{8.18}$$

As indicated in Eq. (8.18), the dynamic reliability concerning first-passage failure is equivalent to the reliability of a series system with an infinite (uncountable) number of components. Consequently, to obtain an exact solution, it is imperative to process the infinite-dimensional joint probability distribution of the stochastic process $X(t)$. In contrast, methods based on crossing process theory merely necessitate the joint probability distribution between two-time instants, yet they can only provide approximate results. To address this, Li and Chen (2009) proposed the absorbing boundary condition method and the extreme value distribution method, both capable of yielding exact solutions for structural dynamic reliability. This section will focus on the latter.

When a dynamic system has multiple failure modes, the problem is termed global reliability analysis. In such cases, by constructing an equivalent extreme-value (EEV) event, the global reliability analysis can thus be converted into a one-dimensional integration problem involving the PDF of EEV. For example, if $X_i(i=1,2,\cdots,m)$ represents the inter-story drift angle of each floor in a building, the global reliability can be defined as:

$$R(t)=\Pr\{\cap_{i=1}^{m}[X_i(\tau)\leq b_i],0\leq\tau\leq t\} \tag{8.19}$$

At this point, the EEV can be constructed as:

$$X_{\max}(t)=\max_{1\leq i\leq m}\left\{\max_{0\leq\tau\leq t}[X_i(\tau)-b_i]\right\} \tag{8.20}$$

Thus, the global reliability defined in Eq. (8.19) can be transformed into:

$$R(t)=\Pr\{X_{\max}(t)\leq 0\}=\int_{-\infty}^{0}p_{X_{\max}}(x,t)\mathrm{d}x \tag{8.21}$$

Therefore, once the PDF of the EEV $X_{\max}(t)$ defined in Eq. (8.20) can be obtained, the dynamic reliability analysis can be transformed into a one-dimensional integration problem. It is worth mentioning that no approximations are introduced in this transformation process.

In fact, the EEV of a stochastic process $X(\Theta,t)$ can be obtained by constructing a virtual stochastic process and solving the corresponding GPDEE, which can be defined as (Li and Chen 2009):

$$X_{\max}(t)=\max_{0\leq\tau\leq t}\{X(\Theta,\tau)\}=\phi(\Theta,t) \tag{8.22}$$

Obviously, $X_{max}(t)$ is also a random variable dependent on the source random vector Θ and the time duration $[0, t]$.

Moreover, a virtual stochastic process can be constructed as:

$$Z(\tau) = \psi[\phi(\Theta, t), \tau] \tag{8.23}$$

which satisfies the condition:

$$Z(\tau)|_{\tau=0} = 0, \quad Z(\tau)|_{\tau=\tau_c} = \phi(\Theta, t) = X_{max}(t) \tag{8.24}$$

To obtain the solution, one simple approach is to set $Z(\tau) = \psi[\phi(\Theta, t), \tau] = \tau\phi(\Theta, t)$ and $\tau_c = 1$.

In this way, by performing a probability density evolution analysis on the probability-preserved system formed by $(Z(\tau), \Theta)$, the PDF $p_{X_{max}}(x, t)$ of the EEV $X_{max}(t)$ can be obtained. Thus, we can readily achieve the global reliability of a stochastic dynamic system.

8.2 Random Seismic Response and Reliability Analysis of a Nonlinear Concrete Gravity Dam

8.2.1 Engineering Background Description

The Jin-an-qiao hydropower station is located on the middle reaches of Jinsha River in Yunnan Province, China, and the distance to Lijiang City is 50 km. The central water-retaining structure is a roller-compacted concrete (RCC) gravity dam. The seismic safety evaluation indicates that the precautionary intensity of the dam is 9-degree. The earthquake action on the dam is assessed by considering the parameters of ground motions corresponding to return periods of 5000 and 10,000 years, with the peak horizontal seismic accelerations on the bedrock being 0.399 g and 0.475 g, respectively (Hong et al. 2011). The dam has a crest elevation of 1424.0 m and a maximum structural height of 160 m. Its crest length is 640.0 m and its minimum foundation surface is 1264.0 m. This investigation focuses on Dam Block #5, of which the cross-sectional geometry is presented in Fig. 8.1.

In the analysis, the mass density of the rock is set close to zero to satisfy the assumption of a massless foundation model. The dynamic elastic modulus of both concrete and rock mass is taken as 1.3 times their respective static values, as specified in the "Specifications for Seismic Design of Hydraulic Structures" (DL5073-2000 2000) of China. The finite element model (FEM) of the dam and the foundation is discretized using 4-node plane-strain elements. The foundation mesh extends 179 m (1.6 times the dam height) upstream and downstream, and 224 m (2 times the dam height) vertically downward from the dam-foundation interface. This FEM comprises 6453 nodes and 6240 elements. The FEM mesh for

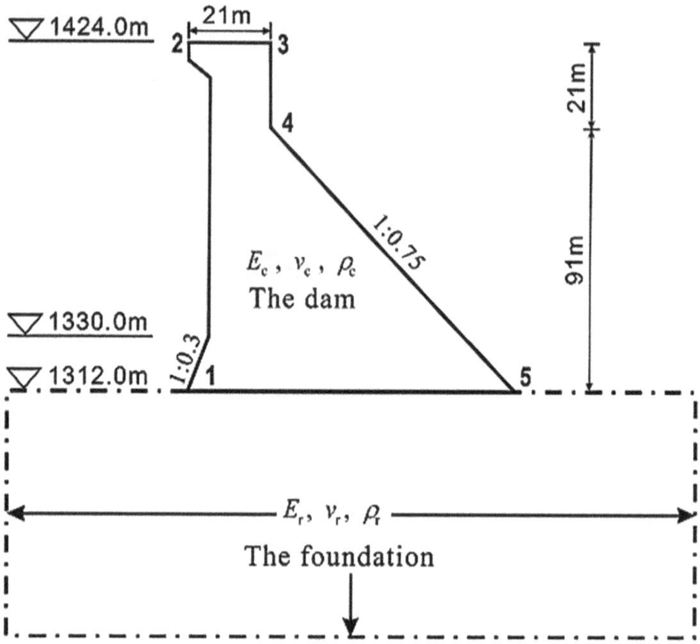

Fig. 8.1 Sectional view of Dam Block #5 (Liu et al. 2018)

the dam-foundation system is depicted in Fig. 8.2. The horizontal displacements at both the upstream and downstream boundaries are restricted, and the horizontal as well as vertical displacements at the bottom boundary are restricted. The constitutive relationship for the concrete is modeled using the plastic-damage model (Lee and Fenves 1998; Omidi and Lotfi 2013). When the reservoir is at the normal water level, the first two natural frequencies of the dam-foundation system are 1.7859 Hz and 4.1025 Hz, respectively. The mechanical parameters of the dam concrete and rock mass are given in Table 8.1.

Under normal operating conditions, the static loads include the self-weight of the dam, hydrostatic pressure on the upstream and downstream surfaces, silt pressure from upstream deposits, and uplift pressure at the dam base. The dynamic loads consist of hydrodynamic pressure and seismic excitation. For the hydrodynamic pressure, water compressibility is neglected, and its effect is simulated as an additional mass distributed over the upstream surface of the dam. Regarding the earthquake action, the horizontal component is exerted along the stream direction, and the vertical component is also taken into account, given that the dam is a typical spatial structure. Specifically, the seismic intensity of vertical component is taken as two-thirds of the horizontal one. The ground motion duration is 15 s with the time interval of 0.01 s. The evolutionary power spectrum density (EPSD) for simulating the non-stationary ground motion employs the Type III model (Eq. (2.73), which combines the intensity modulation function (Eq. (2.60)) and the Kanai-Tajimi

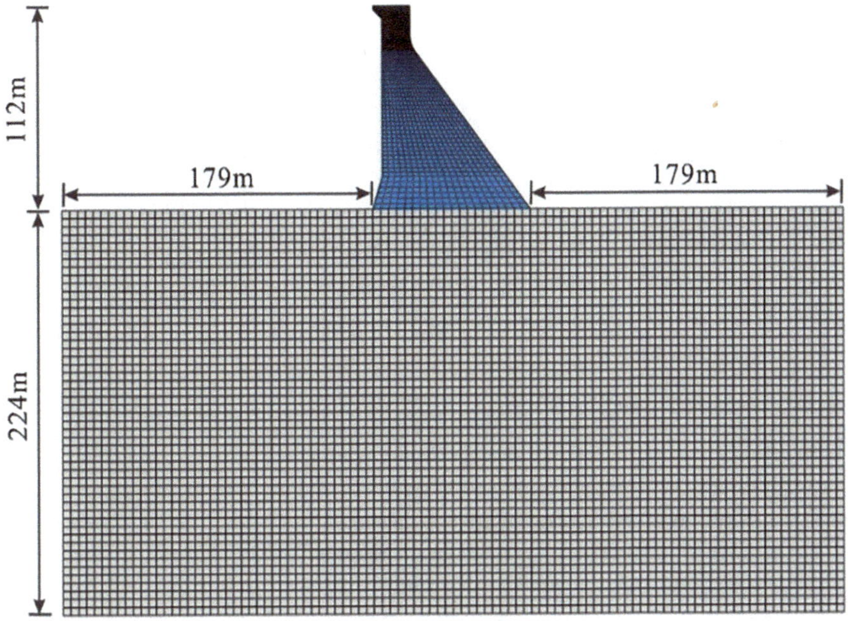

Fig. 8.2 The FEM mesh for the dam-foundation system (Liu et al. 2018)

Table 8.1 Mechanical parameters of the dam concrete and rock mass

Material	Elastic modulus (GPa)	Poisson's ratio	Mass density (kg/m^3)
Concrete	25.5	0.167	2600
Rock mass	15.19	0.25	2700

spectrum with time-varying parameters. The set of ground motion acceleration representative time-histories with response spectrum compatible with the "Specifications for Seismic Design of Hydraulic Structures" (DL5073-2000 2000) of China is generated using the proposed dimension-reduction simulation method. The acceleration representative time-histories are shown in Fig. 8.3. Details can be found in the reference (Liu et al. 2018).

8.2.2 *Random Seismic Response Analysis of the Concrete Gravity Dam*

Research on gravity dam failures (Jiang et al. 2013; Hariri-Ardebili and Saouma 2015) identifies two primary seismic weak areas: the upstream heel, which is susceptible to horizontal penetrating cracks, and the dam-foundation interface, where stress concentrations might occur due to foundation constraints and bond

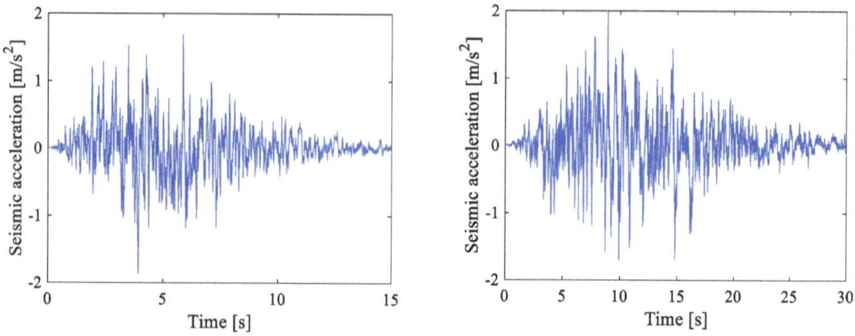

(a) Soil site I (b) Soil site IV

Fig. 8.3 The ground motion acceleration representative time-histories. (**a**) Soil site I (**b**) Soil site IV

strength variability. Consequently, both areas necessitate specific attention in stochastic seismic response and reliability studies.

To achieve a comprehensive damage assessment of the dam under stochastic earthquake actions, three relative displacement angles are defined as follows:

$$\Phi_1(t) = \frac{u_2(t) - u_1(t)}{H_1} \tag{8.25a}$$

$$\Phi_2(t) = \frac{u_3(t) - u_4(t)}{H_2} \tag{8.25b}$$

$$\Phi_3(t) = \frac{u_4(t) - u_5(t)}{H_3} \tag{8.25c}$$

where $\Phi_i(t)$ $(i = 1, 2, 3)$ denote the relative displacement angles. $u_i(t)$ $(i = 1, 2, \cdots, 5)$ denote the horizontal displacement response of point i, as shown in Fig. 8.1, respectively. H_1 denotes the height between Points 1 and 2 of the dam, i.e., $H_1 = 112$ m. H_2 denotes the height between Points 3 and 4 of the dam, i.e., $H_2 = 21$ m. H_3 denotes the height between Points 4 and 5 of the dam, i.e., $H_3 = 91$ m.

According to the definition of Eq. (8.25), $\Phi_1(t)$ reflects the global deformation of the dam structure. A larger absolute value of $\Phi_1(t)$ indicates greater deformation along the dam height and more severe structural damage. $\Phi_2(t)$ characterizes the global deformation of the dam heel region, essentially representing its global inclination and, consequently, the opening of transverse cracks at the slope change area. $\Phi_3(t)$ captures the global deformation of the lower part of the dam structure, approximating its global inclination and indicating the extent of damage at the dam toe.

The location of the dam foundation belongs to soil site I. The seismic input for the analysis is defined by two intensity levels: (a) the design basis ground motion with PGA of 0.399 g (stream-direction) and 0.266 g (vertical-direction); (b) the safety evaluation ground motion with PGA of 0.475 g (stream-direction) and 0.317 g

Table 8.2 Calculation cases

Calculation case	Analysis type	Earthquake action	PGA
Case 1	Linear	Design basis ground motion	0.399 g
Case 2	Nonlinear	Design basis ground motion	0.399 g
Case 3	Nonlinear	Safety evaluation ground motion	0.475 g

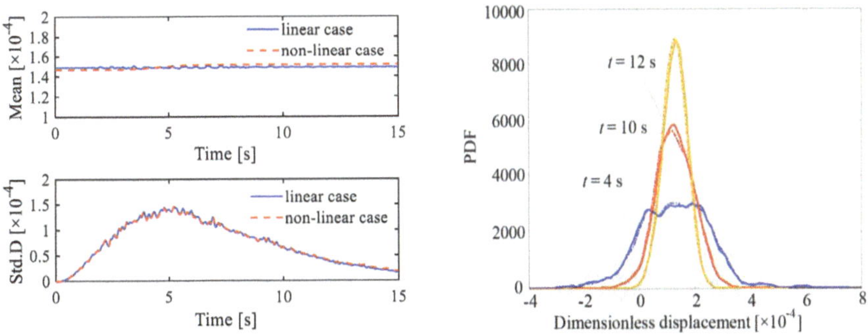

(a) Mean and standard deviation

(b) Typical PDF at different time instants (dotted line denotes linear case; solid line denotes nonlinear case)

Fig. 8.4 Probability information of $\Phi_1(t)$ under the design basis ground motion. (**a**) Mean and standard deviation (**b**) Typical PDF at different time instants (dotted line denotes linear case; solid line denotes nonlinear case)

(vertical-direction). To thoroughly investigate the random seismic response analysis of the dam, three distinct calculation cases are considered in this study, as outlined in Table 8.2.

In fact, for each computational case, the deterministic dynamic time-history analysis of the dam under all the generated seismic representative samples is conducted. Then through combining with the probability density evolution theory, the probabilistic characteristics of random seismic response for the three relative displacement angles $\Phi_1(t)$, $\Phi_2(t)$ and $\Phi_3(t)$ can be achieved. The probabilistic characteristics of stochastic response results for $\Phi_1(t)$ are presented as follows.

The stochastic response statistics for the design basis ground motion (Cases 1 and 2) are summarized in Fig. 8.4. Both the mean and standard deviation (Std. D) (Fig. 8.4a), as well as the PDFs at typical time instants (Fig. 8.4b) exhibit a high degree of fitting trend for both linear and nonlinear cases. This indicates that the probabilistic information for both types of random seismic response is virtually unanimous.

For the safety evaluation ground motion (Case 3), the mean and Std. D are shown in Fig. 8.5a, and the PDFs at three typical time instants are also provided in Fig. 8.5b. In Fig. 8.5c, the temporal evolution of the PDF is vividly visualized in three dimensions, forming a landscape akin to endless mountain ranges. Concurrently, the PDF contour plot in Fig. 8.5d reveals how the probability density transitions over time, resembling the dynamic flow of a whirling fluid.

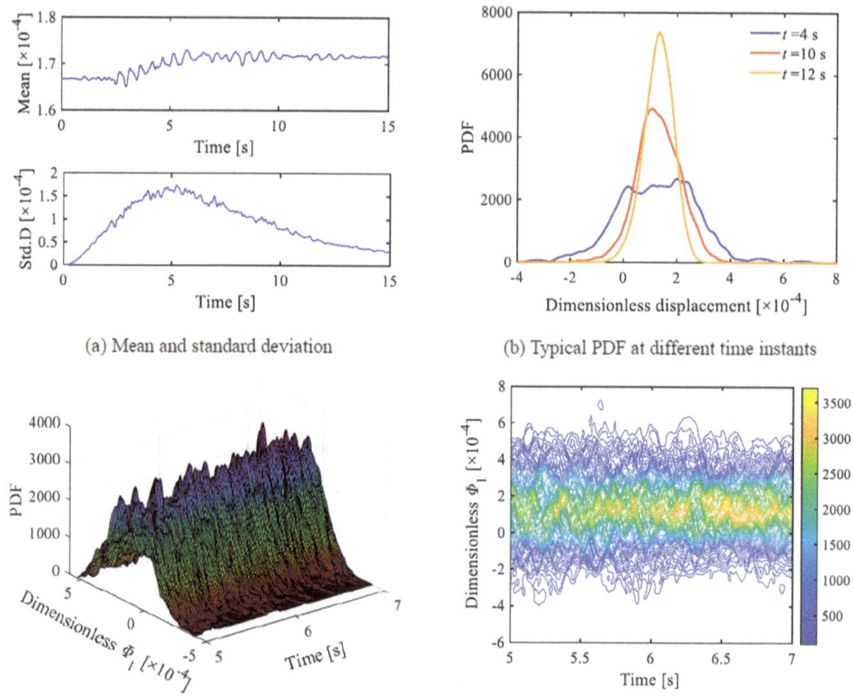

(a) Mean and standard deviation

(b) Typical PDF at different time instants

(c) PDF surface

(d) Contour

Fig. 8.5 Probability information of $\Phi_1(t)$ under the safety evaluation ground motion. (**a**) Mean and standard deviation (**b**) Typical PDF at different time instants (**c**) PDF surface (**d**) Contour

8.2.3 Seismic Reliability Assessment of the Concrete Gravity Dam

The EEV of the relative displacement angle $\Phi_j(t)$, which is a stochastic process, can be defined as (Liu et al. 2018):

$$\Phi_{j,\max} = \max_{t \in [0,T]} \left[\left| \Phi_j(t) \right| \right], \qquad j = 1, 2, 3 \tag{8.26}$$

Clearly, the EEV $\Phi_{j,\max}$ is a random variable.

Then, according to the reliability analysis based on extreme value distribution introduced in Sect. 8.1.2, the seismic reliability of the concrete gravity dam using the relative displacement angles as criterion index can be expressed as:

$$R_{T,j}(t) = \Pr\left\{ \Phi_{j,\max} < \phi_j \right\} = \int_0^{\phi_j} p_{\Phi_{j,\max}}(\phi)\mathrm{d}\phi, \quad j = 1, 2, 3 \tag{8.27}$$

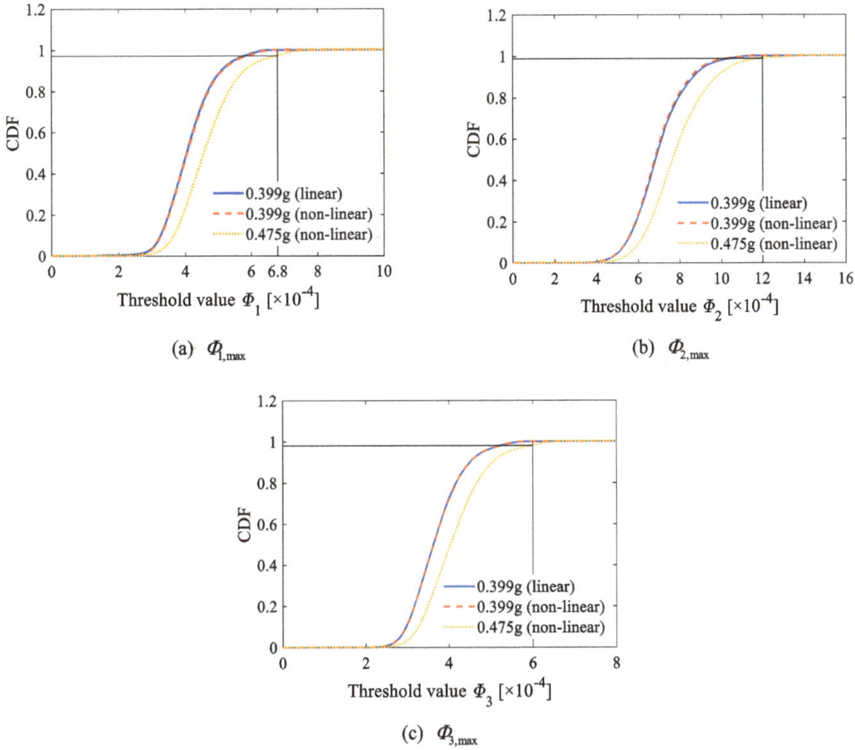

Fig. 8.6 Seismic reliability of the concrete gravity dam. (**a**) $\Phi_{1,\text{ max}}$ (**b**) $\Phi_{2,\text{ max}}$ (**c**) $\Phi_{3,\text{ max}}$

where $p_{\Phi_{j,\max}}(\phi)$ denotes the PDF of the EEV $\Phi_{j,\text{ max}}$, and it can be obtained by constructing a virtual stochastic process and solving the corresponding GPDEE. The cumulative distribution function (CDF) of the EEV $\Phi_{j,\text{ max}}$ then directly yields the seismic reliability $R_{T,j}(t)$. ϕ_j denotes the threshold of the relative displacement angle $\Phi_j(t)$.

Figure 8.6 shows the CDF of the EEV $\Phi_{j,\text{ max}}$, signifying the seismic reliability of the concrete gravity dam. Based on this, we can reach the following remarks.

1. Under the design basis ground motion with a representative horizontal PGA of 0.399 g, the seismic reliabilities for the three relative displacement angles $\Phi_1(t)$, $\Phi_2(t)$ and $\Phi_3(t)$ are nearly identical for both the linear and nonlinear analysis cases. This indicates that the dam effectively remains linear and does not enter a nonlinear state during the design-level event. Consequently, the seismic reliability of the dam can be considered nearly 100%, implying that the structure is probabilistically safe and reliable. The corresponding threshold values for the three relative displacement angles are 6.8×10^{-4},12×10^{-4} and 6.0×10^{-4}, respectively.

2. Considering the safety evaluation ground motion with a representative horizontal PGA of 0.475 g, and the threshold values of 6.8×10^{-4}, 12×10^{-4} and 6.0×10^{-4} for the three relative displacement angles, the resulting dynamic reliabilities are 97.6%, 98.0% and 98.1%, respectively. Since the reliabilities are all greater than 97.5% and very close to each other, it can be concluded that the dam seismic design is both reasonable and safe under the check earthquake scenario from the probabilistic standpoint.
3. When a common threshold value (e.g., 10×10^{-4}) is applied, the relative displacement angle $\Phi_2(t)$ exhibits the lowest seismic reliability. This identifies the dam heel as the most vulnerable region where damage is most likely to initiate. Therefore, the global reliability of the dam can be conservatively represented employing $\Phi_2(t)$ as the failure criterion.

8.3 Performance-Based Global Reliability Assessment of a High-Rise Frame-Core Tube Structure

8.3.1 Generation of Three-Component Ground Motion Compatible with the Design Response Spectrum

Three-component ground motion is inherently a vector-valued process at a specific location. In the context of seismic measurement and structural dynamics, the ground motion vector is conventionally decomposed into three mutually orthogonal components for analysis. The principal axis theory (Penzien and Watabe 1975) postulates the existence of three principal directions for ground motion. The components along these directions are statistically uncorrelated and are characterized by maximum, intermediate, and minimum variances, respectively. To simplify the discussion, the design response spectra and the EPSD for the three principal directions are assumed to be functionally identical, with their distinction lying solely in the assigned parameter values. This section assumes that the three principal axes of seismic motion are aligned with the x-, y-, and z-coordinate axes, where x and y represent horizontal directions and z represents vertical direction.

The design response spectrum in accordance with the "Code for Seismic Design of Buildings" (GB50011-2010 2016) of China is presented in Eq. (2.46) in Sect. 2.3.4. In this part, we just consider the structural system with the damping ratio of 0.05. Li and Liu (2011) indicate that the relationship between the horizontal and vertical seismic response spectrum is not one of simple equivalence, nor can the vertical spectrum be simplistically treated as a fixed proportion of the horizontal spectrum. Significant differences exist between them in aspects such as the peak seismic influence coefficient, the characteristic period, and the decay exponents of the descending branches. For the three-dimensional seismic design response spectrum, Li and Liu (2011) provide recommended values for the spectral parameters. For instance, considering the case that the seismic precautionary intensity of the

Table 8.3 Maximum of the three-component seismic influence coefficient α_{max}

	Earthquake action (seismic precautionary intensity is 8-degree, 0.2 g)		
Seismic component	Frequent ground motion	Basis ground motion	Rare ground motion
x-component	0.17	0.5	1.0
y-component	0.15	0.4	0.8
z-component	0.11	0.3	0.6

Table 8.4 Values of three-component seismic characteristic period T_g (s)

	Soil site II		
Seismic component	The first group	The second group	The third group
x-component	0.35	0.40	0.45
y-component	0.35	0.40	0.45
z-component	0.25	0.30	0.35

Table 8.5 Values of the three-component design response spectrum parameters γ, η_1 and η_2

	Parameters		
Seismic component	γ	η_1	η_2
x-component	0.90	0.02	1
y-component	0.85	0.02	1
z-component	0.80	0.02	1

earthquake action is 8-degree with PGA =0.2 g at soil site II, the values of the parameters for the three-component design response spectrum are given in Tables 8.3, 8.4, and 8.5, respectively.

In Table 8.3, the average values of the maximum seismic influence coefficients for the horizontal components (x- and y-direction) are consistent with the corresponding values in Table 2.6, while the maximum seismic influence coefficient for the vertical component (z-direction) is taken as two-thirds of this average values of the horizontal components. In Table 8.4, the values of the characteristic period T_g for the horizontal components are consistent with the values in Table 2.4, whereas for the vertical component, it is reduced by 0.1 s relative to the base value. Table 8.5 accounts for the differences in the decay exponents of the descending branches in the three-component design response spectrum.

Regarding the EPSD model for non-stationary ground motion processes, this section adopts type III model, i.e., Eq. (2.73), which combines time-varying PSD with intensity modulation function. Specifically, the time-varying PSD based on the Clough-Penzien model, i.e., Eq. (2.74), is employed here (Deodatis 1996):

$$S_v(\omega,t)=\frac{\omega_{g,v}^4(t)+4\zeta_{g,v}^2(t)\omega_{g,v}^2(t)\omega^2}{\left[\omega^2-\omega_{g,v}^2(t)\right]^2+4\zeta_{g,v}^2(t)\omega_{g,v}^2(t)\omega^2}\cdot\frac{\omega^4}{\left[\omega^2-\omega_{f,v}^2(t)\right]^2+4\zeta_{f,v}^2(t)\omega_{f,v}^2(t)\omega^2}\cdot S_{0,v}(t) \tag{8.28}$$

in which

$$\omega_{g,v}(t) = \omega_{0,v} - d\frac{t}{T}, \quad \zeta_{g,v}(t) = \zeta_{0,v} + e\frac{t}{T} \tag{8.29}$$

$$\omega_{0,v} = \overline{\omega}_{g,v} + \frac{d}{2}, \quad \zeta_{0,v} = \overline{\zeta}_{g,v} - \frac{e}{2} \tag{8.30}$$

$$\omega_{f,v}(t) = 0.1\omega_{g,v}(t), \quad \zeta_{f,v}(t) = \zeta_{g,v}(t) \tag{8.31}$$

$$S_{0,v}(t) = \frac{a_{\max}^2}{r_v^2 \pi \omega_{g,v}(t)\left(2\zeta_{g,v}(t) + \frac{1}{2\zeta_{g,v}(t)}\right)} \tag{8.32}$$

where the subscript v represents the different directional components of ground motion, that is $v = x, y, z$. The meanings of the parameters are identical to those in Eqs. (2.74) to (2.78).

In this study, the site soil parameters for the two horizontal components (x- and y-direction) of ground motion are considered identical in the time-varying PSD model. For the vertical component (z-direction), the site soil parameters can be obtained through the following relationship with horizontal components (Xue et al. 2004):

$$\overline{\omega}_{g,z} = 1.58\overline{\omega}_{g,y} = 1.58\overline{\omega}_{g,x} \tag{8.33}$$

$$\overline{\zeta}_{g,z} = \overline{\zeta}_{g,y} = \overline{\zeta}_{g,x} \tag{8.34}$$

Liu and Liu (2017) provide recommended values for site soil parameters of horizontal components, as well as d and e, in accordance with the "Code for Seismic Design of Buildings" (GB50011-2010 2016) of China. Table 8.6 presents specific values corresponding to the first design earthquake group of soil site II for rare ground motion case.

Meanwhile, the values of peak factor r_v for different directional components can be obtained from the three-component seismic design response spectrum. Specifically, the relationship between different directional seismic components are as follows:

$$r_y = \frac{\alpha_{\max,x}}{\alpha_{\max,y}} r_x, \quad r_z = \lambda \frac{\alpha_{\max,x}}{\alpha_{\max,z}} r_x \tag{8.35}$$

where $\alpha_{\max,\ x}$, $\alpha_{\max,\ y}$, $\alpha_{\max,\ z}$ denote the maximum values of seismic influence coefficient in the x-, y- and z-direction, respectively.

Table 8.6 Parameters of time-varying PSD for horizontal seismic components

Parameter	Value
$\overline{\omega}_{g,x}$, $\overline{\omega}_{g,y}$ (rad/s)	15.71
$\overline{\zeta}_{g,x}$, $\overline{\zeta}_{g,y}$	0.72
d (rad/s)	6
e	0.2

The peak factor r_x and vertical correction coefficient λ can be determined through optimal fitting of the seismic design response spectrum. For the case of soil site II and the first design earthquake group, the recommended values are $r_x = 2.75, \lambda = 1.1$ for frequent ground motion, and $r_x = 2.65$, $\lambda = 1.1$ for rare ground motion, respectively.

The three-segment intensity modulation function proposed by Amin and Ang, as presented in Eq. (2.64), has been extensively used for decades to modulate the non-stationary intensity of one-component horizontal ground motion. Its applicability to multi-component ground motion, however, is limited, as the vertical component exhibits distinctly different evolutionary characteristics in rise time, strong-motion duration and decay rate compared to the horizontal components. To better model this multi-component intensity non-stationarity, Li and Liu (2011) developed a refined, continuous four-segment intensity modulation function based on the classical Amin-Ang model, such that:

$$f_v(t) = \begin{cases} 0, & 0 \le t \le t_0 \\ \left(\frac{t - t_0}{t_1 - t_0}\right)^2, & t_0 < t \le t_1 \\ 1, & t_1 < t < t_2 \\ e^{-c(t - t_2)}, & t_2 \le \ t \le T \end{cases} \tag{8.36}$$

where T_0 and t_1 represent the start and end time of the ascending segment, respectively. t_2 denotes the start time of the descent segment. c denotes the attenuation index of the descent segment. Define $T_1 = t_1 - t_0$ as the duration of the ascending segment, and $T_S = t_2 - t_1$ as the duration of the stationary segment, respectively.

Through an extensive statistical analysis of strong motion records, the work of Li and Liu (2011) provided a detailed examination of the directional correlations in multi-component ground motions and furnished parameter values for the proposed four-segment intensity modulation function. These parameters are specified for near-field and far-field conditions at seismic intensity levels of 7-, 8- and 9-degree. To ensure consistency with the “Code for Seismic Design of Buildings” (GB50011-2010 2016) of China, the near- and far-field designations are redefined herein to correspond to the first and third design earthquake groups, respectively, and the parameters for the second design earthquake group are derived as the average of these two. Finally, the parameters for multi-component ground motions for the case of 8-degree and soil site II are explicitly listed in Table 8.7.

In this study, the dimension-reduction representation for univariate non-stationary stochastic process is applied to generate each seismic component process $U_v(t)(v = x, y, z)$, such that (Liu et al. 2022b):

$$U_v(t) = \sum_{n=1}^{N} \sqrt{2S_v(\omega_n, t)\Delta\omega}[X_{n,v} \cos(\omega_n t) + Y_{n,v} \sin(\omega_n t)] \tag{8.37}$$

where $\{X_{n,v}, Y_{n,v}\}$ is a set of standard orthogonal random variables, satisfying the following basic conditions:

$$E[X_{n,v}] = E[Y_{n,v}] = 0, \quad E[X_{n,v}Y_{m,s}] = 0, \quad E[X_{n,v}X_{m,s}] = E[Y_{n,v}Y_{m,s}] = \delta_{mn}\delta_{vs} \tag{8.38}$$

Applying the dimension-reduction representation, $X_{n,v}$ and $Y_{n,v}$ can be defined as the random orthogonal functions of the elementary random variable $\boldsymbol{\Theta} = (\Theta_x, \Theta_y, \Theta_z) = (\Theta_1, \Theta_2, \Theta_3)$ (Liu et al. 2022b):

$$\begin{cases} X_{n,v} = \sqrt{2}\cos(\overline{n}\cdot\Theta_v + \pi/4) \\ Y_{n,v} = \sqrt{2}\sin(\overline{n}\cdot\Theta_v + \pi/4) \end{cases}; \quad v = x, y, z \tag{8.39}$$

where Θ_1, Θ_2 and Θ_3 are used to simulate the acceleration processes of three different directional seismic components, and they are independently and uniformly distributed in the interval $(0, 2\pi]$. The deterministic one-to-one mapping transformation between $\overline{n}(\overline{n} = 1, 2, \cdots, N)$ and $n(n = 1, 2, \cdots, N)$ can be implemented using the method described in Sect. 3.3.

Iterative adjustment of the EPSD is employed herein to ensure a match with the three-component design response spectrum. The corresponding steps are outlined in Sect. 3.4.3. The simulated ground motion considers the seismic intensity of 8-degree and the first earthquake design group at soil site II. The parameters utilized for simulation are: the duration is $T = 30$ s, the time interval is $\Delta t = 0.02$ s, the upper cut-off frequency is $\omega_u = 150$ rad/s, the number of frequency discrete terms is $N = 1000$ and the frequency interval is $\Delta\omega = 0.15$ rad/s. Finally, $n_{sel} = 185$ sets of representative samples for the three-component non-stationary stochastic ground motion with compatible response spectrum are generated.

Figure 8.7 compares the representative acceleration time-histories for different directional components before and after three iterations. The waveforms differ

Table 8.7 Parameter values of the four-segment intensity modulation function

Seismic precautionary intensity	Design earthquake group	Seismic component	Soil site II			
			t_0 (s)	T_1 (s)	T_S (s)	c
8-degree	The first group	x	0.3	4.1	4.1	0.20
		y	0.2	3.9	5.3	0.17
		z	0.0	1.2	10.7	0.16
	The second group	x	0.3	5.6	5.0	0.14
		y	0.2	5.4	6.5	0.12
		z	0.0	2.4	12.9	0.11
	The third group	x	0.3	7.1	6.0	0.08
		y	0.2	6.9	7.8	0.06
		z	0.0	3.6	15.1	0.06

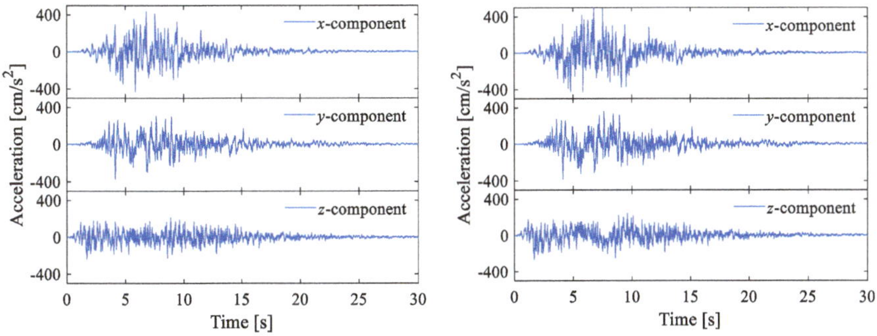

(a) 0th iteration (b) 3rd iteration

Fig. 8.7 Representative time-histories of multi-component ground motion. (**a**) 0th iteration (**b**) 3rd iteration

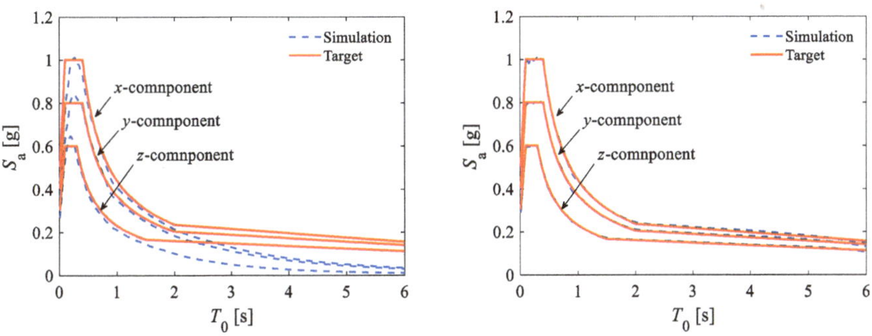

(a) 0th iteration (b) 3rd iteration

Fig. 8.8 Comparison of multi-component response spectra between simulation and design value. (**a**) 0th iteration (**b**) 3rd iteration

markedly, a result originating from the distinct EPSD parameters assigned to each component. Key observations include: the *x*-component has the largest amplitude, slightly greater than the *y*-component, while the *z*-component is smallest; the vertical component also peaks earlier and has a longer strong-motion duration, consistent with the four-stage intensity modulation model. These features collectively confirm the successful simulation of uncorrelated components and the feasibility of the proposed spectral model. Additionally, iteration effectively refines the long-period waveform characteristics, as evident from the adjusted wave patterns after the third iteration.

The comparison between the simulated and design response spectra is presented in Fig. 8.8 for the 0th iteration and 3rd iteration cases. The figures reveal substantial initial discrepancies across all three components, which are effectively eliminated after three iterations. The final simulated response spectra align excellently with the design targets, fully complying with seismic design requirements and thereby validating the iterative scheme.

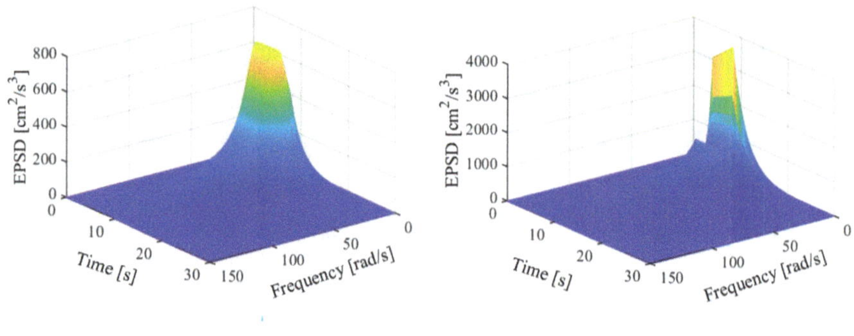

(a) 0th iteration (b) 3rd iteration

Fig. 8.9 Evolutionary power spectrum of the seismic x-component (Liu et al. 2022b). (**a**) 0th iteration (**b**) 3rd iteration

Figure 8.9 presents the comparison of the EPSD for the x-component of the simulated ground motion, illustrating the state before iteration and after the third iteration. The results show a pronounced enhancement in spectral amplitude within the low-frequency band after iteration. This demonstrates that the procedure systematically corrects the low-frequency components, which is physically consistent with the improved fit in the long-period region of the response spectrum. The consistent behavior observed across all three components substantiates the robustness and utility of the iterative scheme.

8.3.2 Engineering Background Description

The dynamic reliability of a 27-story high-rise frame-core tube structure subjected to three- component non-stationary ground motion is investigated as an engineering case study. The structural system employs a box-shaped, closed-section steel-reinforced concrete (SRC) frame combined with a reinforced concrete (RC) core tube, with a total height of 98.1 m. The first floor has a height of 4.5 m, while the rest of the floors have a height of 3.6 m. The structural plane measures 24 m by 24 m, with the layout depicted in Fig. 8.10a. Assuming the location where the structure is situated has a seismic precautionary intensity of 8-degree, falls under the first design earthquake group and soil site II. Further, according to the "Code for Seismic Design of Buildings" (GB50011-2010 2016) of China, the design characteristic period of ground motion is 0.35 s, and the design basis ground motion acceleration is 0.2 g.

Regarding structural materials, C40 concrete is used for floor slabs, frame beams, and SRC frame columns. HRB400 steel reinforcement is specified for shear walls, floor slabs, and frame beams, while Q345 steel sections are adopted for the SRC frame columns. All frame beams have a cross-section of 350 mm (width)×750 mm (height), and all floor slabs are 180 mm thick. The size and materials of structural

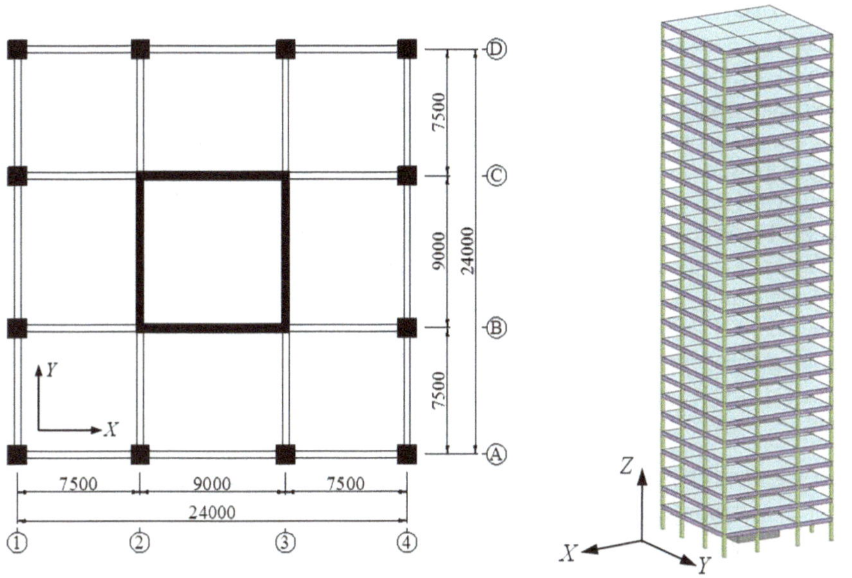

(a) Plane layout (b) Finite element model

Fig. 8.10 Plane layout and finite element model of the high-rise frame-core tube structure (Liu et al. 2022b). (**a**) Plane layout (**b**) Finite element model

Table 8.8 Size and materials of structural members (Liu et al. 2022b)

Story	Size of SRC columns: $b \times h$ (mm)	Size of shape steels: $b_s \times h_s \times t_w \times t'_{w1} \times t'_{w2}$ (mm)	Thickness of shear-walls (mm)	Concrete strength of shear-walls
1–5	1100×1100	550×550×30×30×30	500	C50
6–10	1000×1000	500×500×30×30×30	500	C50
11–20	1000×1000	500×500×30×30×30	500	C40
21–27	800×800	400×400×25×25×25	400	C40

members are provided in Table 8.8 in detail. Additionally, a uniform dead load of 6.0 kN/m^2 and a live load of 2.0 kN/m^2 are applied to all floors, including the roof.

The FEM of the steel-reinforced concrete (SRC) frame-reinforced concrete (RC) core tube structure is implemented, considering nonlinear dynamic response analysis. The FEM comprises 324 column elements, 540 beam elements, 108 wall-column elements, and 243 slab elements, as depicted in Fig. 8.10b. The self-weight of the members is automatically applied by the finite element program. In this model, vertical loads include dead load, live load, and self-weight, while horizontal actions mainly consider the external seismic excitations. The base of the structure is assumed to be rigidly connected to the foundation.

For concrete material, the uniaxial compressive stress-strain relationship specified in the "Code for Design of Concrete Structures" (GB50010-2010 2016) of China is adopted as the constitutive model. A simplified bilinear model is used for the reinforced steel material. Nonlinear moment-rotation relationship is considered at the ends of both beams and columns. The hysteretic behavior of both beam and column plastic hinges is modeled using the modified Takeda three-linear model. Shear walls are simulated using nonlinear shear wall elements based on the fiber model. The P-delta effect is considered for column element groups.

For seismic dynamic analysis, Rayleigh damping is used to represent the structural damping. Based on the fundamental periods of the first two modes, $T_{s1} = T_{s2} = 1.848$ s, and an assumed damping ratio of 5%, the mass and stiffness damping constants are determined as 0.17 and 0.0147, respectively.

8.3.3 Performance-Based Reliability Assessment of the High-Rise Building

The widely applied performance-based seismic design customizes performance objectives according to the specific functions and importance of structural systems, placing emphasis on the global reliability of the structure throughout its whole service life. For the performance indicators, this study employs the deformation-based failure criterion due to its conceptual clarity, operational simplicity and compatibility with current codes. Define $\Phi_{\max}$ as the EEV of the inter-story drift angle response of the whole structure, such that (Liu et al. 2022b):

$$\Phi_{\max} = \max_{j\in[1,N_r]}\left\{\Phi_{j,\max}\right\},\quad j=1,2,\cdots,N_r \tag{8.40}$$

where N_r denotes the story number of the structure. $\Phi_{j,\max}$ represents the EEV of the inter-story drift angle of the j-th story, which can be defined as:

$$\Phi_{j,\max} = \max_{x,y}\left\{\max_{t\in[0,T]}\left|\Phi_{j,x}(t)\right|,\ \max_{t\in[0,T]}\left|\Phi_{j,y}(t)\right|\right\},\quad j=1,2,\cdots,N_r \tag{8.41}$$

where $|\cdot|$ signifies taking the absolute value. $\Phi_{j,x}(t)$ and $\Phi_{j,y}(t)$ denote the inter-story drift angle time-history of the j-th story along x- and y-direction, respectively.

According to the previous theory, the single-story reliability $R_j(\phi_B)$ and structural global reliability $R(\phi_B)$ can be presented in the following form:

$$R_j(\phi_B)=\Pr\left\{\Phi_{j,\max}<\phi_B\right\}=\int_0^{\phi_B} p_{\Phi_{j,\max}}(\phi)\mathrm{d}\phi,\quad j=1,2,\cdots,N_r \tag{8.42}$$

$$R(\phi_B) = \Pr\{\Phi_{\max} < \phi_B\} = \int_0^{\phi_B} p_{\Phi_{\max}}(\phi)\mathrm{d}\phi \tag{8.43}$$

where $p_{\Phi_{j,\max}}$ and $p_{\Phi_{\max}}$ denote the PDFs of $\Phi_{j,\max}$ and $\Phi_{j,\max}$, respectively. Accordingly, the CDFs of $\Phi_{j,\max}$ and $\Phi_{j,\max}$ are the values of dynamic reliability $R_j(\phi_B)$ and $R(\phi_B)$. ϕ_B denotes the allowable (threshold) inter-story drift angle, respectively.

Considering the seismic precautionary intensity of 8-degree and the scenario of rare ground motion, the representative acceleration time-histories generated through the proposed dimension-reduction method for the three-component ground motion serve as the external excitation for the high-rise building. Four input conditions are considered: *X*-excitation, *X*+*Y*-excitation, *X*+*Z*-excitation, and *X*+*Y*+*Z*-excitation. The structural component reliability for each story is determined based on the elastic-plastic stage, say $\phi_B = 1/100$, as detailed in Table 8.9. It should be noted that the long-period range of the design response spectrum is conservatively

Table 8.9 Component reliabilities of the structure under multi-dimensional ground motion

Story No.	*X*-excitation	*X*+*Z*-excitation	*X*+*Y*-excitation	*X*+*Y*+*Z*-excitation
1	100.00%	100.00%	100.00%	100.00%
2	100.00%	100.00%	100.00%	100.00%
3	99.95%	99.88%	99.50%	99.47%
4	99.14%	99.10%	98.51%	98.45%
5	97.69%	97.63%	95.12%	95.20%
6	96.04%	95.98%	92.58%	93.07%
7	94.56%	94.48%	90.83%	91.29%
8	93.31%	93.23%	89.49%	89.79%
9	92.56%	92.47%	88.60%	88.72%
10	92.17%	92.07%	87.97%	88.00%
11	91.81%	91.69%	87.44%	87.43%
12	91.50%	91.37%	86.96%	86.93%
13	91.22%	91.07%	86.49%	86.44%
14	90.98%	90.83%	86.05%	86.00%
15	90.76%	90.64%	85.66%	85.62%
16	90.60%	90.49%	85.34%	85.30%
17	90.47%	90.36%	85.12%	85.09%
18	90.39%	90.27%	84.97%	84.93%
19	90.35%	90.22%	84.90%	84.85%
20	90.35%	90.23%	84.91%	84.86%
21	90.28%	90.15%	84.79%	84.75%
22	90.34%	90.22%	84.95%	84.90%
23	90.41%	90.29%	85.12%	85.07%
24	90.52%	90.41%	85.36%	85.29%
25	90.66%	90.55%	85.62%	85.56%
26	90.81%	90.71%	85.91%	85.85%
27	91.08%	90.99%	86.41%	86.35%

increased to ensure structural safety. Therefore, since the seismic input is compatible with the design response spectrum, the resulting reliability assessment in this study essentially represents a form of conditional reliability.

As evident from the table, under all seismic excitation conditions, the component reliabilities (i.e., single-story reliabilities) of all stories gradually diminish as the structural height increases, reaching the minimum value at the 21st floor. This could be attributed to a sudden reduction in the cross-sectional dimensions of the vertical structural members at that floor, after which the component reliabilities gradually increase again. Analyzing different excitation conditions, the component reliabilities under X+Z-excitation are 0.05% to 0.14% lower than those under single X-excitation. The component reliabilities under X+Y-excitation are 0.45% to 5.54% lower than those under single X-excitation. The component reliabilities under X+Y+Z-excitation are lower than those under X+Y-excitation, except for stories 5–10, and the component reliabilities under X+Y+Z-excitation are 0.41% to 5.44% lower than those under X+Z-excitation. These findings fully demonstrate that the structural reliability subjected to multi-component seismic input is generally lower than that under unidirectional excitation. Especially, bidirectional horizontal excitations would lead to a significant decrease in the structural reliability, while the impact of vertical seismic component on reliability changes is relatively limited.

Performance-based seismic design, which emphasizes multi-level fortification criteria and performance objectives, allows for the targeted strengthening of critical and vulnerable structural components during the design process. Structures designed following this philosophy are better able to satisfy specified seismic performance requirements during future earthquake events. As previously stated, the performance level in this study is defined based on structural deformation which can effectively reflect the overall structural behavior and damage conditions. According to the "Code for Seismic Design of Buildings" (GB50011-2010 2016) of China, the specific quantitative limits for inter-story drift angles under rare ground motion are applied to define these performance levels, as summarized in Table 8.10. Generally, Level 1 corresponds to "Immediate Occupancy", Levels 2 and 3 correspond to

Table 8.10 Performance levels and corresponding inter-story drift angle thresholds

Performance level	Macro damage condition	Description of deformation	Thresholds of inter-story drift angle
Level 1	Basically intact	Slight exceedance of elastic limit.	1/800
Level 2	Light to medium damage	Less than twice the elastic limit.	1/400
Level 3	Medium damage	Less than four times the elastic limit.	1/200
Level 4	Not serious damage	Deformation within 90% of ultimate capacity.	1/111
Level 5	Serious damage	Deformation reaches the plastic deformation limit.	1/100

Table 8.11 Structural global failure probability at different performance levels

	8-degree					9-degree
Earthquake condition	Level 1	Level 2	Level 3	Level 4	Level 5	Level 5
X-excitation	100.00%	98.66%	70.45%	15.68%	9.73%	66.32%
X+*Z*-excitation	100.00%	98.71%	70.41%	15.66%	9.87%	66.71%
X+*Y*-excitation	100.00%	99.81%	90.21%	25.89%	15.27%	85.83%
X+*Y*+*Z*-excitation	100.00%	99.85%	90.12%	25.84%	15.31%	86.03%

"Damage Control", and Levels 4 and 5 correspond to "Life Safety" and "Collapse Prevention", respectively.

Employing the performance levels and the corresponding thresholds of inter-story drift angles listed in Table 8.10 for global reliability assessment of the high-rise frame-core tube structure under severe seismic excitation, the results are summarized in Table 8.11 in detail. In this context, the global failure probability equals one minus the global reliability. For example, under 8-degree seismic excitation in the *X*-direction, the global failure probability at performance level 5 is 9.73%, corresponding to a global reliability of 90.27%. For comparison, Table 8.11 also includes the global failure probabilities of the structure under rare ground motion of 9-degree for performance level 5.

A detailed examination of Table 8.11 yields the following interpreted remarks:

1. A clear inverse correlation exists between the stringency of the performance level (from 5 to 1) and the global structural reliability, evidenced by a progressive increase in failure probability. The failure probabilities for Levels 1 and 2 exceed 95% across all input conditions, showing that these performance states unattainable under 8-degree rare ground motion. The structural compliance with a performance objective under a 10% allowable failure probability is limited to Level 5 for *X* and *X*+*Z*-excitation. Relaxing this allowance to 20% permits compliance for all excitations at Level 5 and for *X* and *X*+*Z*-excitation at level 4.
2. The influence of the vertical seismic component on global reliability is statistically insignificant, as confirmed by failure probability differences of less than 0.15% when comparing excitations with and without *z*-component. Obviously, the employment of two horizontal seismic components induces a substantial increase in failure probability than just one directional excitation, as evidenced by rises of approximately 20%, 10%, and 5% at Performance Levels 3, 4, and 5, respectively. The trends observed under 8-degree excitation remain valid under 9-degree rare ground motion. This suggests that when implementing performance-based seismic design for structures in regions with high seismic precautionary intensity, it is essential not only to enhance the structural seismic resilience but also to reasonably consider the impacts of multi-component earthquake actions, particularly bidirectional horizontal excitation.

8.4 Global Reliability Evaluation of a High-Pier Long-Span Reinforced Concrete Rigid Frame Bridge

8.4.1 Engineering Background Description

This section focuses on a high-pier, long-span RC continuous rigid frame bridge, the main structure of Yilihe Highway Bridge in Yunnan Province, China (Liu et al. 2023). The bridge has a total length of 464.7 m. Its superstructure is a (73 +135 +73) m variable-depth, prestressed concrete continuous rigid frame, supported by a substructure comprising four piers, designated as Piers #1 to #4. Piers #2 and 3 serve as the main piers, featuring rigid connections to the main beam. Piers # 1 and 4, on the other hand, are transition piers connected to the main beam via bearings, indicating a hinged joint. The structural configuration, size, and materials of major components are listed in Table 8.12. The layout of the main bridge type and the cross-section of major components are shown in Fig. 8.11.

The FEM of this high-pier long-span RC rigid-frame bridge utilized the following element types. The variable-depth main girder, variable-cross-section main piers, and transition piers are modeled using BEAM189 elements. The pier caps, pile caps, and piles are modeled using BEAM44 elements. Rigid arms are simulated with BEAM4 elements. LNR bearings and the soil springs for the pile foundations are represented by COMBIN14 spring elements. The approach spans at both ends of the main bridge are considered as additional mass, modeled with MASS21 mass elements. In terms of boundary conditions, full translational and rotational constraints are applied to the nodes at the bottom of the piles and the fixed ends of the soil springs. Only vertical restraints are applied to the two ends of the main girder. For the pile-soil interaction, the effect is simulated using equivalent, layered soil springs. The piles are discretized into segments at 2 m intervals from top to bottom, with soil

Table 8.12 Structural configuration, size and materials of major bridge components

Structural components	Structural configuration	Structure size	Concrete grade
Main girder	Single-box single-chamber variable-section beam	The height from the end of 3 m to the bottom of 8.3 m follows a quadratic parabolic curve.	C55
Main pier	Box-shaped hollow piers with variable cross-sections	Both heights of Piers #2 and #3 are 123 m. cross-sectional dimension from top to bottom is 5.5×6.0 m – 8.5×6.0 m (transverse × longitudinal direction). The wall thickness is 0.8 m.	C40
Transition pier	Box-type hollow piers	The heights of Piers #1 and 4 are 69.5 m and 62.2 m, respectively. Cross-sectional dimension is 6.0×3.5 m. the wall thickness is 0.7 m.	C40
Foundation and cushion cap	Cast-in-place pile foundation	The diameters of the main pier and transition pier pile foundation are 2.2 m and 2.0 m, respectively.	C30

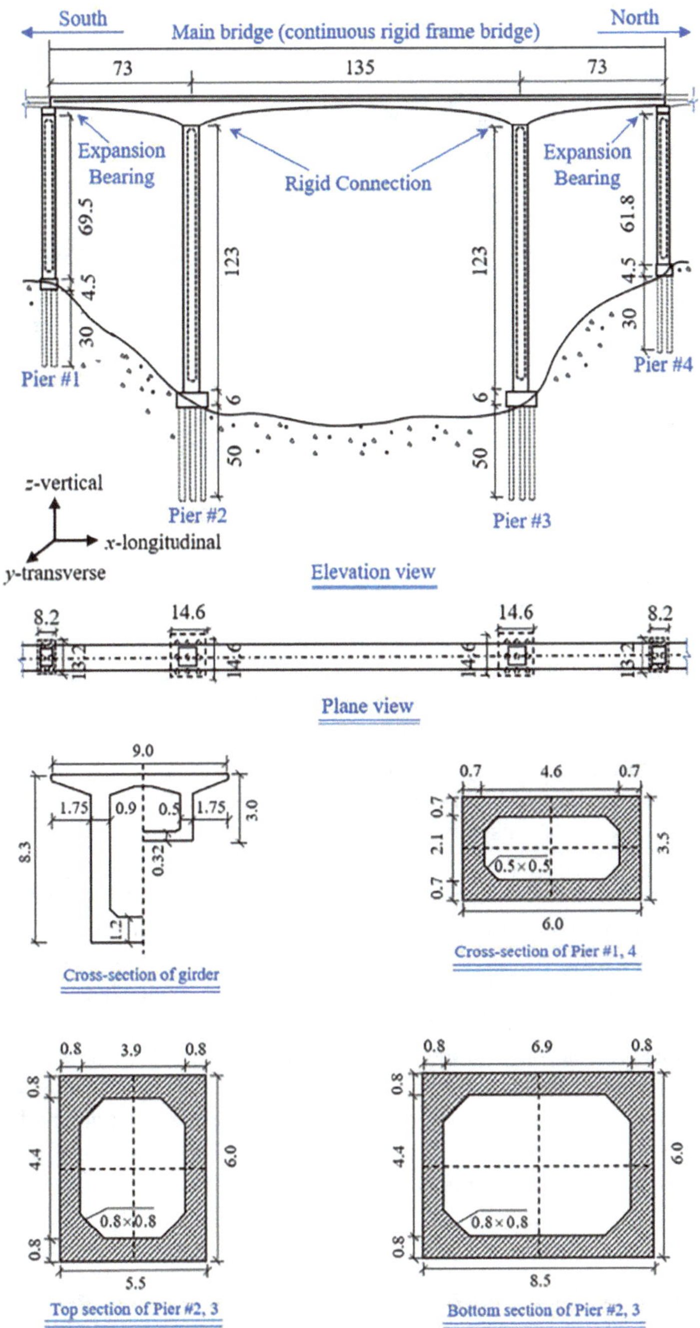

Fig. 8.11 General layout of the bridge and typical cross-sections of major components (Unit: m) (Liu et al. 2023)

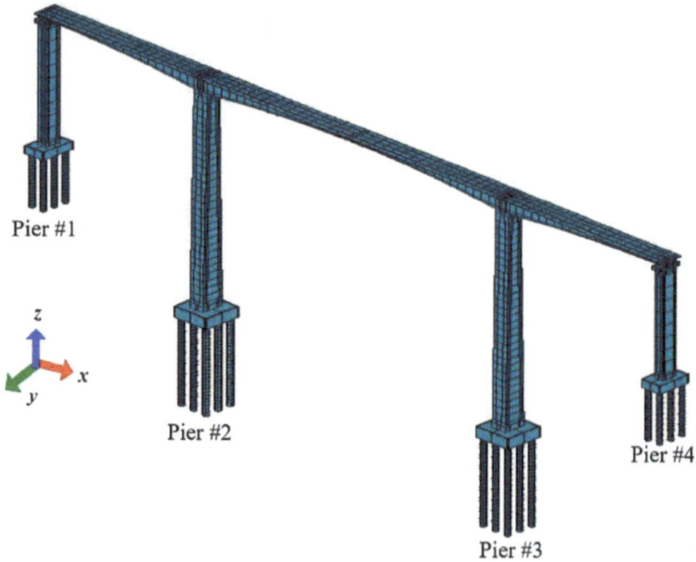

Fig. 8.12 3D FEM of the high-pier long-span RC rigid frame bridge (Liu et al. 2023)

spring elements attached at each node in two horizontal directions. The stiffness of these equivalent soil springs is calculated according to the "m-method" specified in the Chinese code (GB 50011-2010 2016, JTG/T-2008 2008) for analyzing the horizontal displacement and internal forces of elastic piles.

Overall, the established FEM of the high-pier long-span RC rigid frame bridge is depicted in Fig. 8.12. Its first twenty modes of vibration, along with the corresponding natural frequencies and periods, are summarized in Table 8.13.

The inherent dynamic properties of structural natural vibrations fundamentally govern the characteristics of seismic response. Based on detailed examination of frequencies, periods, and mode shapes of the first twenty vibration modes, the following remarks can be drawn. (1) The fundamental mode shape is transverse bending, with the corresponding natural frequency of 0.17 Hz. This indicates quite minimal transverse stiffness in the main span of the bridge. (2) Due to the bearing connections between the transition piers and the main girder, an individual transition pier can exhibit independent longitudinal or vertical bending in certain mode shapes. (3) The formation of the bridge structure is denser and uniformly distributed, indicating that this bridge type is favorable for seismic resistance.

8.4.2 Seismic Dynamic Response Analysis of the High-Pier Long-Span Bridge

The original design requires the high-pier long-span bridge to remain elastic under rare ground motion with intensity of 8-degree. Given its inherent flexibility, the

Table 8.13 Formation characteristics of the first twenty orders for the bridge FEM

Formation No.	Natural frequency (Hz)	Natural vibration period (s)	Main characteristics of formation
1	0.17	5.88	System first-order transverse bending
2	0.22	4.55	System longitudinal deformation
3	0.25	4.00	System second-order transverse bending
4	0.42	2.38	Transition pier #1 longitudinal bending
5	0.48	2.08	Transition pier #4 longitudinal bending
6	0.51	1.96	System first-order transverse bending
7	0.62	1.61	System second-order transverse bending
8	0.71	1.41	System first-order transverse bending
9	0.95	1.05	System third-order vertical bending
10	1.00	1.00	System second-order transverse bending
11	1.18	0.85	System third-order transverse bending
12	1.32	0.76	System fourth-order vertical bending
13	1.35	0.74	System second-order transverse bending
14	1.60	0.63	System third-order vertical bending
15	2.03	0.49	System third-order transverse bending
16	2.05	0.49	Transition pier #1 vertical bending
17	2.30	0.43	System fourth-order vertical bending
18	2.47	0.40	Transition pier #4 vertical bending
19	2.65	0.38	System fifth-order vertical bending
20	3.21	0.31	System sixth-order vertical bending

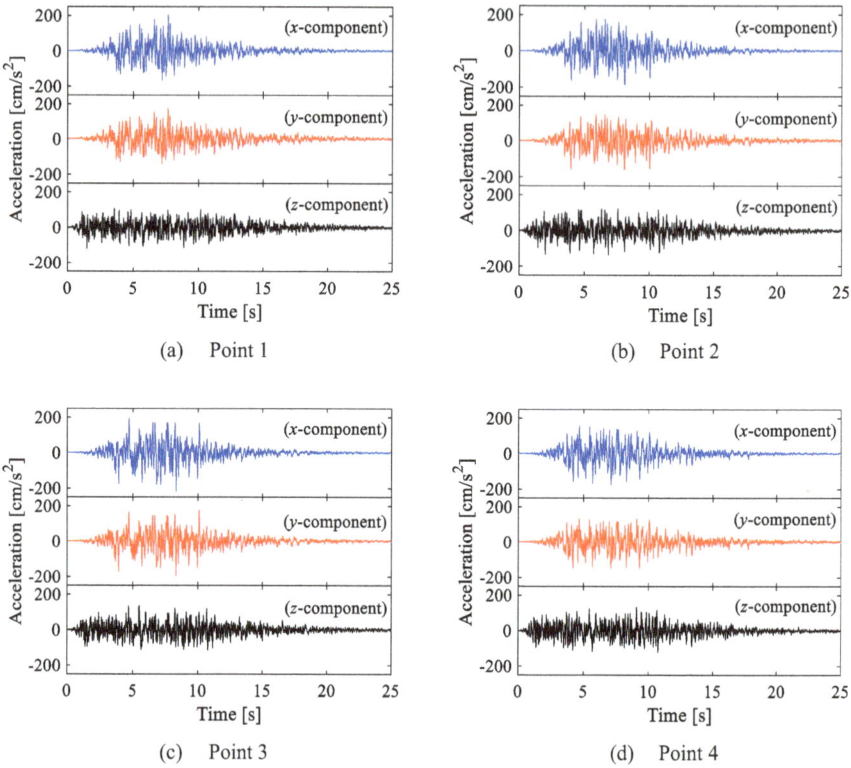

Fig. 8.13 Representative time-histories of multi-point and multi-component stochastic ground motion simulated by dimension-reduction method. (**a**) Point 1 (**b**) Point 2 (**c**) Point 3 (**d**) Point 4

structure experiences considerable horizontal displacements under seismic or wind loads, the impact of which on the structural internal forces should be accounted for. Consequently, the nonlinear dynamic time-history analysis is implemented to capture the geometric nonlinearity, thereby enabling more precise seismic response evaluation. Since the dynamic analysis assumes large displacements with small strains, just the geometric nonlinear effects (e.g., large deformation) are taken into account while the material model remains linear.

Regarding the seismic input for the bridge, considering the spatial variability and multi-component characteristic of ground motion, 266 sets of seismic acceleration representative time-histories are generated by employing the dimension-reduction representation for multi-point and multi-component stochastic ground motion described in Sect. 5.5. In line with the conditions at the bridge soil site, the generated ground motion corresponds to rare ground motion with intensity of 8-degree, featuring PGA of 0.4 g and soil site I. The simulated representative time-histories of multi-point and multi-component stochastic ground motion are shown in Fig. 8.13.

It should also be noted that seismic excitation in a specific direction predominantly exerts a substantial influence solely on the structural dynamic response in the same direction, while its impact on other directions remains relatively minor.

Table 8.14 Load cases of the bridge structural dynamic analysis

Load case	Seismic excitation	PGA
Case 1	*x*-excitation	0.4 g
Case 2	*x*+*y*-excitation	0.4 g
Case 3	*x*+*z*-excitation	0.4 g
Case 4	*x*+*y*+*z*-excitation	0.4 g

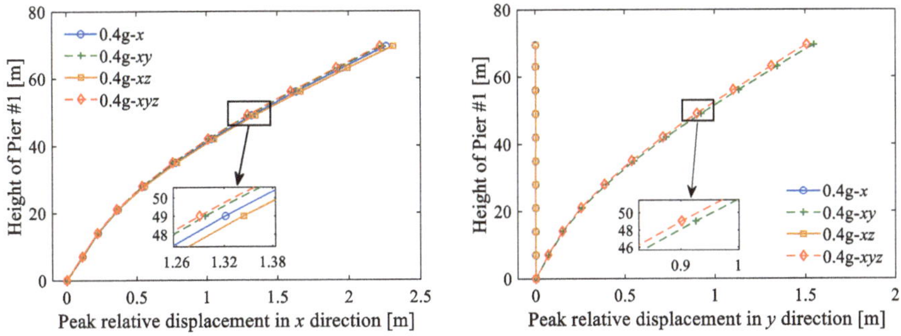

(a) Pier #1 (transition pier)

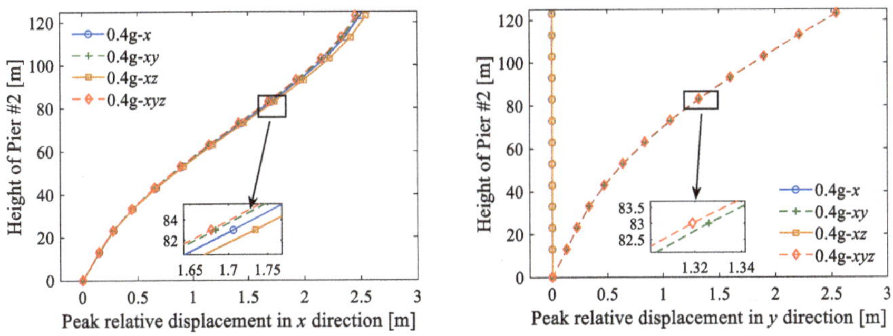

(b) Pier #2 (main pier)

Fig. 8.14 Height wise variation of peak relative displacement in longitudinal (*x*) and transverse (*y*) directions for Piers #1 and #2. (**a**) Pier #1 (transition pier) (**b**) Pier #2 (main pier)

Consequently, for such bridge structures, the longitudinal seismic excitation should be the primary consideration. Therefore, all the subsequently defined seismic loading scenarios include longitudinal excitation. The seismic load cases are specified in Table 8.14, where *x*, *y* and *z* denote longitudinal, transverse and vertical directions, respectively.

Figure 8.14 presents the mean peak relative displacements in the longitudinal (*x*) and transverse (*y*) directions for Piers #1 (transition pier) and #2 (main pier). The absence of *y*-direction displacement in Cases 1 and 3 is attributed to the lack of transverse seismic input. Under identical conditions, it is evident that Pier #2 exhibits larger displacement than Pier #1, a trend that is especially significant in the *y*-

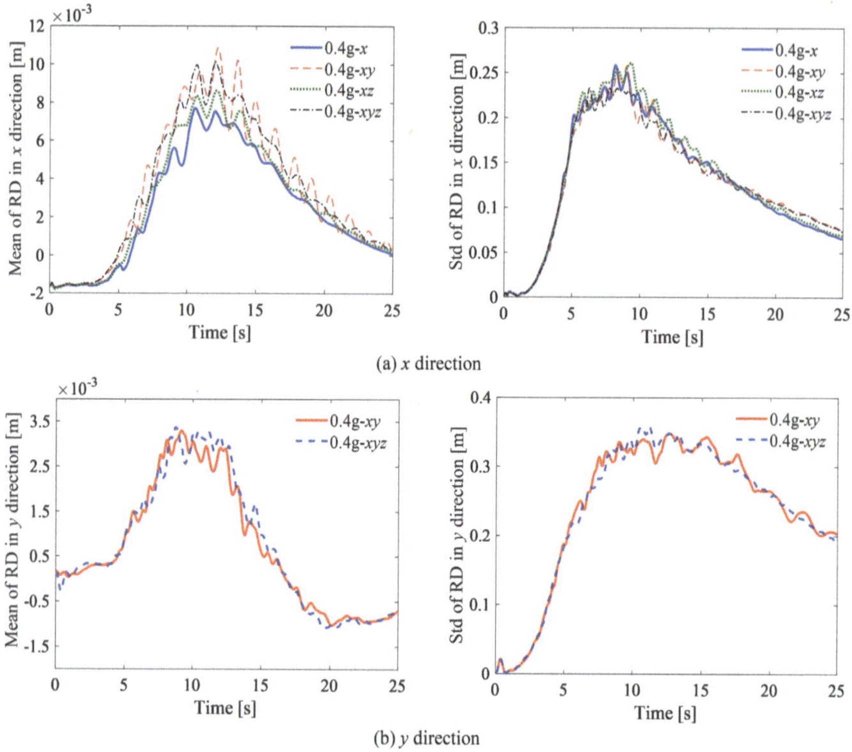

Fig. 8.15 Time-histories of the mean and standard deviation of the relative displacement between the upper and lower ends of the bearing over Pier #1. (**a**) x direction (**b**) y direction

direction. Displacement patterns also differ between the piers: Pier #1 experiences dominantly longitudinal (x) displacement, whereas Pier #2 shows slightly larger transverse (y) than longitudinal displacement, underscoring the importance of transverse seismic effects in such flexible, high-pier structures. Additionally, for any given pier and direction, the peak displacement is largely consistent across all cases where that direction is excited, confirming that the vertical seismic component has a minimal influence on horizontal displacement responses.

To further investigate the vulnerability of the bearings on the transition piers, the relative displacement between their upper and lower ends in the longitudinal (x) and transverse (y) directions under different working conditions is analyzed. Figure 8.15 presents the mean and standard deviation time-histories for the bearing on Pier #1. As shown, the time-history curves of the relative displacement exhibit an initial increase followed by a decrease, consistent with the trend of the input seismic accelerations. In this bridge, LNR horizontal force-dispersing rubber bearings are installed on the transition piers. These bearings have an effective rubber thickness of 174 mm, with an allowable shear displacement of 261 mm (1.5 times the rubber thickness) under frequent and basis ground motion, and an ultimate shear

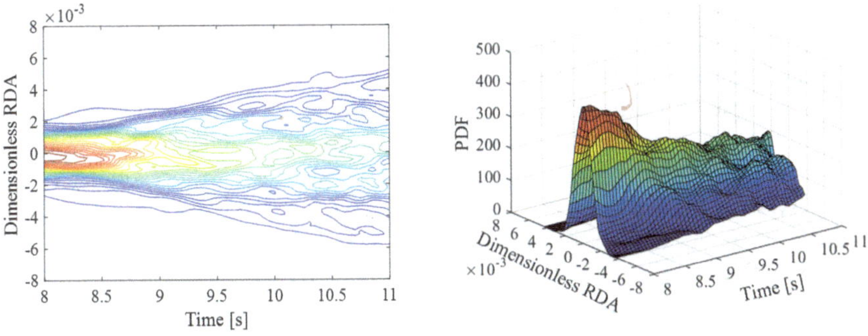

(a) Contour (b) PDF surface

Fig. 8.16 Probability information of relative displacement angle (RDA) of Pier #2 in x direction. (**a**) Contour (**b**) PDF surface

displacement of 435 mm (2.5 times the rubber thickness) under rare ground motion (Xie 2006). The calculated results demonstrate that the maximum relative displacement under each 0.4 g condition remains below the ultimate shear displacement, confirming that the bearings on the transition piers remain undamaged under the 8-degree rare ground motion.

A significant advantage of combining the dimension-reduction representation with the PDEM is its capability to determine the probability density evolution of various randomly-excited structural responses, such as deformation and internal forces. Figure 8.16 presents the probability characteristics of the relative displacement angle (RDA) at the top of Pier #2 from $t = 8$ s to 11 s under operating case 1, obtained via PDEM. The PDF contour in Fig. 8.16a shows the temporal progression of probability, while the 3D surface in Fig. 8.16b depicts its evolution over time. This methodology enables the straightforward acquisition of probability density information for any physical quantity, at any time and location within the structure, conclusively validating the practical power of the dimension-reduction representation method.

8.4.3 *Global Seismic Reliability Evaluation of the High-Pier Long-Span Bridge*

Damage to the substructure is a primary cause of failure in high-pier bridges during severe earthquakes (Sun et al. 2019). Therefore, the seismic reliability of such bridges can be approximated by assessing the reliability of the high piers, typically through nonlinear analysis based on displacement or curvature ductility. The "Specifications for Seismic Design of Highway Bridges" (JTG/T 2231-01-2020 2020) of China mandates a deformation check for piers under rare ground motion, which involves controlling the top displacement or the plastic rotation in the potential hinge region.

However, the deformation indicator in the specification faces the following limitations. It is specifically calibrated for regular bridges under rare seismic events. More significantly, the deformation criteria depend on the unique characteristics of each bridge pier, including height, yield curvature, hinge length, and section size, which differ from one pier to another. Consequently, this method can only evaluate the seismic performance of a single bridge pier but not the global reliability of the bridge system.

Inspired by the concept of the inter-story drift angle in the "Code for Seismic Design of Buildings" (GB50011-2010 2016) of China, we propose a novel metric for bridge piers. In this approach, a pier is subdivided into several segments, and the relative displacement angle (RDA) for each segment is defined as follows (Liu et al. 2023):

$$\Phi_j^i(t) = \frac{\left|u_{j,\mathrm{t}}^i(t) - u_{j,\mathrm{b}}^i(t)\right|}{H_j^i} \tag{8.44}$$

where $i = 1, 2, 3, 4;\ j = 1, 2, \cdots, n_s^i$. The superscript i denotes the number of piers, and the subscript j denotes the number of pier segments. $\Phi_j^i(t)$ presents the RDA of the j-th segment in the i-th pier. $u_{j,\mathrm{t}}^i(t)$ and $u_{j,\mathrm{b}}^i(t)$ relatively signify the displacements at the top and bottom of the j-th segment in the i-th pier. H_j^i presents the height of the j-th segment in the i-th pier, and n_s^i presents the number of segments in the i-th pier, respectively. The schematic diagram illustrating the division of bridge pier segments is presented in Fig. 8.17.

Then, the EEV of the RDA associated with the pier segment can be presented as (Liu et al. 2023):

$$\Phi_{j,\max}^i = \max_{t \in [0,T]} \left\{ \left| \Phi_j^i(t) \right| \right\} \tag{8.45}$$

Furthermore, the EEV of each bridge pier can be defined as

$$\Phi_{\max}^i = \max_{j \in [1, n_s^i]} \left\{ \Phi_{j,\max}^i \right\} \tag{8.46}$$

Thus, the EEV in terms of the whole bridge system can be defined as:

$$\Phi_{\max} = \max_{i \in [1,4]} \left\{ \Phi_{\max}^i \right\} \tag{8.47}$$

Obviously, the dynamic reliability of the pier segments $R_{T,j}^i(\varphi_s)$, the dynamic reliability of the piers $R_T^i(\varphi_s)$, and the global dynamic reliability of the bridge $R_T(\varphi_s)$ can be obtained as follows:

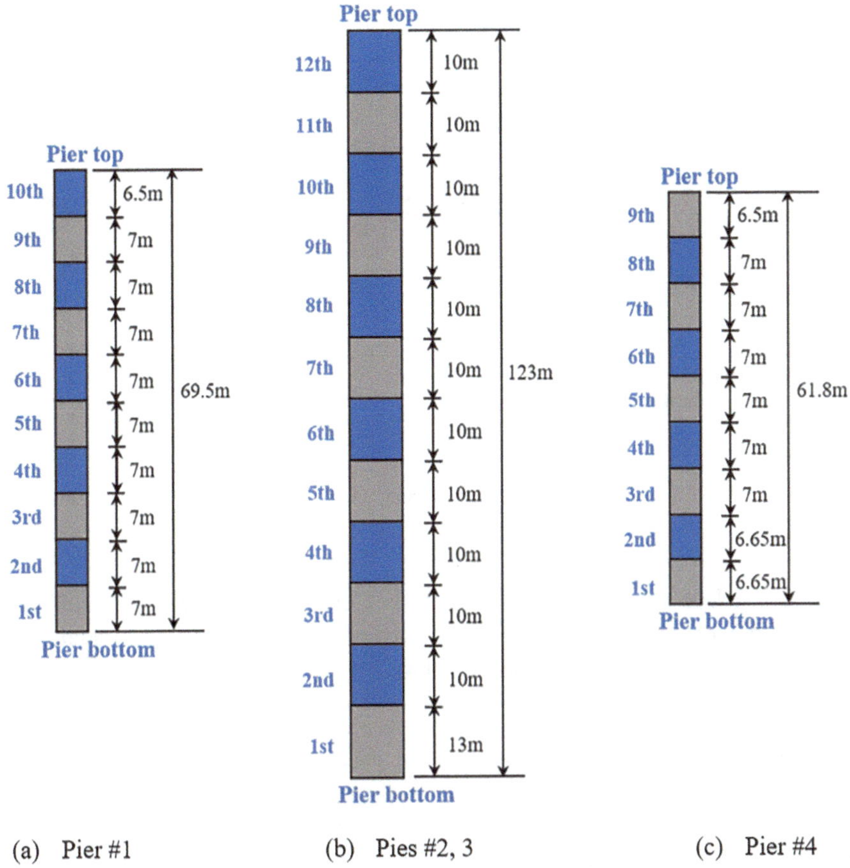

Fig. 8.17 Division of bridge pier segments. (**a**) Pier #1 (**b**) Pies #2, 3 (**c**) Pier #4

$$R^i_{T,j}(\varphi_s) = \Pr\left\{\Phi^i_{j,max} < \varphi_s\right\} = \int_0^{\varphi_s} p_{\Phi^i_{j,max}}(\varphi)\mathrm{d}\varphi \tag{8.48}$$

$$R^i_T(\varphi_s) = \Pr\{\Phi^i_{\max} < \varphi_s\} = \int_0^{\varphi_s} p_{\Phi^i_{\max}}(\varphi)\mathrm{d}\varphi \tag{8.49}$$

$$R_T(\varphi_s) = \Pr\{\Phi_{\max} < \varphi_s\} = \int_0^{\varphi_s} p_{\Phi_{\max}}(\varphi)\mathrm{d}\varphi \tag{8.50}$$

where $p_{\Phi^i_{j,\max}}$, $p_{\Phi^i_{\max}}$ and $p_{\Phi_{\max}}$ are the PDFs of the EEVs $\Phi^i_{j,\max}$, $\Phi^i_{\max}$ and $\Phi_{\max}$, respectively. Similarly, the CDFs of the EEVs are the corresponding dynamic reliabilities, which can be readily calculated by the PDEM.

From the above, with the help of the dimensionless RDA, a direct quantitative evaluation of the structural global reliability can be implemented, effectively accommodating multiple failure modes where the heights of multiple piers differ and

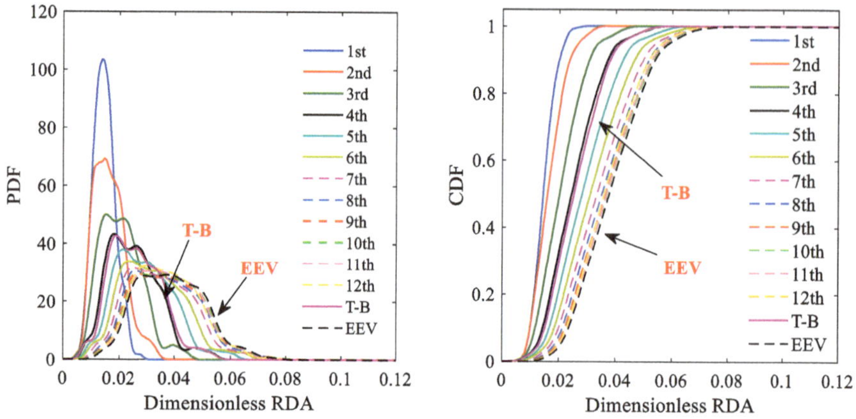

(a) PDF (b) CDF

Fig. 8.18 Reliability of Pier #2 under the three-component seismic action (0.4 g-*xyz*). (**a**) PDF (**b**) CDF

applicable to both linear and nonlinear analysis. Consequently, this approach offers a robust and overarching perspective on the seismic safety of the bridge system.

The seismic reliability of Pier #2 under the earthquake action of case 4 in Table 8.14 is presented in Fig. 8.18. The results indicate that for a single bridge pier, the PDF curves become shorter and more dispersed with increasing height, while their peaks shift toward larger values. This trend reflects an increase in the magnitude of structural dynamic responses as the height of the bridge pier rises. Consequently, the seismic reliability, represented by the CDF curves, decreases progressively from the bottom to the top of the pier. Further observation reveals that the reliability evaluated using the TB-RDA (RDA of the top and bottom of the pier) is approximately equivalent to that of a pier segment located between one-third and one-half of the height from the bottom. In contrast, the reliability assessed by the EEV of the RDA (Eq. (8.49)) is the lowest, indicating that piers designed against this criterion would possess superior seismic resistance.

Figure 8.19a reveals that the PDF curves for the main piers (Piers #2 and #3) are virtually identical, while those for the transition piers (Piers #1 and #4) are similar to each other but lower and more scattered. The global-EEV PDF is further right-shifted. These PDF characteristics directly correspond to the reliability trends in Fig. 8.19b, where the main piers exhibit superior reliability compared to the transition piers at any given RDA threshold. This distinction directly results from the strategic reinforcement of the main piers, thereby confirming the soundness of seismic design for the bridge.

Figure 8.20 serves to quantify the impact of multi-component seismic input on the dynamic reliability of the bridge by comparing four excitation scenarios. The analysis reveals that the 0.4 g-*xyz* condition produces the largest dynamic response (Fig. 8.20a) and the relative conservative global reliability (Fig. 8.20b), unequivocally demonstrating the significant influence of multi-directional excitation and

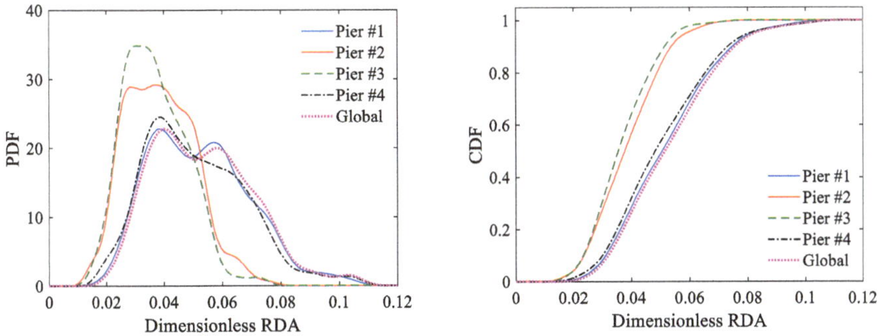

(a) PDF (b) CDF

Fig. 8.19 Reliabilities of the piers and global reliability of the bridge under the three-component seismic action (0.4 g-*xyz*). (**a**) PDF (**b**) CDF

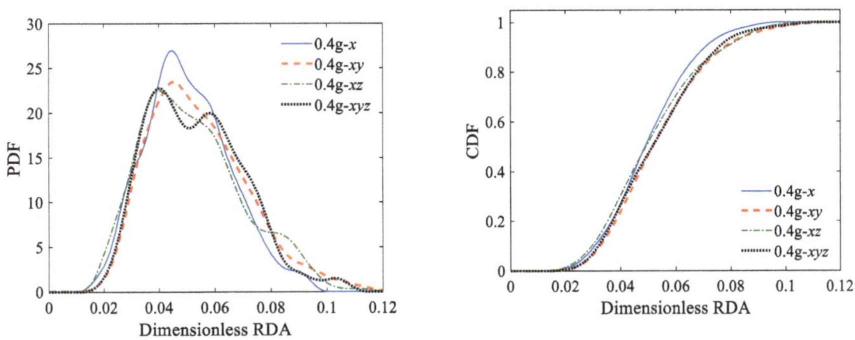

(a) PDF (b) CDF

Fig. 8.20 Global reliability of the bridge under the four seismic excitation conditions. (**a**) PDF (**b**) CDF

justifying its inclusion in seismic design. The proposed method can readily yield specific reliability indices for any pier or the global system under a given threshold. For example, with $\varphi_s = 0.08$ under 0.4 g-*xyz* condition, the pier reliabilities are 96.58%, 100%, 100%, and 96.12%, with the global reliability of 95.23%, considering the geometric nonlinearity. These values are lower than the linear counterparts, underscoring the necessity of nonlinear analysis.

In reality, there are presently no established regulations concerning the RDA failure criteria for bridges. Nevertheless, we can determine the allowable RDA threshold based on the target global reliability level, thereby achieving performance-based design by controlling the deformation of bridge piers to satisfy the threshold requirements. This framework can also guide practice: retrofitting less reliable piers to enhance global safety in existing bridges, while minimizing reliability differences among piers to optimize new designs. Future work should focus on establishing definitive RDA thresholds for various limit states to enable widespread engineering application.

8.5 Conclusion Remarks

This chapter introduces the principles of probability density evolution analysis and dynamic reliability assessment using the PDEM. The applications of the dimension-reduction simulation method for seismic ground motions are illustrated through three engineering case studies. The main conclusions are as follows.

1. Numerical examples demonstrate that the stochastic ground motions generated through the proposed dimension-reduction simulation method can be effectively integrated with the probability density evolution theory, enabling refined nonlinear dynamic response analysis as well as component and global dynamic reliability assessment of complex engineering structures, including dams, high-rise buildings, and high-pier long-span rigid-frame bridges, subjected to complex stochastic earthquake actions.
2. The refined reliability assessment of the structure enables the identification of vulnerable parts within both the structure and its components. For instance, in the case of the gravity dam, the relative displacement angles at various locations serve as failure criteria, revealing that the dam heel exhibits the lowest reliability, thus indicating it as a weak part. Similarly, for high-rise buildings and high-pier long-span bridges, the vulnerable parts can be pinpointed by identifying the stories and pier segments with the lowest reliability. Ultimately, by implementing appropriate measures to strengthen these weak parts, the overall seismic resilience of the structure can be enhanced.
3. Engineering examples have verified that the influence of multi-component seismic action cannot be ignored. The dynamic response of structures under multi-component ground motion is significantly greater than that under just one component earthquake excitation. In particular, the bidirectional horizontal seismic components play a pivotal role, whereas the vertical component has minimal impact on seismic performance indicators such as structural deformation, except for its influence on axial forces.
4. Further research should be conducted on the criteria for assessing the seismic reliability of complex engineering structures. Currently, apart from the "Code for Seismic Design of Buildings" (GB50011-2010 2016) of China, which explicitly specifies the threshold values of inter-story drift angles for buildings at various performance levels as direct criteria for evaluating their seismic reliability, there is no unified standard for determining the seismic reliability of complex structures like dams and high-pier bridges. This study proposes a general approach: the thresholds of relevant indicators can be deduced from the target global reliability of the structure, thereby guiding both the seismic reinforcement of existing structures and the seismic design of new projects. However, this approach necessitates extensive analytical efforts and represents a promising avenue for future research.

References

Chen JB, Li J (2007) The extreme value distribution and dynamic reliability analysis of nonlinear structures with uncertain parameters. Struct Saf 29(2):77–93

Chen JB, Wan ZQ (2019) A compatible probabilistic framework for quantification of simultaneous aleatory and epistemic uncertainty of basic parameters of structures by synthesizing the change of measure and change of random variables. Struct Saf 78:76–87

Deodatis G (1996) Non-stationary stochastic vector processes: seismic ground motion applications. Probabilistic Eng Mech 11(3):149–167

DL5073-2000 (2000) Specifications for seismic design of hydraulic structures. China Water & Power Press, Beijing. (in Chinese)

GB50011-2010 (2016) Code for seismic design of buildings. China Architecture & Building press, Beijing. (in Chinese)

Hariri-Ardebili MA, Saouma V (2015) Quantitative failure metric for gravity dams. Earthq Eng Struct Dyn 44(3):461–480

Hong YW, Du CB, Jiang SY (2011) Innovative design and construction of a high RCC gravity dam in high seismic intensity region. Pract Period Struct Des Constr 16(2):67–72

Jiang SY, Du CB, Hong YW (2013) Failure analysis of a cracked concrete gravity dam under earthquake. Eng Fail Anal 33(5):265–280

JTG/T 2231-01-2020 (2020) Specifications for seismic design of highway bridges. China Communications Press, Beijing. (in Chinese)

JTG/T B02-01-2008 (2008) Guidelines for seismic design of highway bridges. China Communications Press, Beijing. (in Chinese)

Lee J, Fenves GL (1998) A plastic-damage concrete model for earthquake analysis of dams. Earthq Eng Struct Dyn 27(9):937–956

Li J (2016) Probability density evolution method: background, significance and recent developments. Probabilistic Eng Mech 44:111–117

Li J, Chen JB (2009) Stochastic dynamics of structures. John Wiley & Sons, Singapore

Li YM, Liu LP (2011) Design ground motion of engineering structures. Science Press, Beijing. (in Chinese)

Li J, Chen JB, Fan WL (2007) The equivalent extreme-value event and evaluation of the structural system reliability. Struct Saf 29(2):112–131

Liu ZJ, Liu ZX (2017) Simulation of fully non-stationary ground motion based on seismic design response spectrum. J Vib Eng 30(3):457–465. (in Chinese)

Liu ZJ, Liu ZX, Chen DH (2018) Probability density evolution of a nonlinear concrete gravity dam subjected to nonstationary seismic ground motion. J Eng Mech 144(1):04017157

Liu ZJ, Chen JB, Peng YB (2022a) Dynamics of structures. China Architecture & Building Press, Beijing. (in Chinese)

Liu ZJ, Ruan XX, Liu ZX (2022b) Performance-based global reliability assessment of a high-rise frame-core tube structure subjected to multi-dimensional stochastic earthquakes. Earthq Eng Eng Vib 21(2):395–415

Liu ZX, Liu ZJ, Ai QH et al (2023) Global reliability evaluation of a high-pier long-span continuous RC rigid frame bridge subjected to multi-point and multi-component stochastic ground motions. Soil Dyn Earthq Eng 164:107623

Omidi O, Lotfi V (2013) Earthquake response of concrete arch dams: a plastic-damage approach. Earthq Eng Struct Dyn 42(14):2129–2149

Penzien J, Watabe M (1975) Characteristic of 3-dimensional earthquake ground motions. Earthq Eng Struct Dyn 3(4):365–373

Sun ZG, Wang DS, Wang T et al (2019) Investigation on seismic behavior of bridge piers with thin-walled rectangular hollow section using quasi-static cyclic tests. Eng Struct 200:109708

Xie X (2006) Seismic response and earthquake resistant design of bridges. China Communications Press, Beijing. (in Chinese)

Xue SD, Wang XS, Cao Z (2004) An efficient algorithm for multi-dimensional and multi-support random seismic analysis of spatial reticulated structures. World Earthq Eng 20(3):43–49. (in Chinese)

Appendix A: Fourier Transform Pairs

In this book, the Fourier transform pairs between the PSD function $S(\omega)$ and the correlation function $R(\tau)$ of stationary stochastic processes, say the Wiener-Khinchin formula, have been introduced, such that:

$$\begin{cases} S(\omega) = \frac{1}{2\pi}\int_{-\infty}^{\infty} R(\tau)\mathrm{e}^{-\mathrm{i}\omega\tau}\mathrm{d}\tau \\ R(\tau) = \int_{-\infty}^{\infty} S(\omega)\mathrm{e}^{\mathrm{i}\omega\tau}\mathrm{d}\omega \end{cases} \tag{A.1}$$

or

$$\begin{cases} S(\omega) = \int_{-\infty}^{\infty} R(\tau)\mathrm{e}^{-\mathrm{i}\omega\tau}\mathrm{d}\tau \\ R(\tau) = \frac{1}{2\pi}\int_{-\infty}^{\infty} S(\omega)\mathrm{e}^{\mathrm{i}\omega\tau}\mathrm{d}\omega \end{cases} \tag{A.2}$$

In fact, the aforementioned formulas provide two typical Fourier transform pairs. Moreover, numerous other Fourier transform pairs can be found in literature, as illustrated in Table A.1. The coefficients listed in the table should be used in corresponding pairs.

In the aforementioned Fourier transform Eqs. (A.1) and (A.2), frequency is denoted by the circular frequency ω, which has dimension of rad/s or 1/s or s^{-1}. It should be noted that 1rad is equivalent to the real number 1. However, in engineering practice, frequency f is also widely employed, with dimension measured in Hz or 1 cycle/s. Consequently, it is sometimes necessary to convert the PSD function $S(\omega)$ into the spectral density $S(f)$ expressed in terms of f. Here, it is worth mentioning that the energy corresponding to the frequency intervals $\mathrm{d}\omega$ and $\mathrm{d}f$ remains the same, that is:

Z. Liu et al., *Dimension-reduction Representation of Stochastic Ground Motion and Engineering Applications*, https://doi.org/10.1007/978-981-92-0168-6

Table A.1 The coefficients of Fourier transform pairs

Coefficients	1	2	3	4	5
Coefficients of $S(\omega)$ function	$\frac{1}{2\pi}$	1	$\frac{1}{2}$	$\frac{1}{\pi}$	$\frac{1}{\sqrt{2\pi}}$
Coefficients of $R(\tau)$ function	1	$\frac{1}{2\pi}$	$\frac{1}{\pi}$	$\frac{1}{2}$	$\frac{1}{\sqrt{2\pi}}$

$$S(\omega)\mathrm{d}\omega = S(f)\mathrm{d}f \tag{A.3}$$

Meanwhile, note that:

$$\omega = 2\pi f, \quad \mathrm{d}\omega = 2\pi \mathrm{d}f \tag{A.4}$$

Thus, we have:

$$S(f) = 2\pi S(\omega) = 2\pi S(2\pi f) \tag{A.5}$$

This demonstrates that by replacing ω in $S(\omega)$ with $2\pi f$ and multiplying the PSD function by 2π yields the PSD expressed in terms of $S(f)$.

For the Fourier transform Eq. (A.1), the Wiener-Khinchin formula expressed in terms of f is as follows:

$$\begin{cases} S(f) = \int_{-\infty}^{\infty} R(\tau)\mathrm{e}^{-\mathrm{i}2\pi f\tau}\mathrm{d}\tau \\ R(\tau) = \int_{-\infty}^{\infty} S(f)\mathrm{e}^{\mathrm{i}2\pi f\tau}\mathrm{d}f \end{cases} \tag{A.6}$$

Appendix B: Representative Point Sets Applying the Number Theoretical Method

In Chap. 3, the number theoretic method (NTM) of representative point selection for two elementary random variables was discussed. Here, we further present the NTM of representative point selection for s-dimensional ($s \geq 3$) elementary random variables. As a matter of fact, a uniform point set $\{\boldsymbol{\vartheta}_k = (\vartheta_{1,k}, \vartheta_{2,k}, \cdots, \vartheta_{s,k});\ k = 1, 2, \cdots, n_{\mathrm{sel}}\}$ over the hypercube $[0, 1]^s$ can be generated by the following expressions (Hua and Wang 1981):

$$\vartheta_{i,k} = \frac{h_i k}{n_{\mathrm{sel}}} - \mathrm{int}\left(\frac{h_i k}{n_{\mathrm{sel}}}\right), \quad i = 1, 2, \ldots, s; \quad k = 1, 2, \cdots, n_{\mathrm{sel}} \tag{B.1}$$

or (Fang and Wang 1996; Li and Chen 2009):

$$\vartheta_{i,k} = \frac{2h_i k - 1}{2n_{\mathrm{sel}}} - \mathrm{int}\left(\frac{2h_i k - 1}{2n_{\mathrm{sel}}}\right), \quad i = 1, 2, \ldots, s; \quad k = 1, 2, \cdots, n_{\mathrm{sel}} \tag{B.2}$$

Here, n_{sel} denotes the number of representative points, and $\boldsymbol{h} = (h_1, h_2, \cdots, h_s)$ is called the generator vector, respectively.

For ease of application, the literature (Hua and Wang 1981; Fang and Wang 1996) suggests the number of representative points and values for the generator vector $\boldsymbol{h}$ when $s = 2 \sim 18$. Tables B.1, B.2, B.3 and B.4 respectively present the main data in cases of $s = 2, 3, 4, 5$, encompassing the number of representative points n_{sel} and the corresponding values of the generator vector $\boldsymbol{h}$. Further details can be consulted in the literature (Hua and Wang 1981; Fang and Wang 1996).

Table B.1 Number of representative points n_{sel} and values of generator vector $\boldsymbol{h}$ ($s = 2,\ h_1 = 1$)

n_{sel}	34	55	89	144	233	377	610	987
h_2	21	34	55	89	144	233	377	610

Z. Liu et al., *Dimension-reduction Representation of Stochastic Ground Motion and Engineering Applications*, https://doi.org/10.1007/978-981-92-0168-6

Table B.2 Number of representative points n_{sel} and values of generator vector $\boldsymbol{h}$ ($s = 3,\ h_1 = 1$)

n_{sel}	35	101	135	185	266	418	597	828
h_2	11	40	29	26	27	90	63	285
h_3	16	85	42	64	69	130	169	358

Table B.3 Number of representative points n_{sel} and values of generator vector $\boldsymbol{h}$ ($s = 4,\ h_1 = 1$)

n_{sel}	307	562	701	1019	2129	3001	4001	5003
h_2	42	53	82	71	766	174	113	792
h_3	229	89	415	765	1281	266	766	1889
h_4	101	221	382	865	1906	1269	2537	191

Table B.4 Number of representative points n_{sel} and values of generator vector $\boldsymbol{h}$ ($s = 5,\ h_1 = 1$)

n_{sel}	1069	1543	2129	3001	4001	5003	6007	8191
h_2	63	58	618	408	1534	840	509	1386
h_3	762	278	833	1409	568	117	780	4302
h_4	970	694	1705	1681	3095	3593	558	7715
h_5	177	134	1964	1620	2544	1311	1693	3735

Appendix C: Dirac Delta Function

In physics, it is often necessary to study the density of a physical quantity distributed in space or time, such as mass density, charge density and so forth. However, abstract models such as particles, point charges, and instantaneous forces are also commonly used in physics. They are not continuously distributed in space or time, but concentrated at a certain point in space or a certain moment in time. Therefore, how can their density be represented? the Dirac Delta function serves as a potent tool for describing them.

C.1 Definition

The First Definition

If a function $f(x)$ can satisfy the following two conditions:

$$f(x) = \begin{cases} \infty, & x = x_0 \\ 0, & \text{otherwise} \end{cases} \tag{C.1}$$

$$\int_{-\infty}^{\infty} f(x)\mathrm{d}x = 1 \tag{C.2}$$

then $f(x)$ is called the Dirac Delta function, which is usually denoted by $\delta(\cdot)$, such that:

$$f(x) = \delta(x - x_0) \tag{C.3}$$

The Dirac Delta function is a type of generalized function, with its graph depicted in Fig. C.1.

Z. Liu et al., *Dimension-reduction Representation of Stochastic Ground Motion and Engineering Applications*, https://doi.org/10.1007/978-981-92-0168-6

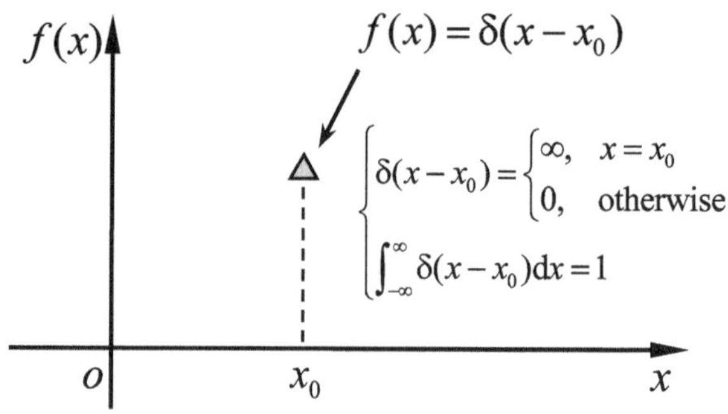

Fig. C.1 Dirac Delta function

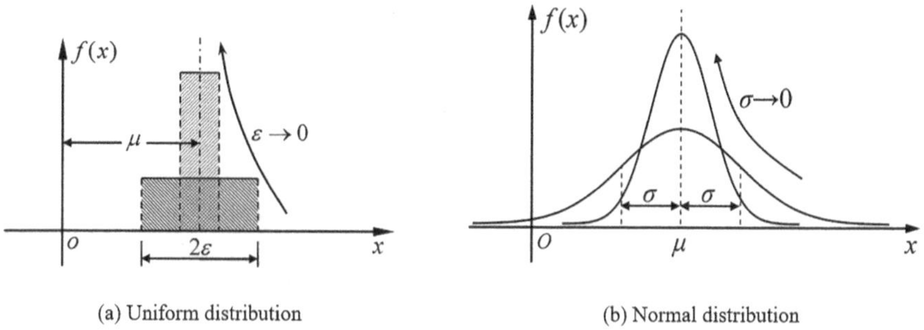

Fig. C.2 The Dirac Delta function as the limit case of probability density function

According to the definitions of the Dirac Delta function provided in Eqs. (C.1) and (C.2), the Dirac Delta function can be viewed as the PDF of a uniformly distributed random variable over the interval $[\mu - \varepsilon, \mu + \varepsilon]$, taken in the limit as the interval length tends to zero, as shown in Fig. C.2a, such that:

$$p_{\text{Uniform}}(x;\varepsilon) = \begin{cases} \frac{1}{2\varepsilon}, & \mu - \varepsilon \leq x \leq \mu + \varepsilon \\ 0, & \text{otherwise} \end{cases} \tag{C.4}$$

Obviously, we have:

$$\lim_{\varepsilon \to 0} p_{\text{Uniform}}(x;\varepsilon) = \delta(x - \mu) \tag{C.5}$$

Similarly, the Dirac Delta function can also be regarded as the PDF of a normally distributed random variable in the limiting case where its standard deviation tends to zero, as presented in Fig. C.2b, such that:

$$p_{\text{Normal}}(x;\sigma) = \frac{1}{\sqrt{2\pi}\sigma} e^{-\frac{(x-\mu)^2}{2\sigma^2}} \tag{C.6}$$

when $\sigma \to 0$, we can get the Dirac Delta function:

$$\lim_{\sigma \to 0} p_{\text{Normal}}(x; \sigma) = \delta(x - \mu) \tag{C.7}$$

The Second Definition

The Dirac Delta function can also be regarded as the derivative of the unit step function. Indeed, based on the definition of the unit step function, also known as the Heaviside function, given by:

$$u(t - t_0) = \begin{cases} 1, & t \geq t_0 \\ 0, & \text{otherwise} \end{cases} \tag{C.8}$$

If we define $\dot{u}(t - t_0) = \mathrm{d}u(t - t_0)/\mathrm{d}t$, then the following can be obtained:

$$\int_{-\infty}^{\infty} \dot{u}(t - t_0)\mathrm{d}t = u(\infty) - u(-\infty) = 1 \tag{C.9}$$

$$\dot{u}(t - t_0) = \begin{cases} \infty, & t = t_0 \\ 0, & \text{otherwise} \end{cases} \tag{C.10}$$

Compare Eqs. (C.1) and (C.2), it is clear that (Li and Chen 2009):

$$\dot{u}(t - t_0) = \delta(t - t_0) \tag{C.11}$$

Thus, the Dirac Delta function can be regarded as the derivative of the unit step function.

C.2 Basic properties

According to the conditions in Eqs. (C.1) and (C.2), for a continuous function $g(x)$, the following holds:

$$\int_{-\infty}^{\infty} g(x)\delta(x - x_0)\mathrm{d}x = g(x_0) \tag{C.12}$$

This property is known as the sifting property of the Dirac Delta function. Specifically, when $x_0 = 0$, we have:

$$\int_{-\infty}^{\infty} g(x)\delta(x)\mathrm{d}x = g(0) \tag{C.13}$$

In practical applications, a time function $x(t)$ typically holds meaning when $t \geq 0$, thus the aforementioned sifting property can be reformulated as:

$$\int_0^t x(\tau)\delta(\tau - t_0)\mathrm{d}\tau = x(t_0) \tag{C.14}$$

where $0 \leq t_0 \leq t$.

According to the sifting property of the Dirac Delta function, the Fourier transform pair of the Dirac Delta function can be readily derived by:

$$\mathcal{F}[\delta(t - t_0)] = \int_{-\infty}^{\infty} \delta(t - t_0)\mathrm{e}^{-\mathrm{i}\omega t}\mathrm{d}t = \mathrm{e}^{-\mathrm{i}\omega t_0} \tag{C.15}$$

$$\mathcal{F}^{-1}\left[\mathrm{e}^{-\mathrm{i}\omega t_0}\right] = \frac{1}{2\pi}\int_{-\infty}^{\infty} \mathrm{e}^{-\mathrm{i}\omega t_0}\mathrm{e}^{\mathrm{i}\omega t}\mathrm{d}\omega = \frac{1}{2\pi}\int_{-\infty}^{\infty} \mathrm{e}^{\mathrm{i}\omega(t - t_0)}\mathrm{d}\omega = \delta(t - t_0) \tag{C.16}$$

In particular, when $t_0 = 0$, we derive:

$$\int_{-\infty}^{\infty} \delta(t)\mathrm{e}^{-\mathrm{i}\omega t}\mathrm{d}t = 1 \tag{C.17}$$

$$\delta(t) = \frac{1}{2\pi}\int_{-\infty}^{\infty} \mathrm{e}^{\mathrm{i}\omega t}\mathrm{d}\omega \tag{C.18}$$

Similarly, we can also obtain:

$$\delta(\omega) = \frac{1}{2\pi}\int_{-\infty}^{\infty} \mathrm{e}^{-\mathrm{i}\omega t}\mathrm{d}t \tag{C.19}$$

Next, let's consider the integral of the Dirac Delta function involving compound functions, for instance:

$$\mathcal{I}[g(x)] = \int_{-\infty}^{\infty} g(x)\delta[\varphi(x) - x_0]\mathrm{d}x \tag{C.20}$$

If the integral variable x is changed to $\varphi^{-1}(y)$, then Eq. (C.20) becomes:

$$\mathcal{I}[g(x)] = \int_{-\infty}^{\infty} |J|\, g\left[\varphi^{-1}(y)\right] \delta(y - x_0)\mathrm{d}y = |J|\, g\left[\varphi^{-1}(x_0)\right] \tag{C.21}$$

where $J = \frac{\mathrm{d}\varphi^{-1}}{\mathrm{d}y}$ is the Jacobi quantity and $|J|$ is its modulus, respectively.

Finally, present the sifting properties of the derivative for the Dirac Delta function. If we define:

$$\delta'(x) = \frac{\mathrm{d}\delta(x)}{\mathrm{d}x} = \lim_{\varepsilon \to 0} \frac{\delta(x+\varepsilon) - \delta(x)}{\varepsilon} \tag{C.22}$$

It is easy to prove that for any continuous function $g(x)$, there is:

$$\int_0^l g(x)\,\delta'(x-t)\mathrm{d}x = -g'(t) \tag{C.23}$$

More generally, there exist:

$$\int_0^l g(x)\,\delta^{(n)}(x-t)\mathrm{d}x = (-1)^n g^{(n)}(t) \tag{C.24}$$

where n is a positive integer.

C.3 Relationship Between the Kronecker Delta and the Dirac Delta Function

Now, consider the integral of the Dirac Delta function $\delta(x - x_0)$ when x_0 is confined to a tiny neighborhood interval $[j - \varepsilon, j + \varepsilon]$, then its integral value is (Li and Chen 2009):

$$\mathcal{I}_{ij} = \lim_{\varepsilon \to 0} \int_{j-\varepsilon}^{j+\varepsilon} \delta(x-i)\mathrm{d}x = \lim_{\varepsilon \to 0} \int_{i-\varepsilon}^{i+\varepsilon} \delta(x-j)\mathrm{d}x = \begin{cases} 1, & i=j \\ 0, & \text{otherwise} \end{cases} \tag{C.25}$$

In general, when i and j are integers, the integral $\mathcal{I}_{ij}$ is called the Kronecker delta, that is:

$$\delta_{ij} = \begin{cases} 1, & i=j \\ 0, & \text{otherwise} \end{cases} \tag{C.26}$$

Hence, the Dirac Delta function can be regarded as the continuous version of the Kronecker delta, while the Kronecker delta can also be seen as the discretized version of the Dirac Delta function.

In fact, we can extend the Kronecker delta to any two real numbers x and x', that is:

$$\delta_{xx'} = \begin{cases} 1, & x=x' \\ 0, & \text{otherwise} \end{cases} \tag{C.27}$$

Thus, we can further obtain another relationship between the Kronecker delta and the Dirac delta function, such that (Liu et al. 2022):

$$\delta(x - x')\mathrm{d}x = \delta_{xx'} \tag{C.28}$$

Appendix D: Power Spectrum of Filtered White Noise Model

D.1 Power Spectrum of Response for SDOF Systems Subjected to Stochastic Excitation

Consider the vibration of a single degree of freedom (SDOF) linear system subjected to stationary stochastic excitation is as follows:

$$\ddot{Y}(t) + 2\zeta_0\omega_0\dot{Y}(t) + \omega_0^2 Y(t) = X(t) \tag{D.1}$$

Here, excitation $X(t)$ is a zero mean stationary stochastic process, with its correlation function $R_X(\tau)$ and PSD function $S_X(\omega)$ both known. According to the Wiener-Kinchin formula introduced in Chap. 4, we have:

$$S_X(\omega) = \int_{-\infty}^{\infty} R_X(\tau)\mathrm{e}^{-\mathrm{i}\omega\tau}\mathrm{d}\tau \tag{D.2}$$

$$R_X(\tau) = \frac{1}{2\pi}\int_{-\infty}^{\infty} S_X(\omega)\mathrm{e}^{\mathrm{i}\omega\tau}\mathrm{d}\omega \tag{D.3}$$

In order to establish the PSD function transfer relationship between input and output stochastic processes, the relationship between sample finite Fourier transform and PSD function is utilized, such that (Li and Chen 2009; Liu et al. 2022):

$$S_X(\omega) = \lim_{T\to\infty}\frac{1}{2T}E\left[F_X(\omega,T)F_X^*(\omega,T)\right] \tag{D.4}$$

where the sample finite Fourier transform is:

Z. Liu et al., *Dimension-reduction Representation of Stochastic Ground Motion and Engineering Applications*, https://doi.org/10.1007/978-981-92-0168-6

$$F_X(\omega, T) = \int_{-T}^{T} X(t) e^{-i\omega t} dt \tag{D.5}$$

It is noted that the finite Fourier transform of $\ddot{Y}(t)$ is as follows:

$$\begin{aligned} F_{\ddot{Y}}(\omega, T) &= \int_{-T}^{T} \ddot{Y}(t) e^{-i\omega t} dt \\ &= \dot{Y}(t) e^{-i\omega t}\big|_{-T}^{T} + (i\omega) \int_{-T}^{T} \dot{Y}(t) e^{-i\omega t} dt \\ &= \dot{Y}(t) e^{-i\omega t}\big|_{-T}^{T} + (i\omega) Y(t) e^{-i\omega t}\big|_{-T}^{T} + (i\omega)^2 \int_{-T}^{T} Y(t) e^{-i\omega t} dt \\ &= (i\omega)^2 F_Y(\omega, T) + \psi_{\ddot{Y}}(\omega, T) \end{aligned} \tag{D.6}$$

Similarly, the finite Fourier transform of $\dot{Y}(t)$ is given by:

$$F_{\dot{Y}}(\omega, T) = (i\omega) F_Y(\omega, T) + \psi_{\dot{Y}}(\omega, T) \tag{D.7}$$

in which

$$\psi_{\ddot{Y}}(\omega, T) = \left[\dot{Y}(t) + (i\omega) Y(t)\right] e^{-i\omega t}\big|_{-T}^{T} \tag{D.8}$$

$$\psi_{\dot{Y}}(\omega, T) = Y(t) e^{-i\omega t}\big|_{-T}^{T} \tag{D.9}$$

By performing finite Fourier transform on both sides of Eq. (D.1), we can obtain the following:

$$\psi_{\ddot{Y}}(\omega, T) + 2\zeta_0\omega_0\psi_{\dot{Y}}(\omega, T) + \left[\left(\omega_0^2 - \omega^2\right) + 2i\zeta_0\omega_0\omega\right] F_Y(\omega, T) = F_X(\omega, T) \tag{D.10}$$

Multiply both sides of the above formula by self-conjugation, divide by $2T$, and let $T \to \infty$, note that:

$$\begin{aligned} &\lim_{T\to\infty} \frac{1}{2T} \psi_{\ddot{Y}}(\omega, T) = 0, \quad \lim_{T\to\infty} \frac{1}{2T} \psi_{\dot{Y}}(\omega, T) = 0 \\ &\lim_{T\to\infty} \frac{1}{2T} \left|\psi_{\ddot{Y}}(\omega, T)\right|^2 = 0, \quad \lim_{T\to\infty} \frac{1}{2T} \left|\psi_{\dot{Y}}(\omega, T)\right|^2 = 0 \end{aligned} \tag{D.11}$$

Then, we have the following:

$$\lim_{T\to\infty} \frac{1}{2T} \left[F_Y(\omega, T) F_Y^*(\omega, T)\right] = H(\omega) H^*(\omega) \lim_{T\to\infty} \frac{1}{2T} \left[F_X(\omega, T) F_X^*(\omega, T)\right] \tag{D.12}$$

where

$$H(\omega) = \frac{1}{\omega_0^2 - \omega^2 + 2\mathrm{i}\zeta_0\omega_0\omega} \tag{D.13}$$

denotes the frequency response function. For the sake of convenience, $|H(\omega)|^2$ is generally referred to as the transfer function.

From the above derivation process, it can be seen that the additional term $\psi_{\dot{Y}}(\omega, T) + 2\zeta_0\omega_0\psi_Y(\omega, T)$ in Eq. (D.10) can be directly omitted without altering the outcome. The simplified Eq. (D.10) is identical to the result obtained from the Fourier transform. This simplification treatment greatly facilitates practical engineering applications.

Taking the mathematical expectation on both sides of Eq. (D.12) and considering Eq. (D.4), we have:

$$S_Y(\omega) = |H(\omega)|^2 S_X(\omega) \tag{D.14}$$

It can thus be observed that the output PSD function of a linear system represents the product of its input PSD function and the transfer function, with the latter being entirely determined by the characteristics of the physical system, say the frequency response function.

D.2 Power Spectrum of Double Filtered White Noise Models

Clough-Penzien Spectrum

Chapter 4 has provided the typical motion equations for the double filtered white noise process model, followed by:

$$\begin{cases} \overline{X}_2(t) = \ddot{X}_g(t) + \ddot{X}_f(t) \\ \ddot{X}_g(t) + 2\zeta_g\omega_g\dot{X}_g(t) + \omega_g^2 X_g(t) = -\ddot{X}_f(t) \\ \ddot{X}_f(t) + 2\zeta_f\omega_f\dot{X}_f(t) + \omega_f^2 X_f(t) = -w(t) \end{cases} \tag{D.15}$$

where $w(t)$ represents a stationary white noise process with the two-sided PSD, say $S_w(\omega) = S_0$. $\overline{X}_2(t)$ is the stationary seismic acceleration process on the ground surface, and relative to the bottom of bedrock.

Then, let's derive the PSD for the stationary seismic acceleration process $\overline{X}_2(t)$, which is known as the Clough-Penzien spectrum.

Firstly, applying the finite Fourier transform to both sides of the second formula in Eq. (D.15) and neglecting the additional term, we can derive the following:

$$F_{X_g}(\omega, T) = -H_g(\omega) F_{\ddot{X}_f}(\omega, T) \tag{D.16}$$

where the expression for $H_g(\omega)$ is analogous to that in Eq. (D.13), such that:

$$H_g(\omega) = \frac{1}{\omega_g^2 - \omega^2 + 2i\zeta_g \omega_g \omega} \tag{D.17}$$

Meanwhile, applying the finite Fourier transform to both sides of the third formula in Eq. (D.15) and neglecting the additional term, we can obtain the following:

$$F_{X_f}(\omega, T) = -H_f(\omega) F_w(\omega, T) \tag{D.18}$$

$$H_f(\omega) = \frac{1}{\omega_f^2 - \omega^2 + 2i\zeta_f \omega_f \omega} \tag{D.19}$$

Furthermore, by applying the finite Fourier transform to both sides of the first formula in Eq. (D.15), we have:

$$\begin{aligned} F_{\ddot{X}_2}(\omega, T) &= F_{\ddot{X}_g}(\omega, T) + F_{\ddot{X}_f}(\omega, T) \\ &= (i\omega)^2 F_{X_g}(\omega, T) + F_{\ddot{X}_f}(\omega, T) \\ &= \left[1 + \omega^2 H_g(\omega)\right] F_{\ddot{X}_f}(\omega, T) \\ &= \left[1 + \omega^2 H_g(\omega)\right] (i\omega)^2 F_{X_f}(\omega, T) \\ &= \left[1 + \omega^2 H_g(\omega)\right] \left[\omega^2 H_f(\omega)\right] F_w(\omega, T) \end{aligned} \tag{D.20}$$

Finally, multiply both sides of Eq. (D.20) by its complex conjugate on the right, divide by $2T$, let $T \to \infty$, and then take the mathematical expectation on both sides, we can derive the result:

$$\begin{aligned} S_{\ddot{X}_2}(\omega) &= \frac{\omega_g^4 + 4\zeta_g^2 \omega_g^2 \omega^2}{\left(\omega_g^2 - \omega^2\right)^2 + 4\zeta_g^2 \omega_g^2 \omega^2} \cdot \frac{\omega^4}{\left(\omega_f^2 - \omega^2\right)^2 + 4\zeta_f^2 \omega_f^2 \omega^2} S_w(\omega) \\ &= \frac{\omega_g^4 + 4\zeta_g^2 \omega_g^2 \omega^2}{\left(\omega_g^2 - \omega^2\right)^2 + 4\zeta_g^2 \omega_g^2 \omega^2} \cdot \frac{\omega^4}{\left(\omega_f^2 - \omega^2\right)^2 + 4\zeta_f^2 \omega_f^2 \omega^2} S_0 \end{aligned} \tag{D.21}$$

This is Eq. (4.35) in Chap. 4, known as the Clough-Penzien spectrum.

Lai-Ye-Li Spectrum

In Chap. 4, we also presented an alternative type of motion equations for the double filtered white noise process model, such that:

$$\begin{cases} \overline{X}_3(t) = \ddot{X}_{\mathrm{g}}(t) + \ddot{X}_{\mathrm{f}}(t) + w(t) \\ \ddot{X}_{\mathrm{g}}(t) + 2\zeta_{\mathrm{g}}\omega_{\mathrm{g}}\dot{X}_{\mathrm{g}}(t) + \omega_{\mathrm{g}}^2 X_{\mathrm{g}}(t) = -\left[\ddot{X}_{\mathrm{f}}(t) + w(t)\right] \\ \ddot{X}_{\mathrm{f}}(t) + 2\zeta_{\mathrm{f}}\omega_{\mathrm{f}}\dot{X}_{\mathrm{f}}(t) + \omega_{\mathrm{f}}^2 X_{\mathrm{f}}(t) = -w(t) \end{cases} \tag{D.22}$$

Similar to the aforementioned derivation, starting with the second formula of Eq. (D.22), we can initially derive:

$$\begin{aligned} F_{X_{\mathrm{g}}}(\omega, T) &= -H_{\mathrm{g}}(\omega)\left[F_{\ddot{X}_{\mathrm{f}}}(\omega, T) + F_w(\omega, T)\right] \\ &= -H_{\mathrm{g}}(\omega)\left[(\mathrm{i}\omega)^2 F_{X_{\mathrm{f}}}(\omega, T) + F_w(\omega, T)\right] \end{aligned} \tag{D.23}$$

Secondly, based on the third formula in Eq. (D.22), we can derive:

$$F_{X_{\mathrm{f}}}(\omega, T) = -H_{\mathrm{f}}(\omega)F_w(\omega, T) \tag{D.24}$$

According to the first formula of Eq. (D.22) and taking into account Eqs. (D.23) and (D.24), we can obtain:

$$\begin{aligned} F_{\overline{X}_3}(\omega, T) &= F_{\ddot{X}_{\mathrm{g}}}(\omega, T) + F_{\ddot{X}_{\mathrm{f}}}(\omega, T) + F_w(\omega, T) \\ &= (\mathrm{i}\omega)^2 F_{X_{\mathrm{g}}}(\omega, T) + (\mathrm{i}\omega)^2 F_{X_{\mathrm{f}}}(\omega, T) + F_w(\omega, T) \\ &= \omega^2 H_{\mathrm{g}}(\omega)\left[-\omega^2 F_{X_{\mathrm{f}}}(\omega, T) + F_w(\omega, T)\right] - \omega^2 F_{X_{\mathrm{f}}}(\omega, T) + F_w(\omega, T) \\ &= -\left(\omega^4 H_{\mathrm{g}}(\omega) + \omega^2\right)F_{X_{\mathrm{f}}}(\omega, T) + \left(\omega^2 H_{\mathrm{g}}(\omega) + 1\right)F_w(\omega, T) \\ &= \left[\omega^4 H_{\mathrm{g}}(\omega)H_{\mathrm{f}}(\omega) + \omega^2 H_{\mathrm{f}}(\omega) + \omega^2 H_{\mathrm{g}}(\omega) + 1\right]F_w(\omega, T) \\ &= \frac{\omega_{\mathrm{g}}^2 + 2\mathrm{i}\zeta_{\mathrm{g}}\omega_{\mathrm{g}}\omega}{\omega_{\mathrm{g}}^2 - \omega^2 + 2\mathrm{i}\zeta_{\mathrm{g}}\omega_{\mathrm{g}}\omega} \cdot \frac{\omega_{\mathrm{f}}^2 + 2\mathrm{i}\zeta_{\mathrm{f}}\omega_{\mathrm{f}}\omega}{\omega_{\mathrm{f}}^2 - \omega^2 + 2\mathrm{i}\zeta_{\mathrm{f}}\omega_{\mathrm{f}}\omega} F_w(\omega, T) \end{aligned} \tag{D.25}$$

Finally, multiply both sides of Eq. (D.25) by its complex conjugate on the right, divide by $2T$, let $T \to \infty$, and then take the mathematical expectation on both sides, we can achieve the following:

$$S_{\overline{X}_3}(\omega) = \frac{\omega_{\mathrm{g}}^4 + 4\zeta_{\mathrm{g}}^2\omega_{\mathrm{g}}^2\omega^2}{\left(\omega_{\mathrm{g}}^2 - \omega^2\right)^2 + 4\zeta_{\mathrm{g}}^2\omega_{\mathrm{g}}^2\omega^2} \cdot \frac{\omega_{\mathrm{f}}^4 + 4\zeta_{\mathrm{f}}^2\omega_{\mathrm{f}}^2\omega^2}{\left(\omega_{\mathrm{f}}^2 - \omega^2\right)^2 + 4\zeta_{\mathrm{f}}^2\omega_{\mathrm{f}}^2\omega^2} S_0 \tag{D.26}$$

which is the PSD presented in Eq. (4.38).

References

Hua LK, Wang Y (1981) Applications of number theory to numerical analysis. Springer, Berlin

Fang KT, Wang Y. Number-theoretic Methods in Statistics. Chapman and Hall, London, 1994; Science Press, Beijing, 1996.

Li J, Chen JB (2009) Stochastic Dynamics of Structures. John Wiley & Sons, Singapore

Liu ZJ, Chen JB, Peng YB (2022) Dynamics of Structures. China Architecture & Building Press, Beijing. (in Chinese)

Z. Liu et al., *Dimension-reduction Representation of Stochastic Ground Motion and Engineering Applications*, https://doi.org/10.1007/978-981-92-0168-6

FSC
www.fsc.org
MIX
Papier | Fördert
gute Waldnutzung
FSC® C083411

Kolibri 360 GmbH
Ferdinand-Jühlke-Straße 7
99095 Erfurt, Deutschland
produktsicherheit@kolibri360.de